Thermal Cycles of Heat Recovery Power Plants

Authored by

Tangellapalli Srinivas
Department of Mechanical Engineering,
B.R. Ambedkar National Institute of Technology Jalandhar,
India

Thermal Cycles of Heat Recovery Power Plants

Author: Tangellapalli Srinivas

ISBN (Online): 978-981-18-0377-2

ISBN (Print): 978-981-18-0375-8

ISBN (Paperback): 978-981-18-0376-5

Published by Bentham Science Publishers Pte. Ltd. Singapore. All Rights Reserved.

need for a court order if at any point you breach any terms of this License Agreement. In no event will any delay or failure by Bentham Science Publishers in enforcing your compliance with this License Agreement constitute a waiver of any of its rights.

3. You acknowledge that you have read this License Agreement, and agree to be bound by its terms and conditions. To the extent that any other terms and conditions presented on any website of Bentham Science Publishers conflict with, or are inconsistent with, the terms and conditions set out in this License Agreement, you acknowledge that the terms and conditions set out in this License Agreement shall prevail.

Bentham Science Publishers Pte. Ltd.
80 Robinson Road #02-00
Singapore 068898
Singapore
Email: subscriptions@benthamscience.net

CONTENTS

FOREWORD 1

It is an honour for me to write the foreword for the book titled "Thermal Cycles of Heat Recovery Power Plants," written by Dr. T Srinivas. The scope of the book is in the area of heat recovery power generation and thermal cycle analysis. The book covers various heat recovery power generation systems and thermodynamic cycle analysis. Given the growing global energy demand, there is a need to improve power generation efficiency and to conserve energy resources. Global warming and the need to reduce greenhouse gas emissions demands higher efficiency fossil fuel-based power plants. The waste heat recovery and utilization will help to reduce greenhouse gas emissions. The cycle analysis plays a dominant role in understanding the heat recovery based power plants, including the performance. The book covers various thermal cycles and their analysis, including the performance of heat recovery based systems.

Srinivas is a well-known researcher in the areas of thermal power generation, combined cycle and cogeneration systems, waste heat recovery, solar energy, and exergy analysis. He has published extensively in journals and conference proceedings as an author and also with research collaborators. We have worked together on research projects in thermal power generation, waste heat recovery, and solar energy and published them in reputed journals and conference proceedings.

The "Thermal Cycles of Heat Recovery Power Plants" book covers the latest advances in heat recovery power generation systems, various thermal cycles, and analysis, including recent advances. The book incorporates recent advances in research and developments in heat recovery based power generation systems. I am confident that the book will be very useful to senior undergraduate level students, graduate-level students, researchers working in the area of thermal power generation and waste heat recovery power generation, and for practicing engineers in the area of thermal power generation, waste heat recovery, and energy management.

Bale V. Reddy
Department of Mechanical and Manufacturing Engineering
Faculty of Engineering and Applied Science
Ontario Tech University (UOIT)
Oshawa, ON, Canada

FOREWORD 2

The power generation from the waste heat recovery is equivalent to the power from the renewable energy sources as it does not generate any new or additional carbon dioxide to the environment. Therefore, these power plants are also eligible to claim the carbon credits as per the policy norms. I wish this book on the power thermal cycle of heat recovery power plants nurtures new ideas of power plants to tap the waste heat recovery. This book deals with the thermodynamic analysis of very important vapour power cycles such as organic Rankine cycle, organic flash cycle, Kalina cycle, steam Rankine cycle, and steam flash cycle from the modelling to the optimization of performance parameters and highlighting the challenges and opportunities. The latest power cycles, including the Kalina cycle, organic flash cycle, and steam flash cycle, are thoroughly analysed with exhaustive models and examples. The book also covers important practical aspects of the power cycles with detailed case studies, which may be very useful for the students. I know that Dr. T. Srinivas is also an author of 'Flexible Kalina Cycle Systems', which is focused on cooling cogeneration cycle based on Kalina cycle working principle. The chapter on comparison of the power cycles based on various thermodynamic characteristics may be very useful to the students and researchers. I hope this book contributes to the understanding of the power cycle's concepts, design, and development of new plants to the students, scholars, faculty, and practicing engineers for innovative developments.

K. Srinivas Reddy
Department of Mechanical Engineering
Indian Institute of Technology Madras
Chennai, India

PREFACE

Worldwide many thermal industries are working without tapping the valuable waste heat into a useful form. Electricity is one of the most extensively used commodities in the world. The existing and futuristic power plant configurations and its characteristics suitable to a waste heat recovery (WHR) are discussed in the book. Novel power plant configurations are developed and elaborated from modelling to the optimization through the simulation. Five different power plants configurations, suitable to heat recovery are presented *viz.* organic Rankine cycle (ORC), organic flash cycle (OFC), Kalina cycle (KC) steam Rankine cycle (SRC) and steam flash cycle (SFC). Out of these power plant layouts, flash cycle (FC) has been recommended because of its adoptability to the heat recovery. The novel flash cycle, which is different from the current geothermal power plant is detailed to augment the heat recovery and power with organic fluid system and steam system. In a power plant, the source temperature may fall below the critical temperature of the fluid or above it. The performance characteristics of these power plants differ with the working fluid and state of heat source, *i.e.*, below or above the critical temperature. Separate performance characteristics and correlations are developed in these two regions for all the selected fluids. The selected working fluids in the heat recovery power plants are R123, R124, R134a, R245fa, R717 and R407C. In FC, the liquid is flashed from high pressure to low pressure at the exit of heat recovery's economizer. The vapour is separated from the process and used in turbine for power augmentation. However, the handling of additional fluid in boiler increases the pump capacity and heat recovery. Therefore, a drop in thermal efficiency has been observed. FC plants are well justified by comparing the existing power plants with its higher production rate. A case study related to cement factory's heat recovery has been presented to understand the power plant nature with heat recovery. The cement factory demands 15 MW for its functioning and the case study showed that the WHR is capable of producing the self-generation to meet the load. A lower heat recovery pressure is suggested for maximum power. Second case study is at a 7.7 MW power plant operating under SFC. The theoretical results are validated with a cement factory's case studies with SRC and SFC. The mathematical simulation has been extended to solve 'n' flashers in SFC. Finally, OFC and SFC are recommended in place of ORC and SRC for maximum output.

Organic flash cycle or steam flash cycle are not reported in the available books in the area of power industry. This book companions the undergraduate and post graduate students of mechanical, electrical and similar streams, power plant engineers, practising engineers, research scholars, faculty and plant trainees in the field of power generation. Latest power plant configurations, selection of working fluids to suit the heat recovery temperature and novel flashing cycle in place of organic Rankine cycle and steam Rankine cycle are the key features of this book.

Tangellapalli Srinivas
Department of Mechanical Engineering
B.R. Ambedkar National Institute of Technology Jalandhar
India

ACKNOWLEDGEMENTS

My first lecture in my teaching profession is on thermal power plants at AANM and VVRSR Polytechnic, Gudlavalleru India. Thanks to the institute for creating such wonderful learning platform through teaching. I had an opportunity to visit Vijayawada Thermal Power Plant (VTPS) with my students from Gudlavalleru Engineering College, India. The plant's staff instructed the practical methods and highlighted the thermal power plant technology. My sincere thanks to the management, all staff members and specially Mr. Muthaiah Chary, Engineer - Maintenance, VTPS for his energetic guidelines and guiding the complete plant processes and components. The industry supports from VTPS, LANCO Power, GMR Energy, India Cements and Sagar Cements enhanced the understanding and nurtured the innovation in power generation technologies with the case studies. I thank all the industrial support for playing a key role in the development of the thermal cycles of heat recovery power plants. I recognize my research scholar, Dr. Pradeep Varma for conducting the valuable case studies at cement factories to formulate the new ideas of power generation through the waste heat recovery.

Thanks to the whole team of VIT University, Vellore, for providing the atmosphere and supporting to shape the fundamental ideas into reality. My sincere thanks to Prof. Lalit Kumar Awasthi, Director, Dr. B.R. Ambedkar, National Institute of Technology Jalandhar for providing the facilities and support to shape this book. My hearty salutations to faculty, non-teaching staff, students and scholars of department of mechanical engineering, NIT Jalandhar for continuous encouragement and providing the things.

My honest gratitude to Dr. P.K. Nag, Professor, IIT Kharagpur for inspiring and advising on thermodynamics applied to various thermal power plants. I am extremely happy to express my deepest gratitude to my PhD guide, Dr. AVSSKS Gupta, Professor, JNT University, Hyderabad for sculpturing me in the field of thermal engineering. It is my fortune to associate with the dynamic and energetic Professor and Guide Dr. BV Reddy, Ontario Tech University, Canada. I wish to express my gratitude to Dr. Reddy for his motivation and backing support. I am happy to convey my thanks to all my research scholars, faculty, staff and students to be a part of my research work. My special thanks to all the staff from CO_2 Research and Green Technologies Centre, VIT University, Vellore for assisting in plant development, erection and testing processes at the laboratory level.

I would like to express my deep sense of gratitude and respect to parents who fashioned my hard work and strong determination since my childhood. My heart felt gratitude to my lovely wife, Kavitha Devi, dearest elder son Rahul and dearest younger son Jignesh; without their regular support and boosting, I cannot do anything.

Finally, thanks to all who were involved in this work, directly or indirectly, in shaping this book to reach its fruitful form.

CONSENT FOR PUBLICATION

Not applicable.

CONFLICT OF INTEREST

The author declares no conflict of interest, financial or otherwise.

Tangellapalli Srinivas
B.R. Ambedkar National Institute of Technology Jalandhar
Department of Mechanical Engineering
India

CHAPTER 1

Introduction on Heat Recovery Power Plants

Abstract: This chapter overviews the heat recovery with power generation plants. The significance of captive power plants has been highlighted. Different heat recovery arrangements as per the category of thermal power cycle have been discussed. The power plant layouts of organic Rankine cycle (ORC), organic flash cycle (OFC), Kalina cycle (KC), steam Rankine cycle (SRC) and steam flash cycle (SFC) are deliberated. The subsequent chapters are focused on the detailed study of thermal cycles of heat recovery power plants.

Keywords: Bottoming cycle, Captive power plant, Energy efficiency, Energy scenario, Thermal power cycles, Topping cycle.

POWER GENERATION FROM WASTE HEAT RECOVERY

Is waste heat recovery (WHR) a renewable energy? The USA framework of climate change declared WHR power as green energy. These projects can claim the carbon credits for earning. The carbon credits are sanctioned for CO_2 reduction but not for renewable energy generation. Energy is capital for any country's development. A lot of waste heat is dumped on earth without dropping its temperature. This book is focused on construction, working, and description of power generation technologies using waste heat recovery. Power generation from fossil fuel causes environmental issues such as carbon dioxide emissions and thermal pollution. The electricity installed capacity of a nation is the sum of utility capacity and captive power capacity. The utility plants are the grid connected power plants. The captive power plants are the decentralized power plants, operating mostly on off-grid mode. The decentralized power generation using renewable technology is one of the promising solutions to address environmental pollution. Apart from the renewable energy sources, power from waste heat recovery is an attractive solution for self-generation of electricity without creating additional carbon dioxide in the environment. Since there is no investment in energy sources such as fuel or renewable energy technologies, the WHR power plants are doing a large business in the power market.

The major consumer of electricity is the industry, followed by domestic, agriculture, commercial, traction, railways, and others.

Since the thermal industries handle heat, they can switch into heat and power, *i.e.*, cogeneration plants. Concrete measures are to be taken for the effective use of waste heat from industries for power generation. Depending on the size of the power plant, these power plants may be operated either captive mode (small capacity) or gird connection mode (high capacity). If the plant capacity is high, the excess amount of electricity can be supplied to the grid. In process industries a lot of hot flue gases are generated from kilns, furnaces and boilers. If effective utilization of those flue gases is done by using proper technology, a considerable amount of energy and money can be saved. The major part of the electricity generation is from conventional sources of energy, *viz.* coal, oil, and natural gas. However, they are exhaustive and harmful to society and the environment. The power from waste heat is one of the opportunity to this challenge. Waste heat recovery is a heat exchanger, which permits the transfer of heat from the waste hot fluid to the working fluid of the power plant. Waste heat recovery units are generally used in cogeneration plants where the outputs consist of power and process heat.

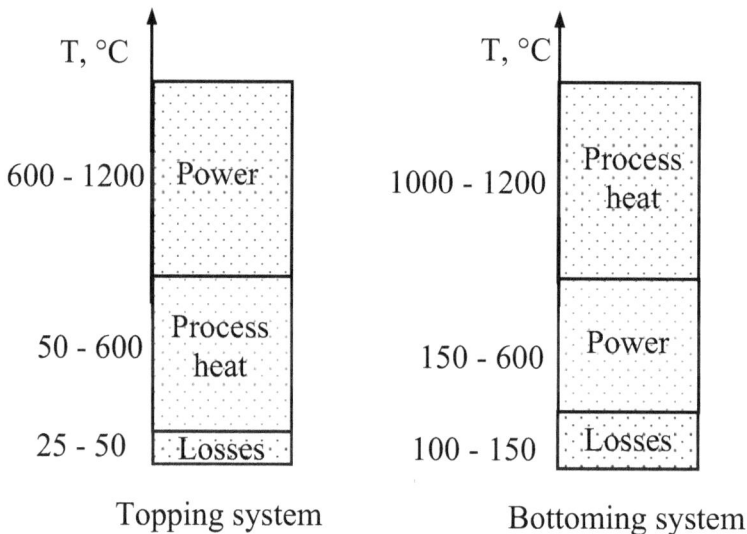

Fig. (1). Captive power through either topping cycle or bottoming cycle operation.

With reference to the position or location of the power cycle in cogeneration, the captive power plants are classified into two *viz.*, topping system and bottoming system. (Fig. **1**) differentiates the topping system and bottoming system with reference to the relative temperature of power and process heat. In a topping system, the high temperature fluid (exhaust gases, steam) drives an engine to produce electricity. In contrast, low temperature heat is used for thermal processes or space heating (or cooling). In a bottoming system, the high temperature heat is first produced for a process (*e.g.*, in a furnace of a steel mill or of glass-works, in a cement kiln). Later the hot gases are used either directly to drive a gas turbine generator if their pressure is adequate, or indirectly to produce steam in a heat recovery boiler, which drives a steam-turbine generator. In the topping system, the fuel is fired mainly to a power plant. Therefore they can not avail the benefit of carbon credits. The bottom system uses waste industrial heat without creating additional emissions, so, these plants can claim the carbon credits.

In this book, waste heat from a cement factory has been selected to develop the power plants. Therefore, the power plant cycles are bottoming systems. The studied power plant cycles are organic Rankine cycle (ORC), organic flash cycle (OFC), Kalina cycle (KC), steam Rankine cycle (SRC), and steam flash cycle (SFC).

The working fluid used in a power plant may be the single fluid system or multi-fluid system. In a single fluid system, a pure substance of working fluid such as water or R123 is used. In a binary fluid system, two working fluids are used together to get the benefit of variable temperature during the phase change. For example, in KC, ammonia and water mixture is used as a working fluid. During the processes of KC, the mixing ratio or concentration of working fluid changes from one state to other state. The mixture of ammonia and water is known as a zeotropic mixture. The zeotropic mixture can be separated by heating and absorbed (mixed) by cooling.

Fig. (**2**) shows the temperature-heat transfer profile of the heat recovery system used in a power plant. In a single fluid plant such as ORC and SRC, the saturation temperature of the working fluid is fixed in the evaporator (Fig. **2a**). The gas temperature is controlled by a constraint called as pinch point (PP). PP ensures the heat transfer from high temperature to low temperature fluid. Approach point (AP) is used between economizer and evaporator to avoid the sudden transition from liquid to vapour. In a binary fluid system, the fluid temperature is variable during the phase change (Fig. **2b**). In KC, the boiling starts at bubble point temperature (BPT) and ends with dew point temperature (DPT) in an evaporator. If the heat recovery is used for steam generation in an SRC or SFC, it is called a heat recovery steam generator (HRSG). If the vapour is generated in a heat exchanger, it is known as a heat recovery vapour generator (HRVG). In ORC, OFC, and KC, the heat exchanger is HRVG. A typical combined cycle power plant use HRSG between the topping cycle (gas power plant) and the bottoming cycle (steam power plant). In the Brayton cycle, the gas turbine inlet temperature is high with the firing of fuel in the gas turbine combustion chamber (GTCC). Therefore, a multi-pressure HRSG is used to suit the high temperature gas. Dual pressure or triple pressure heat recovery is generally used as multi-pressure HRSG in a combined cycle power plant. Fig. (**2c**) shows such a multi-pressure HRSG. The figure shows a dual pressure HRSG where low pressure (LP) and high pressure (HP) sections of heat exchangers are arranged in a sequential order to match the temperature of hot fluid with counter flow arrangement. The counter flow arrangement results in more effectiveness compared to the parallel flow. At low temperature heat recovery, single pressure is sufficient. The multi-pressure heat recovery is complex in nature and suitable for high temperature heat source. The multi-pressure heat recovery is equivalent to binary fluid heat transfer due to temperature glide of the working fluid. Fig. (**2d**) shows the temperature-heat recovery profile of the flash cycle. The flash cycle may be OFC or SFC with organic fluid or steam, respectively. A predetermined amount of liquid at the end of the economizer is used in flasher. In flasher, the high pressure fluid is expanded (throttled or flashed) to a low pressure. The expanded fluid at the low pressure is a mixture of liquid and vapour. The state is nearer to the saturated liquid line. A small quantity of vapour is available for the turbine. A separator is used next to the flasher for vapour collection. The heat recovery is similar the heat recovery used for a pure substance (Fig. **2a**). But the heat load in the economizer is more than the evaporator and superheater due to additional mass in the first part.

ORGANIC RANKINE CYCLE

Majority of power plants at high temperature are combined gas power plant and steam power plant. ORC is suitable to a low temperature source. The rise in price of energy and environmental issues are the major drives to promote energy recovery. The developed countries are successfully extracting the waste heat with ORC. The power through ORC is well established with industrial waste heat recovery (iron, cement and glass), internal combustion (IC) engine, gas turbine, geothermal, solar and biomass supplies.

Very few plants are operating on ORC due to various financial and technical constraints. Currently power plants on ORC are successfully running. Notable cement factories are installed and extended with power generation. Tadipatri, Andhra Pradesh, India is producing 4 MW of power from the waste heat. Thermax India Limited installed 125 kW in Pune with association of Department of Science and Technology (DST). The second plant is a hybrid using solar energy during sunny days and biomass boiler on low radiation. Two more plants are installed purely for experimentation and research purpose by Thermax India limited. ORC is a promising technology for geothermal and biomass but these are not as familiar as solar and wind. There is a huge potential for solar thermal installations and more number of plants running on solar using ORC. Bagasse based plants which are quite common is one more area that ORC can be implemented effectively. A rough estimate has been made and it indicates that 4.4 GW of power can be produced using ORC technology including all sectors. Iron and steel sector potentiality is 148 MW, 35 MW from glass industry, 574 MW of power can be produced from cement sector, 1.4 GW of power can be obtained from solar thermal plants and 2.4 GW of power from biomass.

It becomes difficult to identify the real potential until and unless the total potential of geothermal plants using ORC is assessed properly. It is note worth to mention that if quarter of all the total potential mentioned above can be utilized then the capital cost and environmental impact of building new thermal power plants will be reduced drastically and fossil fuel consumption also reduces. It can be concluded that ORC technology can play a significant role in industrial and renewable energy sector but a detailed analysis has to be carried out in order to know the exact potential of this technology.

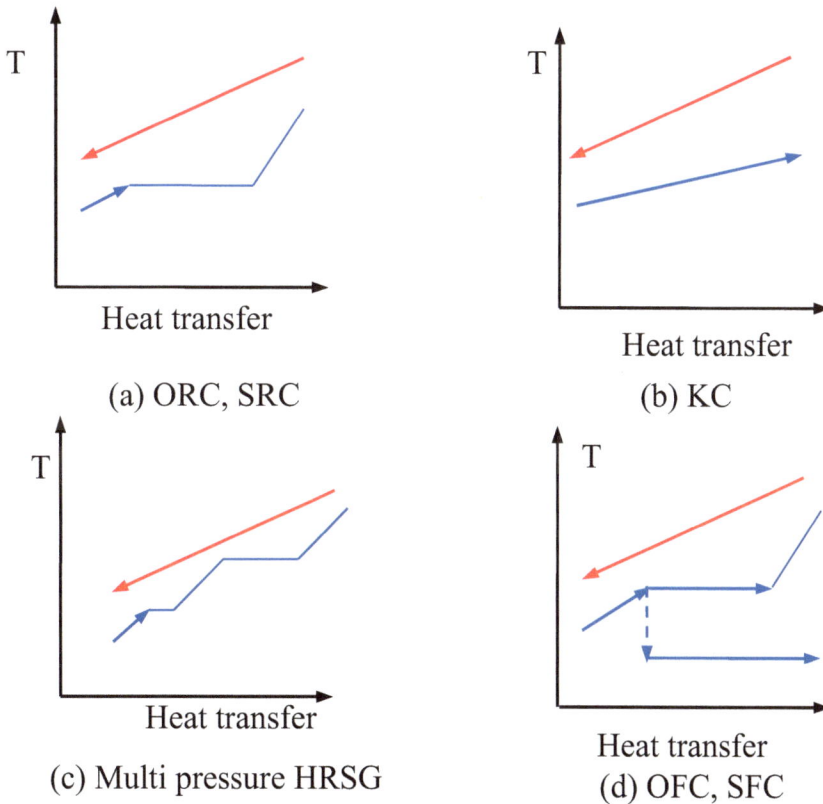

Fig. (2). Heat transfer between hot fluid and cold fluid of heat recovery power plants.

Very few companies and multinational companies (MNC) are working on ORC. Most of the international companies who are expertise in ORC, and many nations are not yet started their full-fledged work at the local level. Different business modules are making the mass scale adoption commercially viable. Working principle of ORC is similar to that of steam Rankine cycle (SRC). The only difference is instead of steam; organic fluid is used. Low boiling point and high vapor pressure of organic fluid makes it suitable for low temperature heat recovery even as low as 65 °C. In ORC, organic fluids such as hydro fluorocarbons, propane, isopentane, isobutene and toluene *etc.* are used. Higher molecular masses of the fluid make it compact, high mass flow rates are possible and turbine efficiencies are in the range of 80-85%. Since the cycle is working at low temperature, the efficiency is low that is having a range of 10-20% depending upon the condenser and evaporator temperature. When compared with SRC (30-40%) the efficiency of ORC (10-20%) is low as SRC works at high temperature. Around 80% of power generated is through SRC which is a matured technology and in near future also it is going to contribute a lot for power production. Frank ofeldt in 1883 used naphtha instead of steam to run the pistons. It triggered the research on organic fluids to run the power engines.

The selection of working fluid plays an important role on performance of thermal power cycle of heat recovery such as Rankine cycle, flash cycle *etc.* The selected fluid should result higher power and thermal efficiency at the chosen source temperature and sink temperature of the power cycle. Isentropic saturation vapour curve or positive slope saturation vapour curve (wet fluid) are suitable to ORC and OFC plants. The negative slope fluid (wet fluid) creates liquid in the turbine and damages the blades. A higher vapour density of the working saves the size of heat exchangers. The low pressure in a condenser increases the volume (low density) and so large size condenser is required which is expensive. The high density fluid save the cost of the equipment and also allows the machines at the lower speeds. This stables the operation of machines such as pumps and turbines. The low viscosity of the fluid in liquid and vapour results high heat transfer and low frictional losses in heat exchangers. Similarly, high thermal conductivity of the fluid also increases the heat transfer coefficient in a heat exchanger. Acceptable vapour pressure is one of the desirable property of the fluid. Too high pressure such as water increases the pumping cost, and other equipment cost. It also increases the complexity and safety measures of the power plant. Therefore, a reasonable pressure to be selected to the plant. The condenser pressure to be positive to avoid the air leakages into the plant. The positive pressure *i.e.* above the atmosphere pressure also avoids the use of vacuum pump. The working fluid should be stable chemically at the high temperature. The high temperature stability permits the plant to operate at higher temperature which favors the performance. The freezing point should be lower than the atmospheric state to avoid the freezing of the working fluid throughout the year. High level of safety in terms of toxicity and flammability is required. The ozone depleting potential (ODP) should be low *i.e.* either zero or close to zero. The fluids having higher ODP are going to phased out by Montreal Protocol. A lower greenhouse warming potential (GWP) is required to the working fluid. GWP is measured with respect to CO_2, as a unity. The cost and availability of the working fluid is one of the important feature to develop the cycle. Currently refrigeration and air conditioning industry is using many organic fluids as working fluids. They are available easily and less expensive in the market.

The main components of ORC are heat exchangers, turbine, generator and pumps. Fig. (**3**) outlines the working principles of simple ORC. Heat recovery from the external source in the form of hot fluid will transfer heat to the organic fluid directly or indirectly to HRVG. The hot organic fluid is impinged on the turbine blades which in turn rotates the turbine shaft. The heat energy is converted to mechanical energy and the mechanical energy is converted into electrical energy in generator. Condenser used to cool the hot organic vapor with the help of water as shown in Fig. (**3**). The hot water from the condenser can be used for process heat which improves the overall efficiency of the plant. The condensed organic fluid is pumped to the boiler and completes the cycle. Thermal efficiency of the ORC can be improved with the internal heater recovery at the exit of vapour turbine, which is called as regenerator.

The various industries that have good potential for waste heat recovery using ORC are cement industry (574.2 MW) followed by iron and steel (148.4MW) and glass industry (35.7MW). A total of 758.3 MW can be produced in India. Nearly 2500 ton per day capacity cement plant can produce 1.6 MW of power using organic Rankine cycle. The present production of cement in India is 327.5 million tons per annum which can produce 574.2 MW of power using ORC technology. Similarly, 500 tons capacity manufacturing glass unit can produce 1 MW power. According to the present production capacity of India a power of 36 MW can be produced. Coming to iron and steel industry 6000 tons per day capacity plant can produce a power of 2.4 MW using ORC. As per the present production scenario of iron and steel in India, a power of 148.4 MW can be generated. The first cement plant in India using ORC technology that was running successfully is Ultratech cement plant, India which is giving 4.8 MW of power. It is 10% of the total power consumption of the plant. This plant registered under CDM (clean development mechanism) and it saved around 80 million India rupee (INR) on the cost of power generation. Due to introduction of ORC in cement plant it decreases the operational cost by minimizing the energy costs.

Based on slope of dry saturated vapour curve on temperature-entropy diagram, the working fluids can be categories into three *viz.* dry, isentropic and wet. The slope of the dry fluids is positive and remains in dry during the expansion in the turbine. Therefore, vapour is not reheated in the power cycle with dry fluids. The slope is zero for isentropic fluids. These fluids also would not demand superheating and reheating of fluid. The slope is negative for wet fluids and needs superheating. The fluid also demands reheating if the power plant is running on high pressure. R123, R113 and R245ca are dry fluids.

Water is the example of wet fluid.

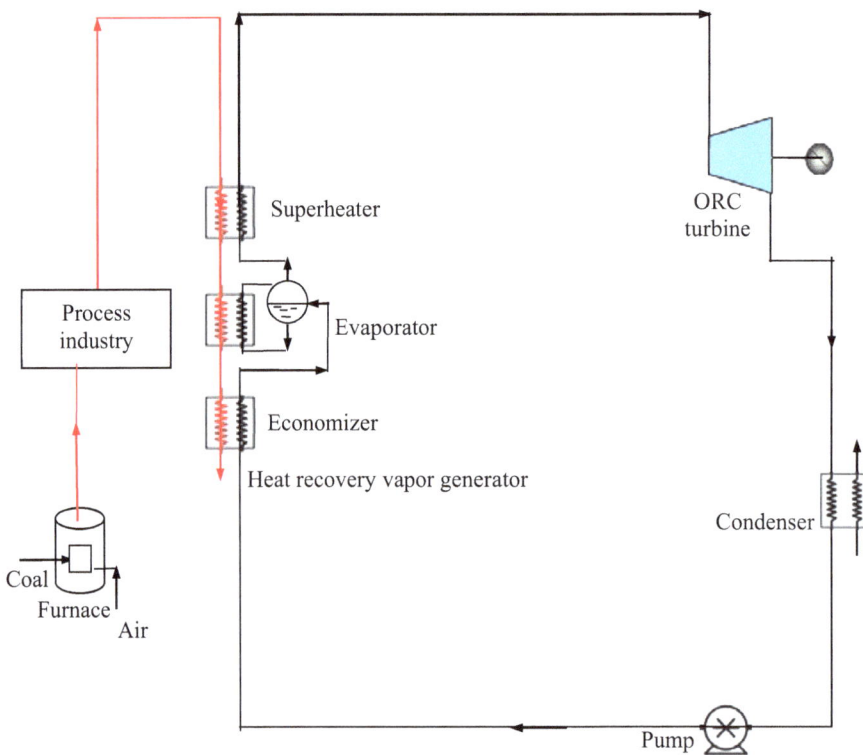

Fig. (3). ORC with heat recovery.

The advantages and disadvantages of ORC are as follows.

The advantages are:

1. The working temperature range varies from as low as 100 °C to as high as 450 °C.

2. Compactness and ORC module standardization makes them easy to install and operate.
3. Size of system ranging from few kW$_e$ to several MW$_e$. This makes them suitable to operate under various thermal sources.
4. Special water treatment plants are not required as in SRC. Simple water treatment is sufficient which reduce the expenses of treated water.
5. The organic fluid molecular mass is more compared with water which rotates the turbine blades at a slower speed. It makes the ORC systems more stable.
6. Long life of the system, more than 20 years is ensured in ORC with a small maintenance. The closed leak proof system ensures no entry of moisture content into the turbine which increases the longevity of turbine blades.
7. Little maintenance is required (erosion of turbine is almost nil) there by low running costs.

The disadvantages are:

1. Low thermal efficiency (10-20%) is the major drawback of ORC as the system is operating at low temperatures. But the overall efficiency can be increased to 95% if hot water after condensation can be utilized for heating and cooling purposes.
2. More stringent measures are to be taken when organic fluids are used because of high flammability and toxicity when compared with steam. But fully concealed systems make them leak proof and flammability chances are almost negligible.
3. Process plants of smaller capacity are not showing interest to implement ORC as the conventional power tariff is not that much higher.
4. This technology is new and only few manufactures are available. Equipment is to be imported. Import and custom duties are making the technology expensive. This is one of the reasons why this technology is not being implemented. Unless proper subsidies and incentives are not declared it becomes difficult to use this technology to the maximum potential.
5. Manufacturer of the equipment is from outside country providing service after sales is costly and the industries are not interested to install the equipment for fear of system break down. So there is a need of large scale capacity building of operators and maintenance technicians.

ORGANIC FLASH CYCLE

Organic flash cycle is a modified version of ORC with added flasher(s). The flashing of fluid is an irreversible throttling process. During the throttling the enthalpy in the process is constant. Therefore, the process is an isenthalpic process. After throttling of high pressure liquid from high pressure to low pressure, liquid and vapour mixture results. The vapour can be separated and used for the power generation in turbine. Therefore, the power generation through the regular processes of preheating in economizer, evaporation and superheating is carried parallel to this flashing process. OFC turbine receives vapour from superheater and flashers. The construction and working of such a simple OFC with single flasher is outlined in Fig. (**4**).

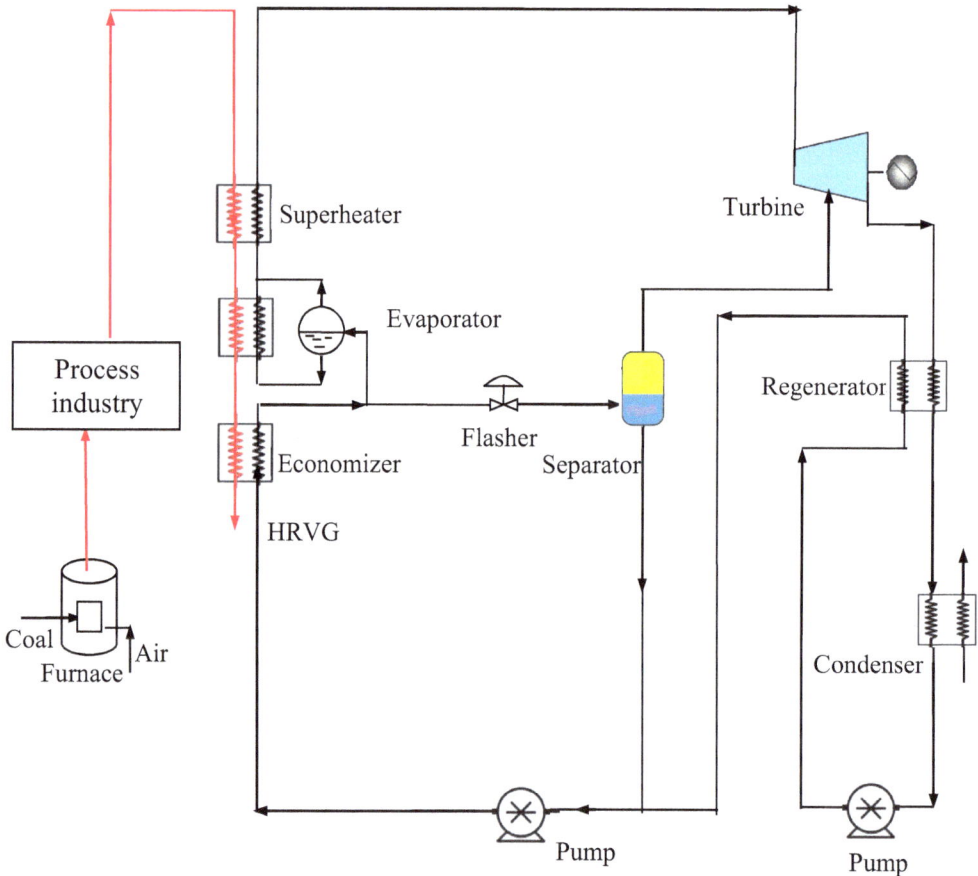

Fig. (4). Organic flash cycle with single flash unit.

KALINA CYCLE

The working of Kalina cycle is similar to a flash cycle. But in KC works on binary fluid system with the advantage of temperature glide with the heat transfer fluid. Similar to OFC, KC also has a regenerator at the exit of turbine. Fig. (**5**) shows the schematic layout of a basic KC with the components of HRVG, separating drum, turbine, absorber and pump.

In basic KC, no regenerator has been shown at the exit of turbine as it consists of main and basic components. The main working principle of KC is absorption (mixing) of two fluids with cooling and separation by heating. In absorber, the vapour absorbs into a weak solution by rejecting heat to the surroundings. The absorber works as a heat sink or condenser.The separation of fluids in heating is the reversal of absorption process. The absorption and separation gives the driving force to the working fluid.The steam Rankine cycle is a well-established technology in power industry that is suitable for high temperatures.

The reversed vapour absorption power cycle is known as vapour absorption refrigeration (VAR) cycle.

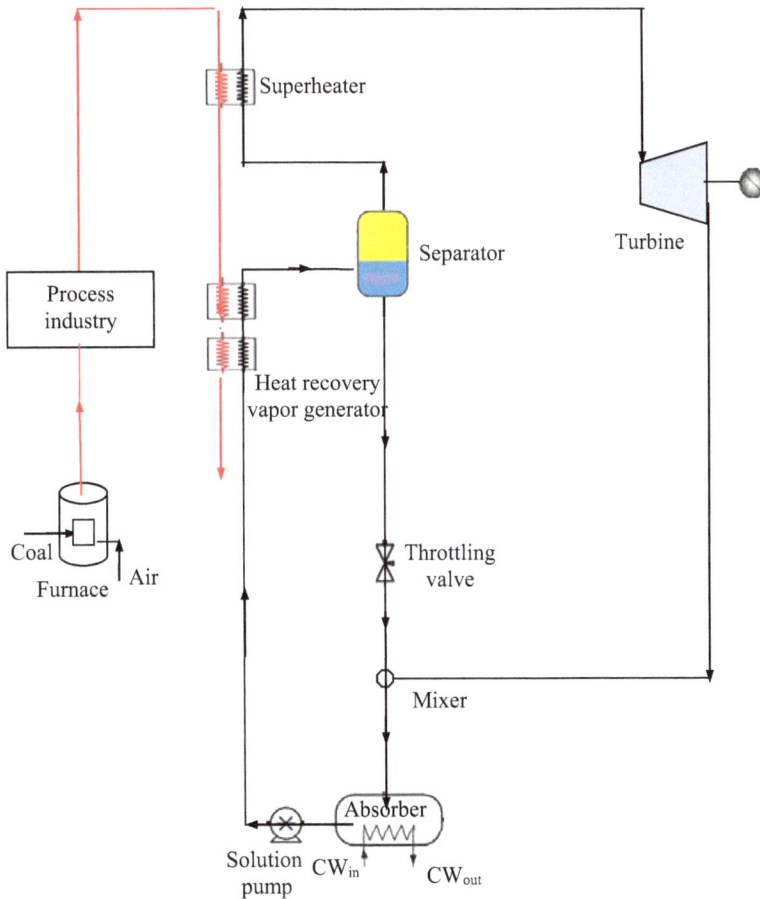

Fig. (5). Basic Kalina cycle operating with heat recovery.

Most of the waste heat recovery systems are designed to run on steam Rankine cycle as shown in Fig. (**5**). If the source temperature exceeds 340 - 370 °C,

Rankine cycle is the best option. The steam from the waste heat boiler is in superheated state. The power from turbine can be used to meet the factory's load and excess can be supplied to grid. Various industrial waste gases that are suitable for SRC are detailed in table **1**.

Table 1. Industrial waste heat for power production with temperature range.

Temperature Range	Source	Temperature, °C	Suitable Power Plant
> 650 °C	Steel electric arc furnace	1370-1650	SRC
	Aluminum furnace	1100-1200	
	Copper refining furnace	760-820	
	Glass melting furnace	1300-1540	
	Coke oven	650-1000	
	Iron cupola	820-980	
	Fume incinerators	650-1430	
	Steel heating furnace	930-1040	
230-650 °C	Gas turbine exhaust	370-450	SRC and ORC
	Cement kiln	450-620	
	Steam boiler exhaust	230-480	
	Reciprocating engine exhaust	320-590	

STEAM RANKINE CYCLE

The challenges that are being faced by using different waste heat gases are corrosive nature, abrasive nature, sticky nature, oily and particulate –laden. We cannot use the exhaust of the various industries directly as it contains sediments, soot, which are difficult to remove. One more constraint is financial burden.

The advantages and disadvantages of SRC are

Advantages

1. Reduction in CO_2 emissions because it uses of waste heat to run plant.
2. Energy efficiency is more.
3. Size of the equipment is reduced.
4. Auxiliary equipment consumes less energy.
5. Fuel used is cheaper when compared with other plants.
6. Generation cost is low.
7. Plant can be installed anywhere if water and fuel are available.

8. It is a matured technology not like other sources solar, nuclear and geothermal. So maintenance of the plant is easy.
9. Space occupation is low compared with hydroelectric plant.

Disadvantages

1. Heat exchangers selection is difficult and exchangers material cost is also high. Because it as to with stand different temperature conditions.
2. Sometimes it may not be economically viable.
3. Cost of the plant will be more.
4. Costly technology is to be used for handling different types of waste gases.
5. Uses large quantity of water.

STEAM FLASH CYCLE

Similar to OFC, SFC also uses flasher(s) to generate the extra vapour for power generation. Since the turbine exit temperature is closed to the sink temperature, the use of regenerator at the exit of turbine is avoided. The construction and working of such a SFC with single flasher is outlined in Fig. (**6**). The working of SFC seems opposite in working of SRC with a deaerator. The deaerator receives the steam from the turbine which favors the efficiency and suppress the power. On other side, SFC turbine receives steam from flasher which favors the power and declines the efficiency.

OBJECTIVES OF THE BOOK

The current book is targeted
- ○ To develop novel configurations to suit the industrial waste heat's temperature.
- ○ To generate the optimum HRSG/HRVG pressure at the given source temperature below and above the critical temperature of fluid.
- ○ To study the performance characteristics of the ORC, OFC, SRC and SFC below and above the critical temperature of the fluid.
- ○ To develop the performance characteristics of KC plant.
- ○ To augment the power by optimizing the process conditions.
- ○ To compare the performance of ORC, OFC, KC, SRC and SFC under the same supply conditions and identify the potential and weakness applicable to heat recovery.
- ○ To conduct the case studies to validate and work the industrial power plant configurations.

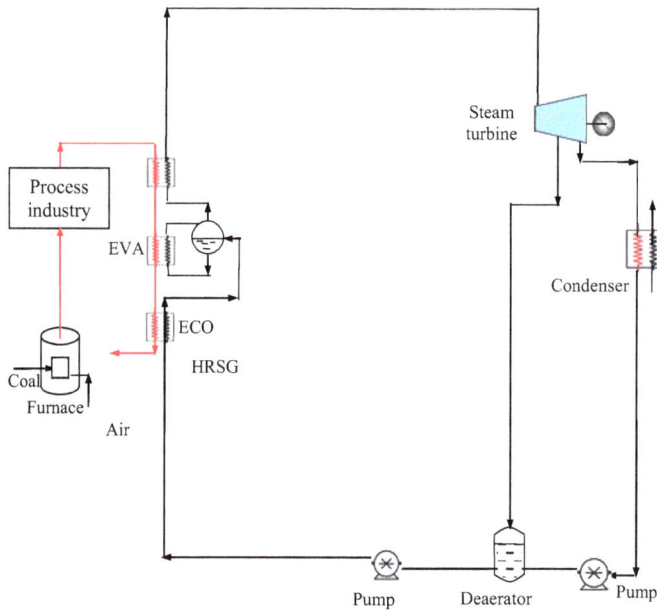

Fig. (6). Steam power plant with a deaerator.

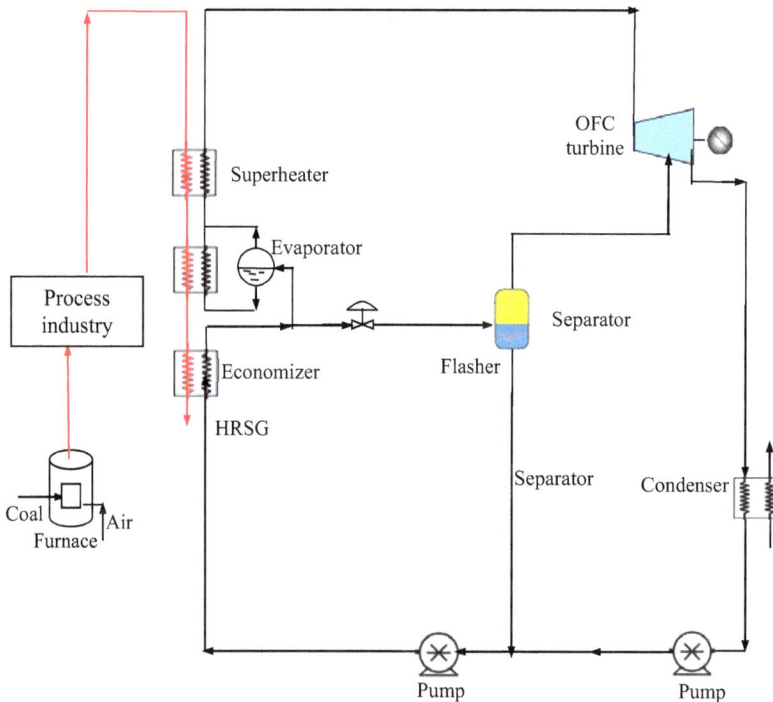

Fig. (7). Steam flash cycle with single flasher.

SCOPE OF THE BOOK

In the current scenario of global emissions, generation of power without producing additional carbon dioxide emissions from the wasting thermal heat has significant weightage in captive power industry. The book dealt with many power plant configurations such as organic Rankine cycle, organic flash cycle, Kalina cycle, steam Rankine cycle and steam flash cycle to tap the waste heat for power production. Out of these thermal power cycles, organic flash cycle and steam flash cycle are not well reported in the literature. Kalina cycle is an advanced thermal power cycle suitable to heat recovery and has potential for cooling cogeneration cycle. This book gives a good reference to the students, researchers and engineering to practice the advanced thermal power cycles for the futuristic implementations and developments.

The waste heat from the cement factory has been selected to generate the power with suitable power plant technology. A novel method of power generation especially suitable to waste heat has been configured and studied to highlight its potential benefits. Nontraditional plants such as liquid flashing into liquid and vapour are compared with the ORC and steam power plants. Recommendations are made to choose the suitable configuration and working fluid to augment the power from the heat recovery. Case studies are conducted at cement factories to link the theoretical work with the industries.

Following are the contributions from the book:

- Mathematical models are developed for OFC and SFC which are not reported in the literature.
- Derived and generalized 'n' number of flashers in a SFC to simplify the tedious numerical work.
- Developed the curve fitting equations for the optimum boiler pressure below and above the critical temperature of the fluid.

CHAPTER 2

History of Heat Recovery Power Plants

Abstract: The present scenario of captive power generation has been outlined with the challenges and the future steps to be implemented. The developments of conventional and the latest power plants are presented. The literature is focused on organic Rankine cycle (ORC), organic flash cycle (OFC), Kalina cycle (KC), steam Rankine cycle (SRC), and steam flash cycle (SFC). The research gap is identified and highlighted the scope for future developments.

Keywords: Flash cycle, Kalina cycle, Organic rankine cycle, Thermal power plant, Waste heat recovery.

INTRODUCTION

World energy consumption is increasing drastically and leading to non-bearable level of carbon dioxide in the atmosphere. Environmental organizations are taking actions to measure and control greenhouse gas emissions. To control these greenhouse gases, the buildings and industries may generate their own electricity to decrease the grid load. Obviously, the technology is to be shifted from the use of conventional energy sources to non-conventional energy sources such as solar, biomass, solar thermal, solar photo-voltaic, geothermal and hydro, *etc.* The use of fossil fuels to be avoided for transportation and space heating and similar. The decentralized power section to be improved [1].

Waste heat recovery (WHR) is a heat exchanger used to transfer the heat from a high temperature fluid to low temperature fluid for the purpose of energy conversion, especially power generation. Captive power plants can be operated with the source of renewable energies, industrial waste, or fuel firing. It can be operated on grid connection or off-grid mode. The waste heat may be taped from the industrial hot gas, diesel engine exhaust, hot water from a process industry such as steel cooling, and steam from cooling towers. WHR may be designed as recuperators, regenerators, heat pipe exchangers, thermal wheels, economizers, or heat pumps. In a conventional power plant, one-third of the fuel energy is converted into electricity, and the rest of two-third is wasted and thrown on the earth. It leads to global warming and environmental disorder.

There is a need to develop energy conversion technologies to tap the waste heat at the low temperature range. Industries handling a considerable amount of heat have a great potential for self-generation power through the recovery of waste heat. The production of electricity using this root is the most economical option compared to the direct fuel-fired plant. To meet the growing energy demand, the available best options are waste heat recovery and renewable energy technologies. Renewable energy technologies are costly and need time for complete marketing. Therefore, heat recovery is the immediately available and feasible solution for power generation without much investment and emission. The review has been conducted to understand the development and identify a suitable technology to tap the potential resources with power plant configurations and working fluids. WHR also can be used for refrigeration and air conditioning (A/C) using vapour absorption refrigeration (VAR) principle. The power and cooling cycles can be combined to form a new cycle for power and cooling. Cooling cogeneration cycle is one such example where power and cooling can be generated from a single cycle.

ORGANIC RANKINE CYCLE

The development of the infra-structure field raised the demand for cement factories. From the cement factories, nearly 40% of the energy is in the form of waste heat [2]. Chandra and Palley [3] suggested organic Rankine cycle (ORC) for the low temperature heat recovery and steam Rankine cycle (SRC) for the high temperature heat recovery of the cement factory after the process heat for cement production. Chen *et al.* [4] studied the ORC with 35 working fluids and concluded the suitable fluids, which are isentropic and dry fluids. In isentropic fluids, the saturated vapour line on the temperature-specific entropy diagram is nearly vertical, and its slope is zero. The slope of the dry fluids is positive, *i.e.*, the saturated vapour line bends towards the right. Water is a wet fluid, and its slope is negative. Bao and Zhao [5] recommended radial inward flow turbine to ORC for the highest isentropic efficiency. They also gave the guidelines to select the prime movers to ORC based on the power plant capacity. They recommended radial flow turbine for large size power plants, positive displacement turbine for medium capacity, and scroll expander for small size power plants. Mago *et al.* [6] worked on regenerative ORC and proved the benefits of this cycle over the ORC without a regenerator. Sun and Li [7] analysed the ORC plant's heat supply temperature with a change in working fluids. The working fluid plays an important role in the process conditions and performance of the ORC plant. The selection of working fluid is a big task in ORC design. Frutiger *et al.* [8] studied 15 working fluids and developed a method to select the suitable working fluid for ORC, Table **1** lists the thermal efficiency of few ORC plants with the source temperature and working fluid. The reported thermal efficiencies are proportional to the supply temperature. Since CO_2 and ethanol plant's temperature is high, the thermal efficiencies are also showing more.

ORGANIC FLASH CYCLE

Organic flash cycle (OFC) is the latest cycle and modified version of the ORC. The OFC presented in this book is different from the geothermal flash cycle. In the conventional flash cycle, complete liquid is subjected to the flashing without undergoing to evaporation and superheating. The reported OFC presented in this book consists of evaporator and superheater and more efficiency and power compared to the traditional flash cycle. Dagdas [21] focused on the flash pressures in a geothermal power plant having the working fluid of steam. Lai and Fischer [22] demonstrated the power augmentation of the flash cycle over ORC. Varma and Srinivas [23] highlighted the power augmentation of OFC over the ORC under the same energy supply conditions. Muhammad *et al.* [24] developed ORC with 1 kW capacity and R245fa fluid. The experiment resulted in maximum thermal efficiency of 5.75%, with 77.74% of the maximum expander's isentropic efficiency. Recently Witanowski *et al.* [25] optimized ORC turbine with toluene fluid with a focus on stator and rotor profile variables, rotor blade twist angle, circumferential lean, and axial sweep angles. The total static efficiency of the ORC turbine is gained by 2.8% from 77.8% to 80.6%.

Table 1. Summary of few ORC plant's performances with working fluid and source temperature.

Authors	Heat Source	Temperature, °C	Thermal Efficiency, %
Yamamoto *et al.* [9]	R123	70	11.0
Lei *et al.* [10]	R123	120	8.0
Lee *et al.* [11]	R245fa	115	6.0
Hu *et al.* [12]	R245fa	95	4.0
White and Sayma [13]	R245fa	105	8.0
Wei *et al.* [14]	R245fa	350	9.5
Yamada *et al.* [15]	R245fa	90	2.0
Yamaguchi *et al.* [16]	CO_2	200	25.0
Chen *et al.* [17]	CO_2	140	9.0
Hsieh *et al.* [18]	R218	100	5.5
Galindo *et al.* [19]	Ethanol	245	22.7
Lu *et al.* [20]	Zeotropic mixture	140	10.5

KALINA CYCLE

Compared to ORC, OFC, SRC, and SFC, the Kalina cycle (KC) has the advantage of temperature glide during evaporation and condensation. Kalina [26], a Russian Engineer, has invented the Kalina cycle with ammonia-water as the working fluid (zeotropic mixture) during 1983. Compared to the Rankine cycle, it consists of a separator located after the vapour generator or in some designs generator itself works as a separator. Mirolli *et al.* [27] proposed thermodynamic evaluation for the Kalina plant located at Husavik, Iceland. Mlcak *et al.* [28] reported the operation of a 2 MW Kalina cycle in Iceland. It has been identified that the Kalina cycle and ORC are suitable power generation technologies for decentralized power generation. Ammonia can be used in thermal power installations in a mixture with water without any problems. It is cheap and readily available, has no corrosive effect on iron and its alloys, and is soluble in water in any concentration. Marcuccilli and Zouaghi [29] recommended radial inflow turbines for binary cycles to get maximum isentropic efficiencies. Lolos and Rogdakis [30] correlated the equations for the performance of solar KC for low temperature heat recovery and suggested 130 °C as a maximum cycle temperature. Condenser pressure and source temperature have been focused on low sink temperature. Dejfors *et al.* [31] suggested that higher maximum pressure can improve the binary mixture cycle. Bai [32] developed models for the components involved in the Kalina cycle using the binary mixture as working fluid. Prisyazhniuk [33] provided the possibilities of reducing the fuel consumption and discharge into the environment in a thermal power plant. Bloomquist [34] proposed KC for the integration of geothermal power projects. Wang and Yu [35] invented and adjustable strong solution concentration in KC and showed the extra power and efficiency. But to justify the added cost of setup with the increased performance, he recommended the thermo-economic study. Zhang *et al.* [36] reviewed the research on the Kalina cycle. They concluded that the Kalina cycle results in better performance than the Rankine cycle and ORC based on energy and exergy efficiency. So far, the only successful application of the Kalina cycle is electrical generation from geothermal plants. It has many cycle configurations suitable for various applications such as low temperature, medium temperature, and high temperature. Kalina cycle is not yet used up to its potential. The efficiency of a power or heat pump cycle largely depends on source and sink temperatures. The exergy analysis completes the thermodynamic analysis by highlighting the weak areas in the thermal systems. The real potential of any thermal system can be diagnosed with the second law analysis as it applies the minimum entropy concept. Esen *et al.* [37] conducted an interesting work on ground-coupled heat pump with horizontal ground heat exchangers to improve the exergy efficiency. Similar to ORC, KC also has greater flexibility to integrate as combined cycle, cooling cogeneration *etc.* with the options of waste heat recovery, solar thermal, geothermal source and similar. Ganesh and Srinivas [38] optimized the working fluid concentration at the components to improve the performance of Kalina cycle. They also focused on KC power plant configurations suitable to the medium temperature of source [39]. Srinivas *et al.* [40] summarized the three configurations of KC at three temperatures of heat source and optimized the process conditions. The KC also can modify into Kalina cooling cogeneration cycle to produce the cooling in addition to power [41, 42].

With turbine as prime mover, a limited number of thermodynamic power cycles are developed with a source temperature in the range of 100 – 150 °C. From the literature, the identified power generation cycles are ORC, OFC and KC operating on vapor turbine. The literature search indicates that there is a thorough study and work on basic ORC, regenerative ORC and KC plants for power generation. OFC, suitable to geothermal power plant (total liquid is flashed into vapor and liquid) is evaluated and analyzed thoroughly. But OFC suitable to waste heat recovery (part of preheated liquid is flashed into vapor and liquid and with evaporator and superheater) is not much focused and evaluated. Common conditions are used to compare the performance of different power generation configurations at low temperature heat recovery. Cement factory has great potential to generate the power from its waste flue gases compared to the other sources. Steam power plant is suitable for high temperature heat recovery in cement factory but ORC, OFC and KC are the more suitable configurations for low temperature heat recovery in cement factory. Hot gas from a cement factory is considered as source for operation of ORC, OFC and KC.

STEAM RANKINE CYCLE

Steam Rankine cycle (SRC) is the most popular and reliable thermal power plant cycle. It has plenty of merits but the only issue is no adoptability to low temperature. Khurana *et al.* [43] worked on SRC operated by the waste heat of a cement factory and showed a low payback period which is less than 2 years. Alexis [44] showed the power generation of 2 MW from SRC and also extended to the space heating and cooling using the bled steam. Srinivas *et al.* [45] formulated the heat recovery solutions for single pressure steam plant, duel pressure plant and triple pressure plant. The work is recommended a multi-pressure HRSG to a combined cycle power plant and single pressure HRSG to a low capacity power plant. Kamate and Gangavati [46] conducted exergy evaluation to a SRC in a sugar factory with two turbines, one is back pressure turbine and another is condensing turbine. The analysis resulted optimum turbine inlet pressure and temperature at 61 bar and 475 °C respectively. Karellas *et al.*, [47] applied energy and exergy method to a SRC and ORC suitable to a cement factory and recommended SRC if the hot gas temperature is greater than 310 °C and ORC if the temperature is less than this. Sen *et al.*, [48] analyzed SRC for a cement factory through thermodynamic evaluation with 1.4 kg/s of steam generation and resulted 38.4% of thermal power cycle efficiency. Sharma and Singh [49] studied on SRC which is the bottoming cycle in a combined cycle power plant and main focus has been given on heat recovery steam generator (HRSG). Recommended to select the correct fin height, density, thickness and steam pressure for good performance. Varma and Srinivas [50] developed a power plant configuration with SRC suitable to a cement factory's waste heat. The work concluded the best operational conditions and gave the feasible solution for self-generation of electricity to meet the factory's load.

STEAM FLASH CYCLE

The literature study shows that there no much work reported on steam flash cycle (SFC) with cement factory's waste heat recovery. In steam power plant, the use of feedwater heaters (regeneration) and use of flashers is quite opposite in function. Srinivas *et al.* [51] optimized the use of feedwater heaters in a thermal power plant with a focus on thermal efficiency. Srinivas and Gupta [52] also generalized the complex numerical solution with multiple number of heaters. In feedwater heating system, the steam flows from turbine to feedwater heater. In the flash cycle, the steam moves from the flasher to turbine. Regenerative Rankine cycle supports the thermal efficiency with a loss in power. Flash cycle supports the power with a loss in thermal efficiency. Wang *et al.* [53] briefed about the single flash steam cycle in cement factory and tabulated the material balance results. Pradeep Varma and Srinivas [54] developed the operational conditions of SFC with two flashers suitable to a waste heat recovery.

SUMMARY

Most of the cement factories are working without adoption of power cogeneration. The reported literature focuses on energy conservation and exergy solutions in cement factory. Many working fluids are developed to generate the power from the waste heat recovery at difference temperature levels. The power augmentation option of fluid flashing which is different compared to geothermal power plant has not been analyzed and reported. The literature search indicates that there is a thorough study and work on basic ORC and regenerative ORC. A flash cycle operating with the source of geothermal energy where the total liquid is flashed into vapor and liquid is evaluated and analyzed. But OFC suitable to waste heat recovery in which a part of preheated liquid, flashed into vapor and liquid and having evaporator and superheater is not been focused much.

CHAPTER 3

Basic Thermodynamics of Heat Recovery Power Plants

Abstract: The basics of thermodynamics required to evaluate a thermal power plant have been summarized. The first law of thermodynamics and the second law of thermodynamics are overviewed to solve the power plant in view of energy analysis and exergy analysis. The chemical reactions with solutions are explained to understand the solid fuel firing in a typical furnace in a power plant.

Keywords: Combustion of fuels, Exergy, First law of thermodynamics, Second law of thermodynamics.

INTRODUCTION

Thermodynamics is the science of energy conversion with framed laws and regulations. In thermodynamics, heat and work are the focused energies. It plays a key role in planning and organizing a thermal system before its making. It also develops the optimum process conditions for the efficient operation of a thermal power plant. Nature has a tremendous amount of energy. The energy always tries to convert from one form to another form. Thermodynamics deals with these energy interactions in a systematic way with reference to certain rules called thermodynamic laws. Sadi Carnot, the father of thermodynamics, developed a benchmark heat engine called Carnot engine and made a goal for a thermal power plant. He also developed many theories that are more relevant to a thermal power plant. The complete thermodynamics study includes four 'E's *viz.* energy, exergy, economics, and environment. The thermodynamics laws *viz.* zero, first, second, and third are designed based on logic and common sense.

THERMODYNAMIC SYSTEM

Thermodynamic system is a prescribed region with finite matter, confined by walls which separate it from the surrondings. A typical thermal power plant consists of four thermodynamic systems *viz.* turbine, condenser, pump, and boiler. The feedwater is heated with a heat source and turned into superheated steam. In the condenser, the vapour is condensed into a saturated liquid state by air circulation or water circulation. The power plant handles various fluid lines such as fuel, air, cooling oil, steam, circulating water, feedwater, and hot gas. Similarly, the systems also involve heat and work transfers. Therefore, the thermodynamic system can be described with mass and energy transactions. To understand the nature of system, it is required to define the terminology used in the system, and they are surroundings, boundary, control surface, control volume, *etc.* The space outside the system is called as surroundings. Boundary is the enclosure that separates the system from the surroundings. The boundary may be real or imaginary. It is a stationary boundary or moving boundary. The system and its surroundings together are called as the universe. Thermodynamic systems can be grouped into an open system, closed system, and isolated system.

In an open system, the mass and energy cross the boundary. In this system, the fixed region in space is the control volume, and the surface of the control volume is called as the control surface. For example, in a steam boiler, feedwater enters into the system and leaves as a superheated steam (mass transfer) by absorbing heat (energy interaction). Compresser, turbine, nozzle, diffuser, steam engine, boiler, *etc.*, are the open systems. If a system allows mass without energy transfer, such as steam flowing in an insulated pipe, that is also an open system. In this case, even though there are no energies crossing the boundary, because of insulated pipe, the fluid carries energy along with the flow, which is called kinetic energy. In addition to this kinetic energy, it also possesses flow work. The frictional resistance in the insulated pipe drops the fluid velocity. So within the control volume, the energy exchange occurs without crossing the control surface or boundary. Therefore a system with mass flow, but without work and heat flow, can be treated as an open system. A system is called a closed system if it does not allow the matter to enter or leave and the energy (heat and work) across its boundary. Examples are gas enclosed in a cylinder, water stored in a container, electronic device, *etc.* In an isolated system, neither mass nor energy transfers across its boundary. Examples of an isolated system are thermos flask and universe. The system and surroundings together form an isolated system.

THERMODYNAMIC PROPERTY

Thermodynamic properties of a system are the measurable characteristics describing the system's nature. It is independent of the nature of the process and depends only on the state or condition. Therefore, the property is a point function. They are classified into two types, *viz.* intensive properties and extensive properties. If the value of the property is independent of the mass of the system, it is an intensive property. It is qualitative in nature. Ex: pressure, temperature, density, velocity, height, viscosity, specific property. The pressure is defined as the force exerted normal to a unit area of the boundary. From the continuum point of view, the pressure at a point is the force per unit area in the limit, where area tends to be very small, *i.e.*, approaches to zero.

Pressure as a result of depth of fluid = $\rho g h$

$$P = \frac{w}{A} = \frac{\rho V g}{A} = \rho g h \qquad (1)$$

Atmospheric pressure is the pressure exerted by the weight of the atmospheric air.

P_{atm} = 760 mm of Hg = 101.325 kN/m^2 = 1.01325 bar =1.01325 × 10^5 Pa = 0.101 MPa = 1.113 kg f/cm^2.

The gauge pressure is measured by the instrument (gauge) such as Bourdon, and the manometer is called the gauge pressure.

The absolute pressure measured from the absolute zero pressure is called absolute pressure. Absolute zero occurs when the molecular momentum of fluid is zero. It happens with the perfect vacuum. If the pressure of fluid is less than the atmospheric pressure, it is known as vacuum, rarefaction, or negative pressure. The pressure measured by the instrument does not consider the velocity of the fluid. The gauge pressure is the static pressure. Pressure due to velocity of fluid = $\rho c^2/2$, N/M^2 . The kinetic head expressed in force per unit area is called dynamic pressure. $KE = 1/2\ mc^2$, $j(or\ Nm)$ and therefore:

$$Kinetic\ pressure = \frac{1}{2}\frac{mc^2}{V} = \frac{\rho c^2}{2},\ \ N/m^2 \qquad (2)$$

Total or stagnation pressure is the sum of static pressure and dynamic pressure. The average density of a system is the ratio of its total mass to its total volume. If the value of the property depends upon the mass of the system, it is known as an extensive property. It is a quantitative property. Ex: volume, surface area, internal energy, P.E. and K.E.

There are five basic or primary thermodynamic properties *viz.* temperature, pressure, volume, entropy, and internal energy. Temperature is a thermal potential and measure of relative hotness or coldness of a system. Pressure is a mechanical potential, normal force per unit area. Volume is the mechanical displacement and the quantity of space possessed by a system. Entropy is the thermal displacement and the quantity of disorder possessed by a system. The kinetic and potential energies of its constituents (atoms and molecules, usually) is the internal energy of a system. The energy possessed by a closed system is internal energy. The secondary properties are enthalpy and specific heat. The total of internal energy and product of pressure-volume is enthalpy. It is the total energy of a system. The energy possessed by an open system is enthalpy. In the open system, the fluid flows with velocity from inlet to outlet. Some drive or work is required for the flow of the working fluid. The energy required for the flow of fluid from inlet to exit of an open system is the flow work. Therefore, in an open system the energy comprises the internal energy with added flow work. Specific heat capacity, c_p or c_v are the energies required to increase the temperature of unit mass of system by one degree.

THERMODYNAMIC STATE, PROCESS AND CYCLE

The combination of states is called a thermodynamic process, and the combination of processes with a closed-loop is called a thermodynamic cycle. A power plant consists of many components and processes with mass and energy interactions. The system may be open or closed for mass transfer. Sometimes the system work as an open system as well as closed system intermittently. For example, in the fuel firing of an engine system, the inlet valve and out valves remain closed and the system works as a closed system. The same engine during suction and exhaust, the inlet and out valves open respectively and open system applies. Thermodynamic cycle consists of states and processes with direction. In a typical thermal power cycle, the cycle is represented in a clock wise direction. In a refrigeration cycle, the same or different cycle may in anti-clock direction. The thermodynamic cycle shows information of the material flow diagram and the processes. P-v, T-s and P-h diagrams are some of the popular diagrams used to represent the cycle. The thermodynamic state can be defined as a condition defined by the properties. Broadly speaking state is the condition of a system at an instant of time described by its properties. A thermodynamic system passing through a series of states constitutes a path. A process is defined as a transition in which a system changes from one initial state to a final state. In a thermodynamic cycle, the system undergoes through a series of processes from one state to other and finally reaches to the initial state. The properties at the beginning and end of the cycle are same.

The processes can be categorized as quasi-static process, reversible process and irreversible process. Process gives information about initial state and final state only. A quasi-static process is one in which the system deviates from one equilibrium state by infinitesimal amounts throughout the entire process. In other words, a process closely approximating to a succession of equilibrium states is known as quasi-static process. A reversible process passes through a series of equilibrium states, which can be restored back by reversal of energy interactions.

It occurs with infinite slow pace without any dissipative effects such as friction, heat transfer with temperature difference, *etc.* Practically all the processes occur with the dissipative effects. A reversible process is a bench mark process to find the degree of deviation from the ideal process. The concept of reversibility can be used to evaluate the effectiveness of a real process to modify or improve the system conditions. The heat will flow with temperature difference and the rotating elements will move with friction. A reversible process can be shown with a continuous line or curve. The reversible processes are motion of bodies without friction, slow expansion or compression without friction, slow isothermal compression/expansion of a gas, electrolysis of water, flow of electric current through inductors and capacitors. Reversible process occurs with infinitely slow manner. It is a slow process and closed to the reversible process. In a fast process more entropy is generated and is an irreversible process.

An irreversible process passes through non-equilibrium states, which cannot be restored back by reversing the energy interactions between system and surroundings. The complete path of irreversible is not defined due to nonexistence of unique state. Actually, all the thermodynamic processes are irreversible. For the sake of simplicity in thermodynamic evaluation and bench mark comparison the reversible processes are introduced. The human life and age is irreversible as the childhood cannot be get back once it is finished. We cannot avoid the friction and potential difference (temperature, pressure, voltage, height, concentration *etc.*) in a process. The degree of irreversibility can be estimated by comparing with the reversible process. An irreversible process is expressed by a dotted line. But to gain the clarity in thermodynamic study, irreversible processes also draw in continuous lines or curves but with definition of degree of irreversibility or conditions. The irreversibilities may be either internal or external. The internal irreversibilities happen within the system with friction and some dissipative effects. The internal irreversibilities are happens in the viscous flow dissipation, free expansion, throttling, mixing, separation, magnetization, hysteresis, heat generation with current resistance *etc.* The external irreversibilities are the heat transfer through a finite temperature differences between hot fluid and cold fluid such as boiler and condenser.

FIRST LAW OF THERMODYNAMICS

Thermodynamics obeys certain rules and regulations, called as thermodynamic laws. These laws provide guide lines to solve, analyze and refine the systems. The first law of thermodynamic considers the energy interactions in a system and gives the base for energy balance. The performance can be solved using first law of thermodynamics.

First law of thermodynamics focuses on energy. It says that energy can be converted from one from to another. It can be destroyed or created and just converts only. So the energy of universe is constant. The first law of thermodynamics provides the base for measurement of energy through the properties. But it has certain limitations as it is not in complete form. It is focused on transfer of energy quantities without dealing the quality transformation. The study on energy conversion direction is missing. It will not say that the process is feasible or not. Since the first law explains the energy conversion, it defines the energy conversion efficiency or first law efficiency of a process or cycle. It also called as thermal efficiency which is the ratio of output to input. Joule conducted an experiment to demonstrate the first law of thermodynamics applicable to a cycle. Let a certain amount of work W_{1-2} be done upon the system by the paddle wheel. The quantity of work can be measured by the product of weight and the vertical height through which the weight descends. The work input to the insulated vessel causes a rise in the temperature of the fluid. For the process 1-2,

$$Q + U_1 = W + U_2 \qquad \qquad (3)$$

He conducted the experiment in a water container with insulation. The temperature of water is increased by the stirrer work. The work done on the system, $U_1 + W = U_2$. The temperature of water increases from T_1 to T_2. The work is used to rise the internal energy of the system *i.e.* $U_2 - U_1$. In second step, the insulation is removed and the water bath is inserted into another water container. The heat is transfer to the water and the system gains its original state.

For the process 2-1,

$$U_2 + Q = U_1 \qquad \qquad (4)$$

Since there is no work, the heat rejection is equal to decrease in internal energy *i.e.* $U_2 - U_1$. Now the system under goes two process one is forward the second the backward and completes a cycle. In the forward process, the water temperature is raised and the energy quantity can be determined from the specific heat of water, fluid mass and temperature rise. Joule conducted many such experiments involving different work interaction in a variety of systems. For example, Joule used work in the form of electrical energy, after measuring the heat transfer by the heating coil, completing the cycle by restoring the initial state.

Joule identified that in all the experiments he conducted; same amount of work is spending to produce the same amount of heat. In all the cases 4186 Joule of work is used to rise the temperature of 1 kg of water through 1 °C rise. He proved that the ratio of W/Q is constant.

It has been found that W_{1-2} is proportional to Q_{1-2}.

$$\oint dW \alpha \oint dQ \qquad (5)$$

$$\oint dW = J \oint dQ \qquad (6)$$

In SI units, J = 1 Nm/J and in MKS units, J = 427 kgfm/kcal. Or it is equal to 4.18 kJ/kcal.

Therefore, first law of thermodynamics says, 'when a system executes a cyclic process, the algebraic sum of work transfers is proportional to algebraic sum of heat transfers'.

$$\text{For SI units, } \oint dQ - \oint dW = 0, \Rightarrow \oint (dQ - dW) = 0 \qquad (7)$$

SECOND LAW OF THERMODYNAMICS

The first law of thermodynamics is essential for the fundamental energy study and performance evaluation. It describes the energy conversions but with few limitations. To overcome these difficulties and get the completeness of energy conversion study, one more dimension to the thermodynamics is required, which is entropy. The entropy (randomness) of system with primary dimensions of pressure, volume and temperature defines the state completely. The second law of thermodynamics adds clarity to energy conversion and makes a significant study. It adds a proper direction to process with the feasibility. The first law of thermodynamics results energy analysis and the second law gives exergy (maximum potential) analysis. Currently many researchers are focusing on second law analysis to refine the system through minimization of entropy generation. Actually the first law together with second law completes the thermodynamic study.

The occurrence of a spontaneous process is due to the finite potential difference. For example, water flows from higher altitude to lower altitude, mass with concentration difference, heat with temperature difference and so on. The reverse of these spontaneous process never happens unless the supply of external agency. The second law limits the direction controls the occurrence of process.

The first law of thermodynamic does not address the following:

1. The direction of heat flow with temperature.
2. The limitations in energy conversion are not described (Practically, it is not

possible to convert all the heat into work).
3. The feasibility of the process is not cleared.

The effectiveness of process is not described in first of thermodynamics, which has significance importance in the design of thermal system. The imperfection of thermodynamic process can be completely scanned using second law of thermodynamics which is not possible with first law. Therefore, second law can be used in the refinement of the thermodynamic system.

In a thermodynamic cycle, all the heat addition cannot be converted into equal amount of work. It shows that it is not possible to convert the low grade energy completely into high grade energy. Some portion of heat must be rejected to the surroundings. Therefore, work is said to be high grade energy and heat is the low grade energy. For example, heat has been supplied to a thermal power plant by burning fuel. This heat is used to generate the steam and the turbine coverts the heat energy into work by expanding the steam. But it is not equal to the heat supply as some heat has been rejected in the condenser. The heat rejection from the plant or the engine is unavoidable phenomena. On other hand if water in an insulated container is heated by an electrical heater, the electrical energy completely converts into heat energy assuming minor electrical losses. Since electricity is high grade energy, the complete conversion into low grade energy is possible. It can be concluded that for low grade applications, use of low grade energy is better than the high grade as it is more expensive. For example, for hot water generation, direct fuel firing or solar thermal energy is a better option compare to electricity use. Kelvin and Max Plank are the two scientists defined the second law of thermodynamics based on heat engine working. The Kelvin-Plank (KP), states that it is impossible to construct a heat engine that executes heat with single TER. If an engine working with single TER, that engine is called perpetual motion machine of second kind (PMM 2). Therefore, PMM 2 is impossible. Clausius conceptualized the second law of thermodynamics based on the heat pump working. As per the Clausius statement, it is impossible to construct a heat pump or refrigerator that removes the heat from a body at lower temperature to a body at high temperature without using work.

CARNOT CYCLE

A standard heat engine or a heat pump is required to estimate the maximum gain from thermal machines. Carnot machine is such an imaginary or hypothetical machine shown as a master piece. Carnot cycle is a reversible cycle consists of four processes as shown in Fig. (**1**).

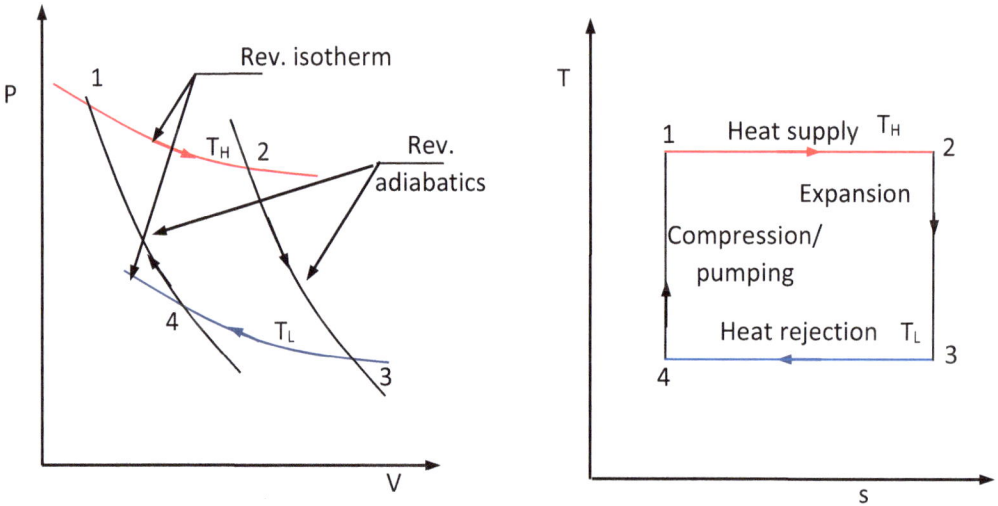

Fig. (1). Representation of Carnot engine on (a) P-V diagram and (b) T-s diagram.

1) Reversible isothermal process (1-2)

Heat is added at constant temperature process.

$$Q_1 = (U_2 - U_1) + W_{1-2} = W_{1-2} \quad \text{(since } U_1 = U_2\text{)} \qquad \textbf{(8)}$$

$$Q_1 = W_{1-2} = mRT_H \ln\left(\frac{V_2}{V_1}\right) \qquad \textbf{(9)}$$

2) Isentropic expansion process (2-3)

$0 = (U_3 - U_2) + W_{2-3}$

3) Isothermal process (3-4)

Heat is rejected at isothermal process

4) Isentropic compression process (4-1)

$0 = (U_1 - U_4) - W_{4-1}$

For Carnot power cycle,

$$\eta_{carnotcycle} = \frac{W_{net}}{Q_1} = \frac{Q_1 - Q_2}{Q_1} = 1 - \frac{Q_2}{Q_1} = 1 - \frac{mRT_L \ln\left(\dfrac{V_3}{V_4}\right)}{mRT_H \ln\left(\dfrac{V_2}{V_1}\right)} = 1 - \frac{T_L}{T_H} \qquad (10)$$

Or,

$$\eta_{carnotcycle} = \frac{W_{net}}{Q_1} = \frac{Q_1 - Q_2}{Q_1} = 1 - \frac{Q_2}{Q_1} = 1 - \frac{T_2(S_3 - S_4)}{T_1(S_2 - S_1)} = 1 - \frac{T_2}{T_1}$$

Similarly, for Carnot refrigeration cycle,

$$COP_R = \frac{Q_2}{W_{net}} = \frac{Q_2}{Q_1 - Q_2} = \frac{T_2(S_3 - S_4)}{T_1(S_2 - S_1) - T_2(S_3 - S_4)} = \frac{T_2}{T_1 - T_2} \qquad (11)$$

ENTROPY

The word entropy was first used by Clausius, taken from the Greek word 'tropee' meaning 'transformation'. Entropy is the fourth property (dimension) in thermodynamics after pressure, volume and temperature. Pressure, volume and temperature are the properties can be measured with the instruments but entropy cannot be measured using instrument and so it a special property. Actually it brings a new look to thermodynamics as it adds the quality to the energy. Entropy is well used to determine or estimate the decay or dissipative effects in the process or cycle to find the degree of reversibility or irreversibility.

The concept of entropy is developed with the statement that two reversible adiabatic paths cannot intersect each other. Assume reverse of this statement. It results a cycle having network, producing by exchanging heat with a single reservoir in isotherm. It is PMM 2. Therefore, two reversible adiabatic paths cannot intersect each other. Therefore, two or more reversible adiabatic processes cannot have a common state. Two constant temperature processes cannot have a common state because of property of matter temperature remaining constant. There must be a property of matter that remains constant during reversible adiabatic process. This new property is called as entropy.

The entropy change can be determined from the P, V and T as follows:

$$S_2 - S_1 = mc_v \ln \frac{T_2}{T_1} + mR \ln \frac{V_2}{V_1} = mc_P \ln \frac{T_2}{T_1} - mR \ln \frac{P_2}{P_1} = mc_v \ln \frac{P_2}{P_1} + mc_p \ln \frac{V_2}{V_1} \quad (12)$$

COMBUSTION OF FUELS

Reactive mixtures involve chemical reaction and this process changes the gas composition before and after the reaction. The difference between the non-reactive mixtures and reactive mixtures is the chemical reaction. In non-reactive mixtures, there is no any chemical reaction or process. This knowledge can be applied in power plant engineering, automobile engineering, aeronautical and similar. A system is said to be in thermodynamic equilibrium if, it is isolated from its surroundings, there would be no macroscopically observable changes. An important requirement for equilibrium is that the temperature be uniform throughout the system or each part of the system in thermal contact. If this condition were not met, spontaneous heat transfer from one location to another could occur when the system was isolated. There must also be no unbalanced forces between parts of the system. These conditions ensure that the system is in thermal and mechanical equilibrium, but there is still the possibility that complete equilibrium does not exist. A process might occur involving a chemical reaction, a transfer of mass between phases, or both. In combustion reactions, rapid oxidation of combustible elements of the fuel results in energy release as combustion products are formed. The three major combustible chemical elements in most common fuels are carbon, hydrogen, and sulfur. Although sulfur is usually a relatively unimportant contributor to the energy released, it can be a significant cause of pollution and corrosion.

Hydrocarbon fuels may be solids, liquids or gases. Liquid hydrocarbon fuels are commonly derived from crude oil through distillation and cracking processes. Examples are gasoline, diesel fuel, kerosene, and other types of fuel oils. The compositions of liquid fuels are commonly given in terms of mass fractions. For simplicity in combustion calculations, gasoline is often considered to be octane, C_8H_{18}, and diesel fuel is considered to be dodecane, $C_{12}H_{26}$. Gaseous hydrocarbon fuels are obtained from natural gas wells or are produced in certain chemical processes. Natural gas normally consists of several different hydrocarbons, with the major constituent being methane, CH_4. The compositions of gaseous fuels are commonly given in terms of mole fractions. Both gaseous and liquid hydrocarbon fuels can be synthesized from coal, oil shale, and tar sands. For combustion calculations, the solid fuel is usually expressed with an ultimate analysis giving the composition on a mass basis in terms of the relative amounts of chemical elements (carbon, sulfur, hydrogen, nitrogen, oxygen) and ash. A fuel is said to have burned completely if all of the carbon present in the fuel is burned to carbon dioxide, all of the hydrogen is burned to water, and all of the sulfur is burned to sulfur dioxide. In practice, these conditions are usually not fulfilled and combustion is incomplete. The presence of carbon monoxide (CO) in the products indicates incomplete combustion. The products of combustion of actual combustion reactions and the relative amounts of the products can be determined with certainty only by experimental means. Among several devices for the experimental determination of the composition of products of combustion are the Orsat analyzer, gas chromatograph, infrared analyzer, and flame ionization detector. Data from these devices can be used to determine the makeup of the gaseous products of combustion. Analyses are frequently reported on a "dry" basis: mole fractions are determined for all gaseous products as if no water vapor were present. Some experimental procedures give an analysis including the water vapor, however. Since water is formed when hydrocarbon fuels are burned, the mole fraction of water vapor in the gaseous products of combustion can be significant. If the gaseous products of combustion are cooled at constant mixture pressure, the dew point temperature is reached when water vapor begins to condense. Corrosion of duct work, mufflers, and other metal parts can occur when water vapor in the combustion products condenses. Oxygen is required in every combustion reaction. Pure oxygen is used only in special applications such as cutting and welding. In most combustion applications, air provides the needed oxygen. Idealizations are often used in combustion calculations involving air: (1) all components of air other than oxygen (O_2) are lumped with nitrogen (N_2). On a molar basis air is then considered to be 21% oxygen and 79% nitrogen. With this idealization the molar ratio of the nitrogen to the oxygen in combustion air is 3.76; (2) the water vapor present in air may be considered in writing the combustion equation or ignored. In the latter case the combustion air is regarded as dry; (3) additional simplicity results by regarding the nitrogen present in the combustion air as inert. However, if high-enough temperatures are attained, nitrogen can form compounds, often termed NO_X, such as nitric oxide and nitrogen dioxide.

The minimum amount of air that supplies sufficient oxygen for the complete combustion of all the combustible chemical elements is the theoretical, or stoichiometric, amount of air. In practice, the amount of air actually supplied may be greater than or less than the theoretical amount, depending on the application. The amount of air is commonly expressed as the percent of theoretical air or the percent excess (or percent deficiency) of air. The air-fuel ratio and its reciprocal the fuel-air ratio, each of which can be expressed on a mass or molar basis, are other ways that fuel-air mixtures are described. Another is the equivalence ratio: the ratio of the actual fuel-air ratio to the fuel-air ratio for complete combustion with the theoretical amount of air. The reactants form a lean mixture when the equivalence ratio is less than unity and a rich mixture when the ratio is greater than unity.

A combustion process involves the burning of fuel with oxygen or a substance containing oxygen such as air. The fuel and oxidation are called reactants and the constituents resulting from the combustion process are called the products. The combustion of carbon, for example involves the simple reaction.

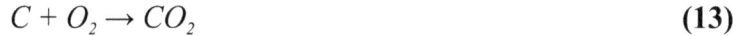

$$C + O_2 \rightarrow CO_2 \tag{13}$$

Let C, H, O, N, S and ash are the results of ultimate analysis of a typical solid fuel.

Conversion of gravimetric to volumetric involves the following steps:

Molar basis,

$$\overline{C} = \frac{C}{12}$$

$$\overline{H} = \frac{H}{1}$$

$$\overline{O} = \frac{O}{16}$$

$$\overline{N} = \frac{N}{14}$$

$$a_1 = \frac{\overline{C}}{\overline{C}}$$

$$a_2 = \frac{\overline{H}}{\overline{C}}$$

$$a_3 = \frac{\overline{O}}{\overline{C}}$$

$$a_4 = \frac{\overline{N}}{\overline{C}}$$

Now the fuel is $C_{a1} H_{a2} O_{a3} N_{a4}$

The combustion reaction is,

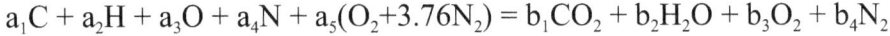

$$a_1C + a_2H + a_3O + a_4N + a_5(O_2+3.76N_2) = b_1CO_2 + b_2H_2O + b_3O_2 + b_4N_2$$

Higher heating value,

$HHV = m_{dry\ fuel}(0.3491C+1.1783H+0.10055S-0.1034O-0.151N-0.0211Ash)1000$, kJ

where *C, H, S, O, N* and ash are in % on mass basis in the dry solid fuel.

Mass of dry fuel per kg mole fuel,

$$m_{dry\ fuel} = a_1 12 + a_2 1 + a_3 16 + a_4 14$$

Enthalpy of formation of fuel,

$$h_{f,\ dry\ fuel} = a_1 h_{fCO2} + \tfrac{1}{2}\ a_2 h_{fH2OL} + HHV$$

The enthalpy of formation of water condensate is considered fuel to higher heating value.

$$a_1C + a_2H + a_3O + a_4N + a_5(O_2+3.76N_2) = b_1CO_2 + b_2H_2O + b_3O_2 + b_4N_2$$

At stoichiometric air fuel supply, the traces of oxygen in products are zero. Therefore, $b_3 = 0$

From the above equation:

C balance: $b_1 = a_1$

H_2 balance: $b_2 = \tfrac{1}{2}\ a_2$

O_2 balance: $\tfrac{1}{2}\ a_3 + a_5 = b_1 + \tfrac{1}{2}\ b_2$

N_2 balance: $b_4 = \tfrac{1}{2}\ a_4 + 3.76\ a_5$

Stoichiometric air fuel ratio = mass of air / mass of fuel

$$= a_5\ (M_{O2} + 3.76\ M_{N2})\ /\ M_{dry\ fuel}$$

Adiabatic flame temperature

$$H_R = H_P$$

Assume adiabatic flame temperature and the iterate the following energy equation to result a final temperature.

$$h_{f,\ fuel} = b_1(h_{f,\ CO2} + h_{CO2,\ T}) + b_2(h_{f,\ H2O} + h_{H2O,\ T}) + b_4\ h_{N2,\ T}$$

It is now fairly easy to see how we may calculate the enthalpies of substances with respect to the zero element state of 298 K and 1 atm. We employ the tables to compute the changes from reference state and add this to the enthalpy of formation.

Thus the total molar enthalpy at temperature T is,

$$\overline{h}_{T,\ total} = \overline{h}_f^{\,0} + \overline{h}_T - \overline{h}_{298} \qquad\qquad (14)$$

Chemical formula of solid fuel, $C_{a1}\ H_{a2}\ O_{a3}\ N_{a4}$

Mass of dry solid fuel per kg mole of fuel, $m_{df} = 12a_1 + a_2 + 16a_3 + 14a_4$

Higher heating value of dry fuel,

$HHV_{df} = m_{df}\ (0.3491C + 1.1783H + 0.10055S - 0.1034O - 0.151N -0.0211Ash)1000$ kJ

where $C,\ H,\ S,\ O,\ N$ and Ash are in % weight basis in the dry solid fuel.

From the definition of HHV, the reactants and products are maintained from and at 25 °C with stoichiometric air fuel ratio.

The energy balance equation to find the enthalpy of formation for dry solid fuel,

$$hf_{fuel\ dry} = a_1\ hf_{CO2} + 0.5a_2\ hf_{H2OL} + HHV_{df}$$

As we have seen before, it is possible to choose different zero levels, or reference states, for tabulations of thermodynamic properties. For the steam tables we choose the saturated liquid state at 0 °C, while for the air tables we choose 0 K and a pressure of 1 atm. For use with above equation tables are employed using a reference temperature of 0 K as well as the standard state temperature of 298 K. The 0 K state is perhaps more common for English units while the 298 K reference is employed for SI tabulations. The modern approach is also to use a reference pressure of 0.1 MPa. instead of 1 atm (0.10132 MPa). This small difference has essentially no effect on the enthalpy or internal energies.

With enthalpy of formation, it is possible to establish a Gibbs function of formation with the zero level set at 298 K and 1 atm for all elements. Values of absolute entropies are set in accordance with the third law of thermodynamics which states that, the entropy of any pure substance in thermodynamic equilibrium approaches zero as the absolute temperature approaches zero. Using this zero level, one may integrate between zero Kelvin and the standard state of 25 °C and 1 atm to obtain the valves of absolute entropy given in table. For those substances which follow ideal-gas behavior, the entropy variation when pressure is given through an application of equation $s_2 - s_1 = \varphi_2 - \varphi_1 - R \ln (P_2/P_1)$ written in terms of molar properties.

$$\bar{s}_2 - \bar{s}_1 = \bar{\phi}_2 - \bar{\phi}_1 - \bar{R} \ln\left(\frac{P_2}{P_1}\right) \tag{15}$$

For ideal gases the absolute entropy at any pressure and temperature can be written as:

$$\bar{s}^0(P,T) = \bar{s}^0 + \bar{\phi}_T - \bar{\phi}_{298} - \bar{R} \ln\left(\frac{P}{P_0}\right) \tag{16}$$

where is taken from table and P_0 is the reference pressure of 1 atm. Absolute values for entropy are required for certain second law analyses of combustion reactions. The important point to remember is that reference states must be carefully specified when dealing with combustion reactions.

The energy liberated in a chemical reaction is given by the following relations.

For constant volume reactions,

$$Q_V = (\Delta U)_V = (U_P - U_R)_V = \bar{u}_{RP} \tag{17}$$

and for constant pressure,

$$Q_P = (\Delta H)_P = (H_P - H_R)_P = \bar{h}_{RP} \tag{18}$$

The quantities given in the above two equations are called the heats of reaction or the internal energy and enthalpy of combustion. If the heats are expressed per mole of fuel, then the symbols and are appropriately used to designate the energies indicated in the equation. The term heating value for a combustion process is in wide use and is a synonym for heat of reaction. When a hydrocarbon fuel is burned in a combustion process, water will appear in the products. The maximum energy release will be obtained when all the water in the products due to combustion is in the liquid state, in such cases the heat of reaction is called a higher heating value. Similarly, a lower heat of reaction will be experienced when all the water in the products due to combustion is in the vapor state, in this case we say that a lower heating value is obtained. In reality, both liquid and vapour will be present in many combustion processes, so that a heating value between the HHV and LHV can be observed.

For dry organic substances contained in solid fuels consisting of *C, H, O* and *N* with a mass ratio of oxygen to carbon less than 0.667, the following expression is obtained in terms of mass ratios [55].

$$\phi_{dry} = 1.0437 + 0.1882 \frac{h}{c} + 0.0610 \frac{o}{c} + 0.0404 \frac{n}{c} \tag{19}$$

where *c, h, o* and *n* are the mass fractions of *C, H, O* and *N* respectively.

For solid fuels with the mass ratio 2.67 > *o/c* > 0.667,

$$\phi_{dry} = \frac{1.0438 + 0.1882 \frac{h}{c} - 0.2509 \left(1 + 0.7256 \frac{h}{c}\right) + 0.0383 \frac{n}{c}}{1 - 0.3035 \frac{o}{c}} \tag{20}$$

For moist fuel, standard chemical exergy, ε

$$\varepsilon^0 \; (kJ/kg) = [LHV^0 \; (kJ \; kg^{-1}) + 2442 \; w] \; \phi_{dry} + 9417 \; S \tag{21}$$

where *w* = mass fraction of moisture in the fuel

h_{fg}^o = 2442 kJ kg^{-1} steam = enthalpy of evaporation of H_2O at standard temperature, T_0

For liquid fuels the effect of sulphur was included in the correlation giving the expression:

$$\phi_{dry} = 1.0401 + 0.1728 \frac{h}{c} + 0.0432 \frac{o}{c} + 0.2169 \frac{s}{c} \left(1 - 2.0628 \frac{h}{c}\right) \tag{22}$$

The accuracy of this expression is estimated to be ± 0.38%.

The chemical exergy of a fuel consisting of a known mixture of gaseous components. However, as the composition of common gaseous fuels varies within relatively narrow limits, once the value of φ has been calculated for a fuel of a typical composition it can be used with reasonable accuracy in other cases. The φ can have values smaller or greater than unity and for most industrial fuels except for peat and wood this value lies within a few percent of unity.

The exergy efficiency is defined as the ratio of maximum obtainable work output from the plant to availability of fuel.

$$\text{Exergy efficiency of plant}, \eta_2 = \left(\frac{\varepsilon^0_{fuel} - i_{total}}{\varepsilon^0_{fuel}} \right) \times 100 \tag{23}$$

Table 3. Typical values of φ for some industrial fuels and other combustible substances.

S. No.	Fuel	$\varphi = \varepsilon^0/(NCV)^0$
1	Coke	1.05
2	Different type of coal	1.06-1.10
3	Peat	1.16
4	Wood	1.15-1.30
5	Different fuel oils and petrol	1.04-1.08
6	Natural gas	1.04 ± 0.5%
7	Coal gas	1.00 ± 1%
8	Blast furnace gas	0.98 ± 1%
9	Hydrogen	0.985
10	Carbon monoxide	0.973
11	Sulphur (rhombic)	2.017

Table 4. Enthalpy of formation and chemical exergy of elements at 298.15 K and 1 atm.

Enthalpy of Formation, kJ/kg mol		Standard Chemical Exergy, kJ/kg mol
-110541		275430
-393546		20140
-241845		11710
-285830		3120
0		3970
0		720
-33098		56220
0		238490
-74873		836510

(Table 4) cont.....

	Enthalpy of Formation, kJ/kg mol		Standard Chemical Exergy, kJ/kg mol
	-83.8E3		1504360
	-104.7E3		2163190
	-134.2E3		2818930
	-125.6E3		2818930
	-179E3		3477050
	-146.8E3		3477050
	-167.2E3		4134590

SUMMARY

The basic thermodynamic and combustion formulation and modeling theories are briefed. Thermodynamics related to power plant's solutions are overviewed to understand the modeling and simulation procedures presented in the subsequent chapters of this book.

CHAPTER 4

Organic Rankine Cycle

Abstract: Basic organic Rankine cycle (ORC) consists of vapor generator (boiler), turbine, condenser and pump. The vapor is generated from the waste heat recovery which is also called as heat recovery vapor generator (HRVG) But due to possibility of internal heat recovery, additionally a regenerator is used between turbine and condenser for internal heat transfer. This chapter highlights the advantage of regenerator over the basic ORC. The performance characteristics of ORC with R123, R124, 134a, R245fa, R717 and R407C are developed. The correlation equations are developed to find the optimum boiler pressure for ORC. The performance has been analyzed with a source temperature below and above the critical temperature of the fluid. The performance characteristics and specifications of ORC with these working fluids are compared.

Keywords: Low temperature heat source, Organic rankine cycle, Regenerator, Thermal power cycle, Working fluid.

INTRODUCTION

A fluid containing carbon based compounds is called as organic fluid. The Organic Rankine Cycle (ORC) uses a fluid with high molecular weight and low boiling point. Organic fluid undergoes chemical deterioration at high temperatures. Therefore, ORC are limited to the low temperature heat sources. The heat rejection from the process industries such as cement factory, steel plant, sponge iron, power plant, kitchen *etc.* is at different temperature levels. In this chapter, low temperature waste heat is used to convert into electricity. The power plant layouts and working fluids are studied. The heat source temperature is changed from 70 °C to 250 °C with the working fluid. The power plant configurations for low temperature heat recovery are ORC, organic flash cycle (OFC) and Kalina cycle (KC). This chapter deals the ORC, its configurations, solutions and performance characteristics. Some working fluids are not suitable to entire range of the temperature. Therefore, from the literature, six working fluids are selected for ORC to study the quantity and quality conversion of waste heat into electricity. Five fluids are single fluids and one is zeotropic mixture. The single fluids are R123, R124 R134a, R245fa and R717. The zeotropic mixture is R407C with fixed concentration. MATLAB computational tool is used for properties development, modeling, processes making, cycle simulation and optimization. The energy balance and exergy balance results are depicted on Sankey diagram and Grassmann diagram respectively. The role of components in cycle is analyzed to operate the plant components for maximum energy conversion. Thermodynamic properties at the optimized conditions are presented on property charts and tables. The plant specifications are developed at these conditions.

ORGANIC RANKINE CYCLE WITH SOURCE TEMPERATURE BELOW THE CRITICAL TEMPERATURE

Fig. (1) shows the power plant layout of a basic ORC. The fluid expands in turbine from state 1 to state 2 and followed by a condensation (2-3). The saturated liquid at state 3 is pumped to HRVG pressure by feed pump (3-4). The pumped fluid (4) is supplied first to economizer (4-5) where sensible heat of the working fluid gains. Later in evaporator (5-6), the latent heat is supplied by the hot gas. Finally, the fluid is superheated (6-1) by hot gas in HRVG. The hot gas temperature is dropped in superheater (10-11), evaporator (11-12) and economizer (12-13) with counter flow arrangement of hot fluid and cold fluid. The condenser is water cooled heat exchanger with circulating water (14-15) as a heat transfer fluid. The hot water from the condenser is cooled in the cooling tower (not shown) for recirculation from cooling tower to condenser. Alternatively, air cooled condenser also can be used in small size plants and water scarcity areas.

In steam Rankine cycle (SRC), the steam condition at the exit of turbine is in wet state and close to the sink temperature. Therefore, the steam is condensed immediately after the turbine in condenser. Fig. (2) shows the temperature-specific entropy plot for the simple ORC. It shows that the exit vapor from the ORC turbine is in superheated state. The exhaust vapor from the turbine is above the saturated temperature. It is to be cooled to condenser temperature and so the condenser load is more with additional sensible heat of in de-superheater. Another side, the condensate needs heating in economizer of HRVG after the pumping. A heat exchanger between these two fluids saves the condenser load and economizer load. Therefore, an internal heat recovery is arranged to transfer the heat from the exhaust vapour to condensate. This heat exchanger is called as regenerator.

Fig. (3) shows the ORC with regenerator. Regenerator is a heat exchanger with a hot fluid and cold fluid are vapor and liquid respectively. The vapor temperature is decreasing from state 2 to state 3 with a cold fluid temperature rise from state 5 to state 6. The capacity of condenser and economizer drops with the association of regenerator in ORC. Regenerator influence on HRSG and thermal efficiency but not on power production. The regenerator shares the economizer's load. It improves the thermal efficiency of the cycle by decreasing heat supply.

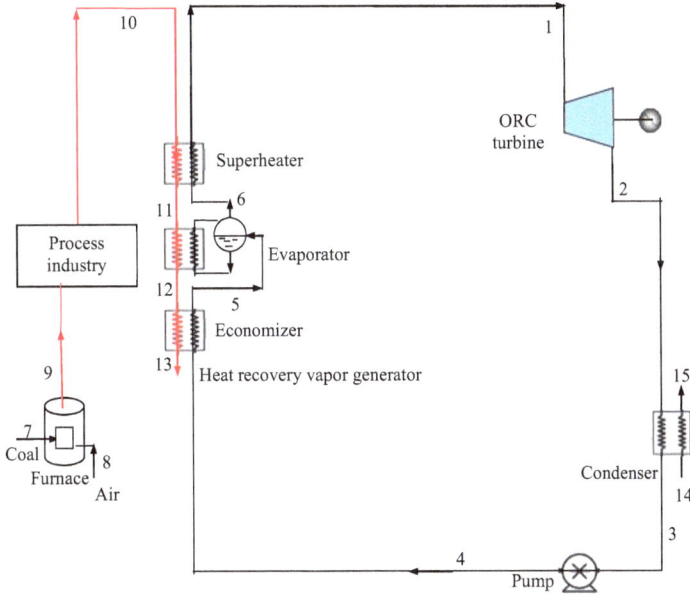

Fig. (1). Basic organic Rankine cycle without regenerator.

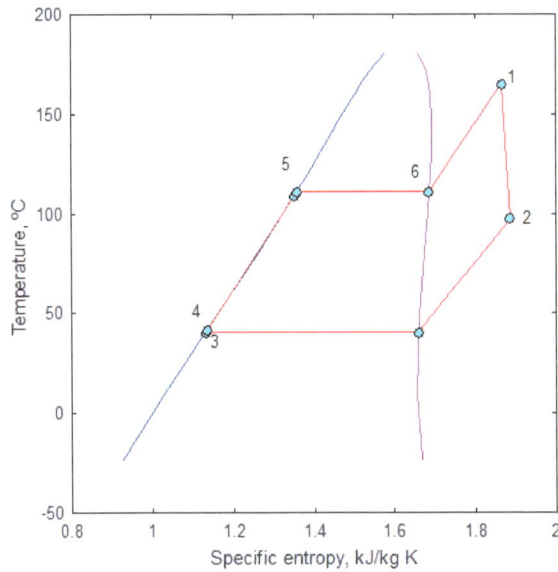

Fig. (2). Temperature-entropy diagram of a basic ORC without regenerator.

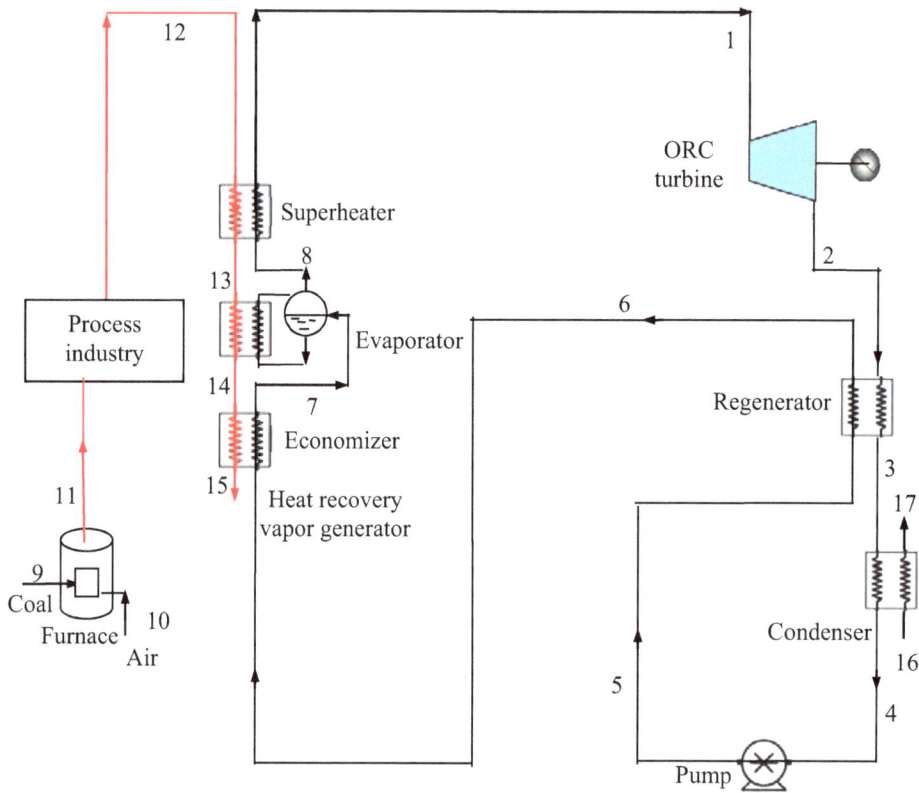

Fig. (3). Organic Rankine cycle with regenerator.

Fig. **(4)** shows the temperature-specific entropy diagram of ORC with regenerator. The supply temperature has been selected to suit the working fluid at sub-critical region. In regenerator, the vapor temperature is decreased from state 2 to state 3 with a rise in temperature of condensate from state 5 to state 6. The temperature, T_6 is less than the temperature, T_2 and can be designed with the effectiveness of heat exchanger or terminal temperature difference (TTD). The degree of internal heat recovery in the regenerator depends on the source temperature. If the source temperature is high, more heat can be recovered and vise-versa. Suppose, the internal heat recovery with R134a and R407C is low due to its lower source temperature. Higher source temperature is possible with R123 and R245fa allows more heat recovery in the regenerator.

The internal heat recovery of regenerative ORC with multi fluid system (zeotrope) is different as the temperature of the fluid changes during the phase change. In the zeotropic fluid system, fluids are selected such that they will mix with each other in cooling and separates on heating. The zeotrope fluid has two saturation temperatures. The beginning of the phase change temperature is called bubble point temperature (BPT) and the end temperature is known as dew point temperature (DPT).

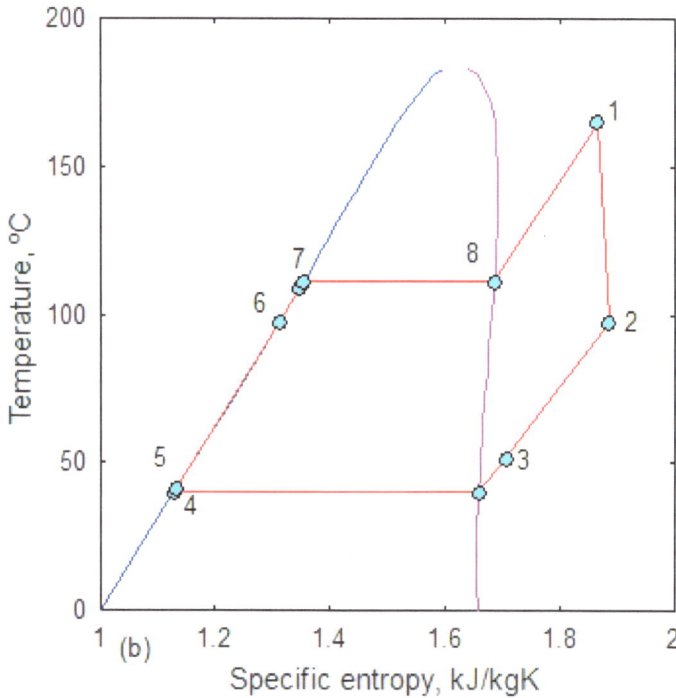

Fig. (4). Temperature-entropy diagram of ORC with regenerator.

Thermodynamic Evaluation of ORC

The source of the power plant is the hot gas from a typical cement factory. The hot gas supply to the power plant is 1000000 Nm^3/h. In a typical process industry, the hot gas flow rate is a few million Nm^3/h. The unit of one million Nm^3/h hot gas supply facilitates the researchers and designers to scale up with the actual flow rate of the hot gas in the power station. The hot gas supply temperature to the ORC is fixed based on the working fluid in ORC.

The solid fuel (coal) state to furnace is 25 °C. The air inlet temperature to furnace is 35 °C. The combustion temperature in the furnace of the industry is 900 °C and maintained below the adiabatic flame temperature.

The gravimetric composition of coal used in the furnace is as follows:

Carbon, $C = 47\%$

Hybrogen, $H = 3.17\%$

Oxygen, $O = 8.7\%$

Nitrogen, $N = 1.5\%$

Sulphur, $S = 0.91\%$

and Ash $= 33\%$

Chemically the solid fuel can be defined in molar form as for easy framing of combustion equations. The carbon atom considered is one hence the coefficient a_1 is unity. Therefore, the other coefficients are equal to H/C, O/C and N/C mole ratios respectively for a_2, a_3 and a_4. The moisture content in coal is 5.72%.

The chemical reaction in the combustion is framed with the coal, air and moisture as follows:

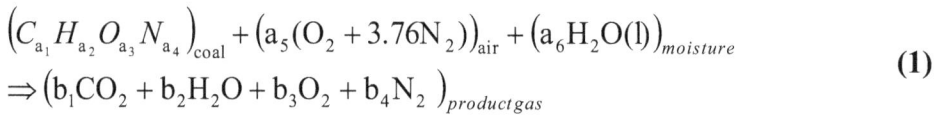

$$\left(C_{a_1}H_{a_2}O_{a_3}N_{a_4}\right)_{coal} + \left(a_5(O_2 + 3.76N_2)\right)_{air} + \left(a_6H_2O(l)\right)_{moisture} \tag{1}$$
$$\Rightarrow \left(b_1CO_2 + b_2H_2O + b_3O_2 + b_4N_2\right)_{product\,gas}$$

In the above equation, the coefficients a_1 to a_4 are known from the gravimetric analysis of coal. The coefficient, a_6 is determined from the moisture content in the fuel. The other coefficients a_6, b_1, b_2, b_3 and b_4 are the five coefficients to be solved. To solve these five unknown coefficients, five equations are framed. The five equations are four mass balance equations (carbon, hydrogen, oxygen and nitrogen) and one energy balance equation. For energy balance equation, the combustion temperature, 900 °C is used. The resulted coefficient, a_5 gives the actual air fuel ratio to get the required combustion temperature as per the assumption. The resulted coefficients b_1 to b_4 gives the gas composition which is the main source to the power plant.

By assuming no excess oxygen in the products of combustion, the minimum air fuel ratio (stoichiometric) and adiabatic flame temperature (AFT) are determined to check the limits with actual air fuel ratio (AFR) and assumed combustion temperature. The minimum AFR and AFT are solved from mass balance and energy balance respectively. The resulted minimum AFR and AFT are 9.22 (minimum) and 1120 °C (maximum) respectively.

The resulted combustion equation at the assumed combustion temperature of 900 °C is given as follows:

$$\left(C\,H_{0.8094}O_{0.1388}N_{0.0274}\right)_{coal} + \left(1.2125(O_2 + 3.76N_2)\right)_{air} + \left(0.086H_2O(l)\right)_{moisture\,content}$$

$$\Rightarrow \left(CO_2 + 0.4857H_2O + 0.2690\,O_2 + 5.29N_2\right)_{product\,gas} \tag{2}$$

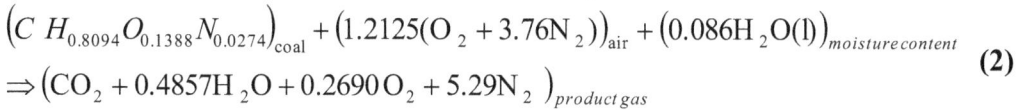

The resulted actual AFR is 11.41 and it is above the minimum air fuel ratio. The combustion temperature assumed (900 °C) is below the adiabatic flame temperature which is the maximum combustion temperature under ideal conditions.

Degree of superheat (DSH) is the temperature difference between the superheated vapor and saturation temperature in the superheater. The impact of superheating in SRC and ORC is different. The DSH is determined from the turbine inlet temperature which is obtained from the source temperature and HRVG saturation temperature.

The TTD at the end of superheater is fixed as 15°C. The supply temperature to HRVG is T_{12}.

From the gas supply temperature and TTD in superheater, turbine inlet temperature is determined as follows:

$$T_1 = T_{12} - TTD_{superheater} \tag{3}$$

The saturation temperature in HRVG is obtained by subtracting the DSH from the turbine inlet temperature. DSH is determined as follows:

$$DSH = T_1 - T_{HRVG,\,saturation} \tag{4}$$

The pinch point (PP) is the minimum temperature difference between hot gas and saturation temperature in HRVG. To ensure the heat transfer from the hot fluid to cold fluid, the hot gas temperature should be kept always above the cold fluid temperature. Therefore, the gas temperature at the evaporator inlet (in the direction of cold fluid) is determined from the PP (Refer Fig. 3 of schematic arrangement).

Fig. (5) shows the temperature-heat transferred diagram of ORC heat source.

$$T_{14} = T_8 + PP \tag{5}$$

The PP in the above equation is taken as 10 °C.

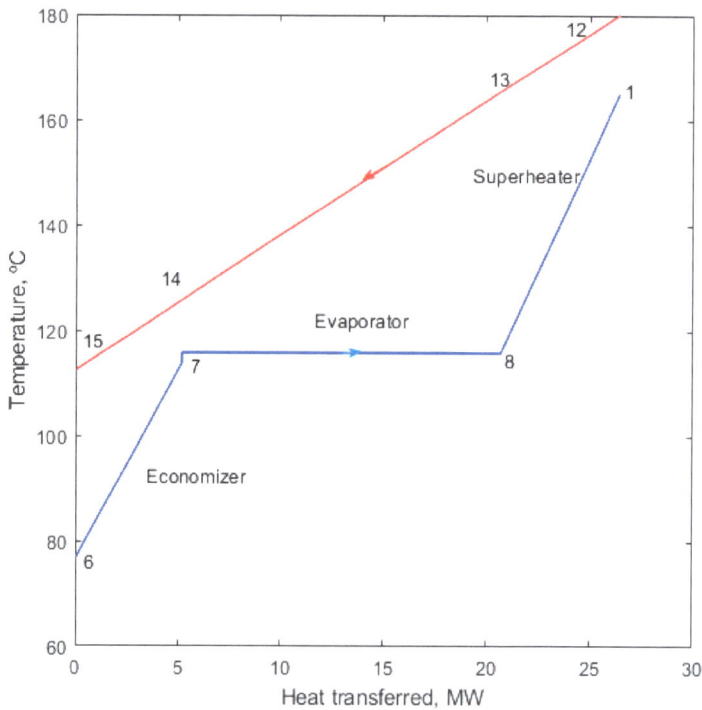

Fig. (5). Temperature-heat transferred diagram of HRVG in ORC plant.

Similar to pinch point, in evaporator section of HRVG, one more design parameter is approach point (AP). At the inlet of evaporator, phase change starts. A sudden shift of fluid from liquid to vapor causes pitting action due to transient condition. It may damage the heat exchanger tubes. Therefore, for the smooth transition from liquid to vapor, subcooled or unsaturated liquid is supplied to the evaporator from the economizer. The evaporator inlet temperature is below its saturation temperature. AP is defined as the temperature difference between the saturation temperature and the liquid supply temperature. In ORC, the AP is considered at 2 °C.

With reference to Figs. (**3** and **4**),

The evaporator inlet temperature,

$$T_7 = T_{HRVG\ saturation} - AP \tag{6}$$

The isentropic efficiencies of solution pump and vapor turbine are assumed as 75% and 80% respectively. The actual condition of fluid at the exit of the pump and turbine is determined from the isentropic expansion and its efficiency.

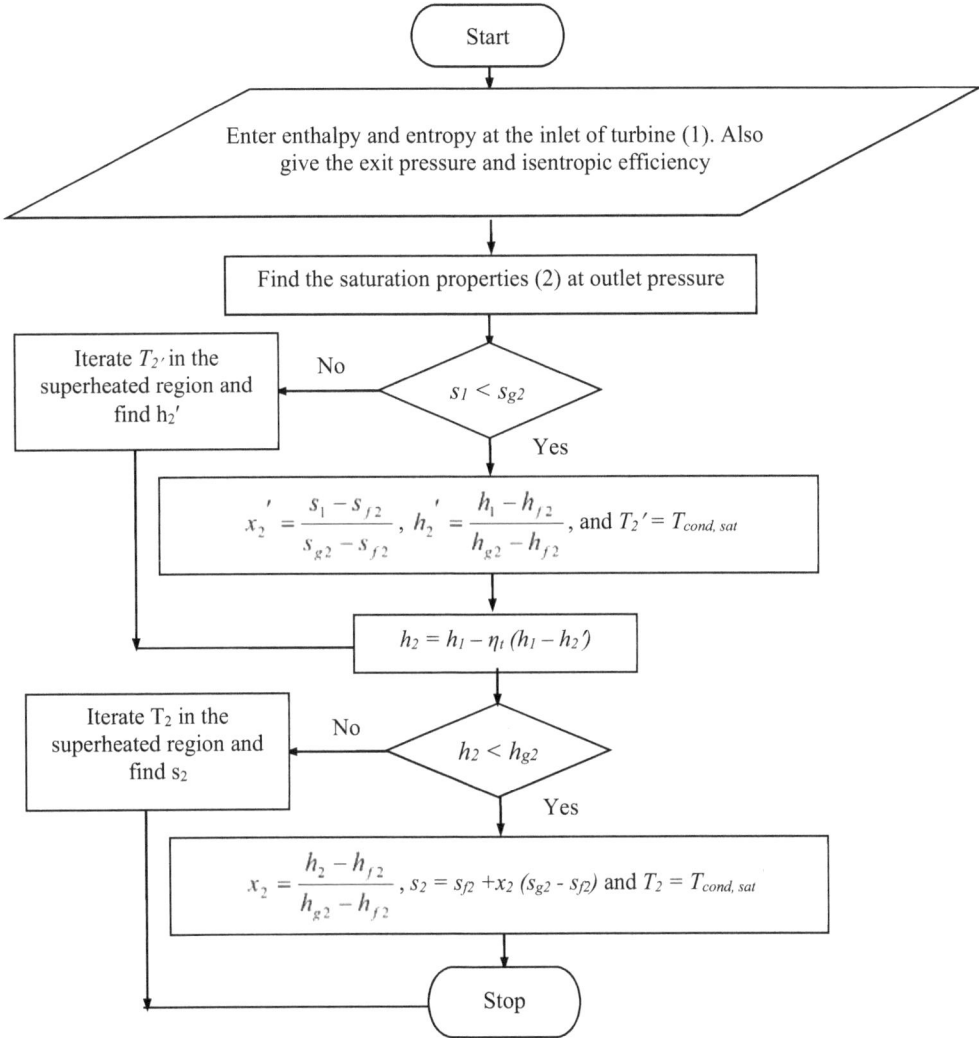

Start

Enter enthalpy and entropy at the inlet of turbine (1). Also give the exit pressure and isentropic efficiency

Find the saturation properties (2) at outlet pressure

Iterate T_2' in the superheated region and find h_2'

No

$s_1 < s_{g2}$

Yes

$$x_2' = \frac{s_1 - s_{f2}}{s_{g2} - s_{f2}}, \ h_2' = \frac{h_1 - h_{f2}}{h_{g2} - h_{f2}}, \text{ and } T_2' = T_{cond,\,sat}$$

$$h_2 = h_1 - \eta_t \,(h_1 - h_2')$$

Iterate T_2 in the superheated region and find s_2

No

$h_2 < h_{g2}$

Yes

$$x_2 = \frac{h_2 - h_{f2}}{h_{g2} - h_{f2}}, \ s_2 = s_{f2} + x_2 \,(s_{g2} - s_{f2}) \text{ and } T_2 = T_{cond,\,sat}$$

Stop

Fig. (6). Simulation of vapor turbine.

The simulation of exit state of vapor turbine is shown in the flow chart Fig. (**6**). After the expansion of vapor in the turbine, it may the saturated vapor, wet vapor or superheated state. To find the actual condition of the vapor at turbine exit, the isentropic expansion to be solved. The isentropic condition is iterated by equating the entropy before and after the expansion. In case, the entry entropy is equal to the saturated vapor entropy at low pressure, the final state is the saturated vapor. If the inlet entropy is less than the saturated vapor entropy of low pressure, the state is wet vapor. In the wet vapor the dryness fraction, enthalpy and entropy can be calculated with the entropy reference. The exit temperature is the saturation temperature at the low pressure (condenser pressure). If the inlet entropy is greater than the saturated vapor's entropy, it is superheated condition. The final temperature after the isentropic expansion is assumed and later iterated till the entropy match. The resulted temperature is the exit temperature after the isentropic expansion. The actual outlet temperature is determined by considering the isentropic efficiency of the turbine. Isentropic efficiency of the turbine is defined as the actual enthalpy drop and isentropic enthalpy drop. The isentropic efficiency of the turbine gives the actual enthalpy of the vapor at the exit. The actual enthalpy of exit vapor is compared with the saturated vapor enthalpy similar to the earlier entropy comparison. Now the actual enthalpy is equal to the saturated vapor enthalpy (saturated vapor), or below the saturated vapor enthalpy (wet vapor) or above the saturated vapor (superheated vapor). Similar to the entropy reference, with reference to enthalpy, the final temperature is iterated. At the resulted temperature and condenser pressure, entropy and other properties are determined.

The iteration of exit state from the solution pump is depicted in flow chart shown in Fig. (7). The turbine is solved in the vapor region and the pump is to be iterated in the liquid region. At the condenser exit condition, *i.e.* saturated liquid condition, the properties, such as enthalpy and entropy can be solved. The steady flow energy equation (SFEE) is applied to the pump and used to find the enthalpy after the isentropic pumping. The actual enthalpy of liquid is determined from the isentropic efficiency of the pump. The resulted enthalpy is used to iterate the exit temperature and other properties.

The exit vapor temperature after the regeneration should be above the saturation temperature. The specific heat of vapor is less than the specific heat of liquid. Therefore, the temperature difference in vapor is more than the temperature difference of liquid. The temperature of vapor is maintained above the saturation by 10 °C as follows:

The vapor temperature at the exit of the regenerator,

$$T_3 = T_4 + 10 \tag{7}$$

The fluid inlet state to HRVG from the heat balance in the regenerator,

$$h_6 = h_5 + h_2 - h_3 \tag{8}$$

The temperature of liquid at the exit of regenerator is iterated from the enthalpy, h_6.

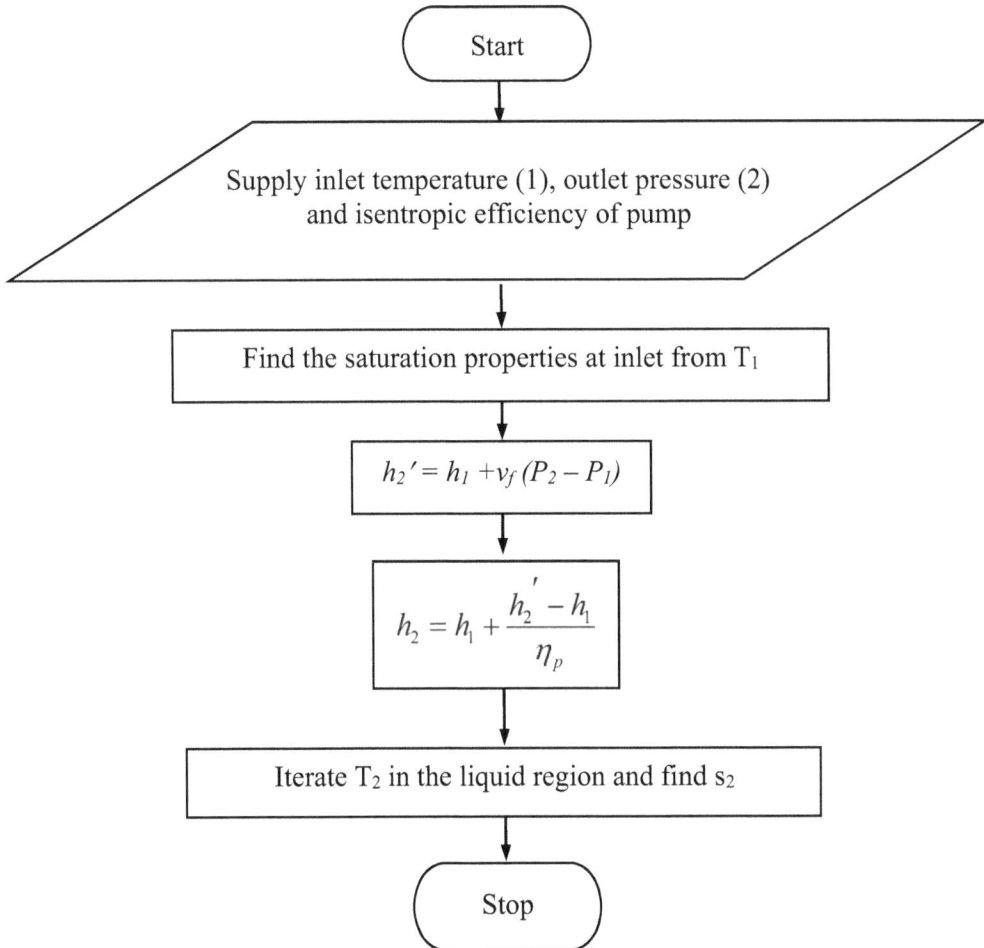

```
                    ┌───────────┐
                    │   Start   │
                    └───────────┘
                          │
                          ▼
      ╱─────────────────────────────────────────────╲
     ╱   Supply inlet temperature (1), outlet pressure (2)  ╲
    ╱          and isentropic efficiency of pump             ╲
   ╱─────────────────────────────────────────────────────────╲
                          │
                          ▼
    ┌─────────────────────────────────────────────────┐
    │   Find the saturation properties at inlet from T₁ │
    └─────────────────────────────────────────────────┘
                          │
                          ▼
          ┌──────────────────────────────┐
          │   h₂' = h₁ + vf (P₂ − P₁)     │
          └──────────────────────────────┘
                          │
                          ▼
          ┌──────────────────────────────┐
          │   h₂ = h₁ + (h₂' − h₁)/ηp     │
          └──────────────────────────────┘
                          │
                          ▼
    ┌─────────────────────────────────────────────────┐
    │   Iterate T₂ in the liquid region and find s₂     │
    └─────────────────────────────────────────────────┘
                          │
                          ▼
                    ┌───────────┐
                    │   Stop    │
                    └───────────┘
```

$$h_2' = h_1 + v_f (P_2 - P_1)$$

$$h_2 = h_1 + \frac{h_2' - h_1}{\eta_p}$$

Fig. (7). Simulation of solution pump.

The shaft power of the turbine is less than the actual enthalpy drops because of frictional losses in the turbomachines. Similarly, more shaft power is required for the actual enthalpy rise in the solution pump. The ratio of shaft power and actual enthalpy drop is called mechanical efficiency of turbine. Similarly, the actual enthalpy rise to shaft power ratio is called as mechanical efficiency of pump. The mechanical efficiency of turbine and pump is assumed as 98%. It indicates 2% frictional losses. Similar to mechanical losses, the electrical losses exist in the electrical machines such as alternator and motor *etc.* The electricity production from the alternator is less than the shaft power from turbine. Therefore, the electrical generator efficiency and the motor are assumed to solve the exact electricity from the electrical machines. The conversion efficiency from shaft power to electricity in turbine and electricity to shaft power in pump is called as electrical generator/pump efficiency of concerned electrical machine. The electrical efficiency of generator and electrical efficiency of pump is assumed as 98%.

The electricity generation from the turbine,

$$W_t = \eta_{mechanical}\, \eta_{generator}\, m_1\, (h_1 - h_2) \tag{9}$$

The electricity required to run the motor connected to the pump,

$$W_p = \frac{m_4(h_5 - h_4)}{\eta_{mechanical}\eta_{motor}} \tag{10}$$

The net electricity generation,

$$W_{net} = W_t - W_p \tag{11}$$

The circulating cooling water supply temperature to the condenser is 25 °C and the temperature rise of water in the condenser is 8 °C.

The cooling water requirement in the condenser with the above listed water inlet temperature,

$$m_{16} = \frac{m_3(h_3 - h_4)}{c_{p,\,w}\,(T_{17} - T_{16})} \tag{12}$$

The heat supply to the cycle,

$$Q_{supply} = m_1\, (h_1 - h_6) \tag{13}$$

The thermal efficiency of the power cycle from the waste heat recovery,

$$\eta_{thermal} = \frac{W_{net}}{Q_{supply}} \times 100 \tag{14}$$

In case of zeotropic mixture, the condensate temperature (T_4) at the exit of condenser is the BPT. Similarly, in the evaporator of HRVG, the boiling starts from BPT and ends with DPT. So the temperature, T_8 is the DPT at the boiler pressure.

The study on exergy decays or irreversibilities plays an imperative role on thermodynamic work. Exergy efficiency of each component can be analyzed to design the effective process in the plant. In this section, component exergy losses or decays are identified and compared to find the major and minor losses caused with entropy generation. It helps to pay the attention on weak areas of the components to make the further modifications or updating the configuration. The resulted irreversibilities also check the correctness of thermodynamic work. The law of entropy increase says that the universe (system and surroundings together) entropy always increases for any process. In case the universe entropy is decreasing or net is negative, the thermodynamic solution of the particular component or process is wrong. The net positive change of entropy or positive irreversibility in the process ensures the correctness of the evaluation.

Since the focus is given on the power but not on the process industry, the hot gas supply to HRVG is the main input to the power plant.

Exergy is the maximum potential of the work can be determined from energy and entropy.

$$\text{The specific exergy, } e = h - T_0\, s \tag{15}$$

The exergy supply to the power plant is the potential of the hot gas to do the maximum work which is called as the exergy of hot gas supply at the inlet of HRVG.

Exergy of hot gas,

$$E_{hot\,gas} = m_{12}\, e_{12} \tag{16}$$

The irreversibility can be determined in two ways one is the exergy balance similar to the heat balance to find the heal losses and the second option is the using entropy change. Since the exergy balance consists of energy and entropy terms in the process, the heat balance cancels the energy quantities leaving the entropy items. Therefore, the irreversibility can be directly determined by counting the entropy change without starting from exergy balance. The following formulation is developed from the both options as per the convenient and suitability of the process.

The irreversibility can be determined from the following exergy balance,

$$\Sigma\, Exergy_{in} = \Sigma\, Exergy_{out} + Irreversibility \tag{17}$$

HRVG is a heat exchanger where the solution is heated and converted into superheated vapor. HRVG consists of three sections viz. economizer, evaporator and superheater. The exergy loss or irreversibility of HRVG includes these three sections.

The irreversitility in HRVG,

$$I_{HRVG} = m_{12}\, e_{12} + m_6\, e_6 - m_{15}\, e_{15} - m_1\, e_1 \tag{18}$$

The irreversibility in turbine,

$$I_t = m_1\, (e_1 - e_2) - W_t \tag{19}$$

Or it may be found as,

$$I_t = T_0\, m_1\, (s_2 - s_1) \tag{20}$$

The irreversibility in the regenerator heat exchanger,

$$I_{reg} = m_2\, (e_2 - e_3) + m_5\, (e_5 - e_5) \tag{21}$$

The irreversibility from the heat transfer in condenser,

$$I_{cond} = m_3\, (e_3 - e_4) + m_{16}\, (e_{16} - e_{17}) \tag{22}$$

Similar to the turbine, the irreversibility in pump also can be determined as follows:

$$I_p = m_4\, (e_4 - e_5) + W_p \tag{23}$$

The cycle irreversibilities and outputs can be balanced with the cycle exergy supply. The cycle exergy supply is the heat transfer fluid connected at the inlet of turbine.

The exit of heat transfer fluid from HRVG carries exergy and it is considered as exergy loss for the cycle.

The hot gas from the HRVG is supplied to stock and it carries an exergy loss.

The irreversibility in exhaust gas from the HRVG,

$$I_{exhaust\ gas} = m_{15}\ e_{15} \tag{24}$$

The hot water from condenser also contributes exergy loss.

The irreversibility from exit hot water of condenser,

$$I_{hw} = m_{17}\ e_{17} \tag{25}$$

The total irreversibility of the cooling cogeneration cycle,

$$I_{total} = I_{HRVG} + I_t + I_{reg} + I_{cond} + I_p + I_{exhaust} + I_{hw} \tag{26}$$

Fig. (**8**) summarizes the methodology adopted in ORC solution. In this metho-dology, the main steps involved are properties generation, processes study, modeling of components by mass balance, energy balance and exergy balance, simulation of the total integrated components, analysis and optimization of system operation conditions and developments of specifications at the optimized results.

Effect of Regeneration on ORC

Fig. (**9**) compares the thermal efficiency of basic ORC and regenerative ORC. The internal heat recovery with regenerator at the turbine exit and boiler inlet saves the heat recovery and improves the thermal efficiency. The heat recovery is constrained by the same pinch point and generates same quantity of fluid. The turbine is not effected by regenerator hence the power will not effect. The comparative result shows that the regenerator is not much effective with low temperature source. The regenerator is recommended with high temperature source. Anyhow, the performance of a thermal power plant increases with increase in source temperature. Therefore, regenerator is recommended to ORC assuming to design it at the high the source temperature.

Optimum Boiler Pressure with Source Temperature

Fig. (**10**) studies the influence of hot gas supply temperature and boiler or HRVG pressure on plant power generation through heat recovery and its cycle thermal efficiency. Suitable temperature range (80 °C to 180 °) and pressure range (2 bar to 25 bar) is selected to operate the plant under sub-critical condition. The HRVG pressure is increased from condenser pressure to maximum boiler pressure below the critical state. The condenser pressure is fixed with the sink temperature. Since the source temperature is changing (increasing), proportionately the upper limit of pressure is increasing. The maximum turbine inlet temperature is selected just below the critical temperature of fluid.

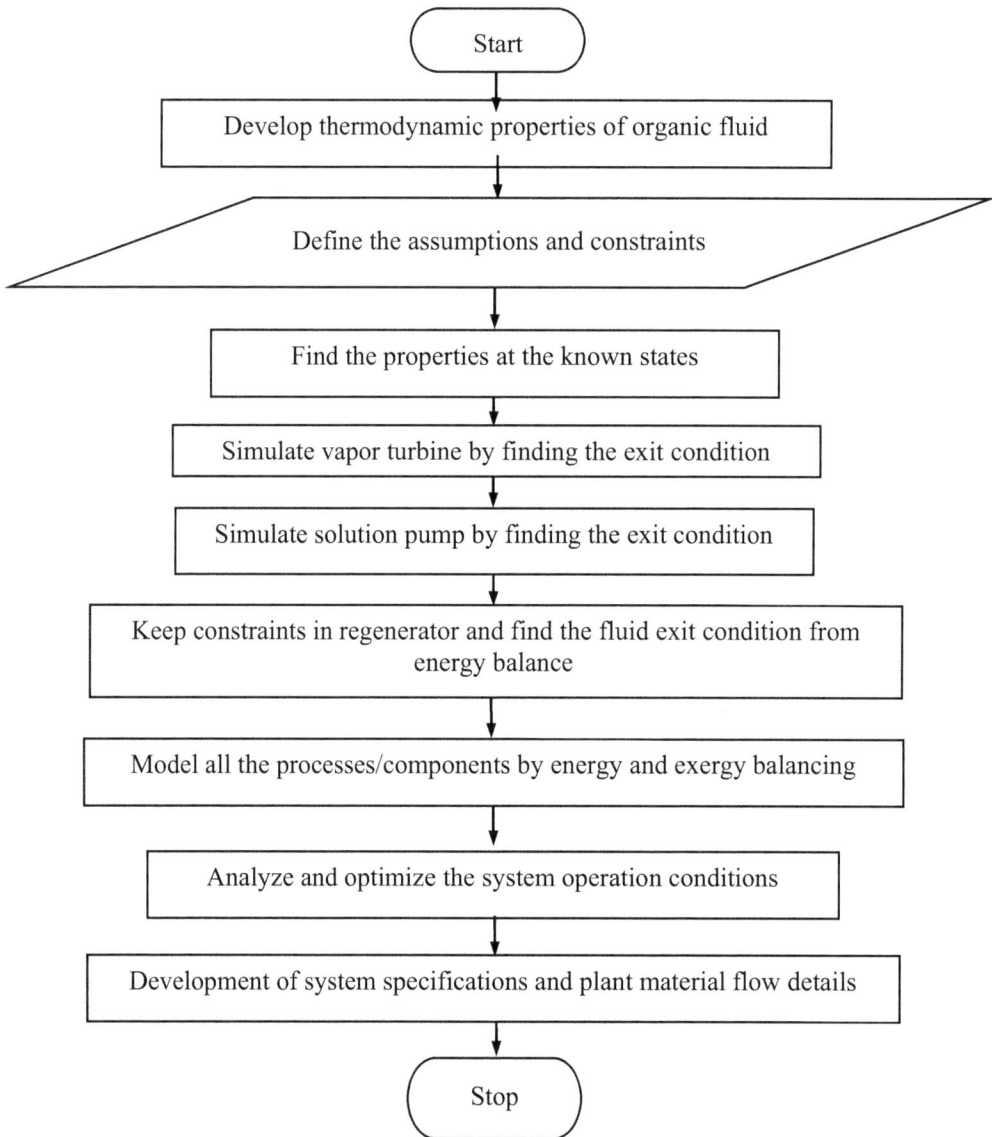

Fig. (8). Steps involved in the study of ORC with heat recovery.

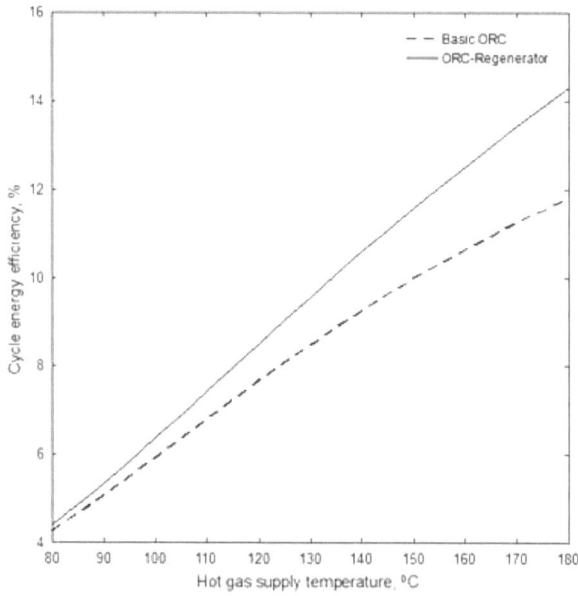

Fig. (9). Comparison of basic ORC and ORC with regenerator.

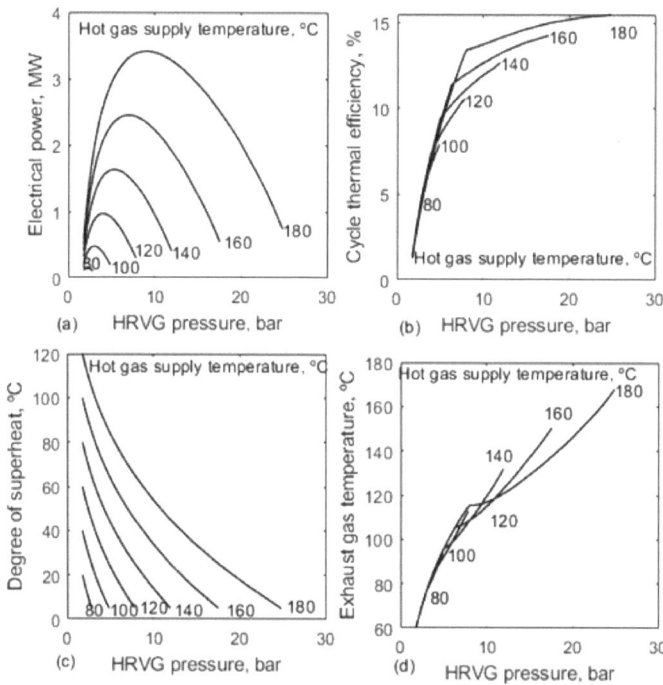

Fig. (10). Influence of source temperature with HRVG pressure on **(a)** net power generation, **(b)** cycle energy conversion efficiency **(c)** degree of superheat and **(d)** exhaust gas temperature with R123 fluid.

Fig. (**10a**) gave some interesting results to choose the HRVG pressure at known source temperature to gain the advantage of higher power production. The optimum HRVG pressure is increasing with increase in source temperature to get maximum power condition. The maximum thermal efficiency condition is demanding high pressure compared to the optimum pressure. The main objective of the heat recovery plant is production of power without minding thermal efficiency. The maximum heat recovery gives maximum power. Therefore, process conditions are selected at the optimum HRVG pressure to augment the power. Fig. (**10c**) is prepared to study the variations in DSH resulted with changes in temperature and pressure. The DSH is increasing with increase in source temperature and decreasing with rise in HRVG pressure. There are two ways to design the HRVG pressure. In the first method, directly the HRVG pressure can be selected at optimum state which lies between condenser pressure and upper limit of pressure. In the second method, the HRVG pressure can be determined by selecting the suitable DSH. The subtraction of TTD and DSH from the source temperature results the saturation temperature of HRVG. The pressure can be determined from saturation temperature. In this work, instead of selecting the pressure from the arbitrary DSH, optimum pressure is developed from the source and sinks temperature. The merging of power generation characteristics with DSH results the optimum DSH to select the pressure from the source temperature as second option. The exhaust gas temperature from the plant is shown in Fig. (**10d**). The optimum pressure increases with increase in source temperature. The exhaust gas temperature also increases with increase in pressure with pinch point limit.

Fig. (**11**) shows the role of source temperature and HRVG pressure on ORC output Fig. (**11a**) and its thermal efficiency Fig. (**10b**) for the working fluid of R124. The changes in DSH with temperature and pressure also included to analyze the role of super heater on performance (Fig. **11c**). As per the earlier results, the optimum boiler pressure is increasing with increase in temperature. With increase in pressure, the DSH decreases as the HRVG saturation temperature increases with the fixed superheat temperature. As per the properties of R124, the source temperature is varied from 70 °C to 120 °C. The corresponding pressure is changed from 7 bar to 23 bar. The increase in exhaust gas temperature can be observed in Fig. (**11d**) with a rise in temperature and pressure.

Fig. (**12**) indicates the (**a**) plant power generation, (**b**) thermal efficiency (**c**) degree of superheat and (**d**) exhaust gas temperature with change in source temperature and boiler pressure for R134a.

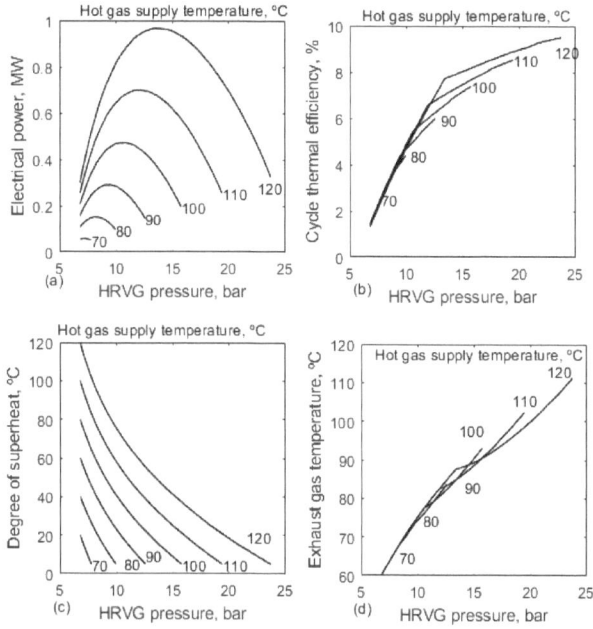

Fig. (11). Influence of source temperature with HRVG pressure on **(a)** net power, **(b)** cycle energy conversion efficiency **(c)** degree of superheat and **(d)** exhaust gas temperature with R124 fluid.

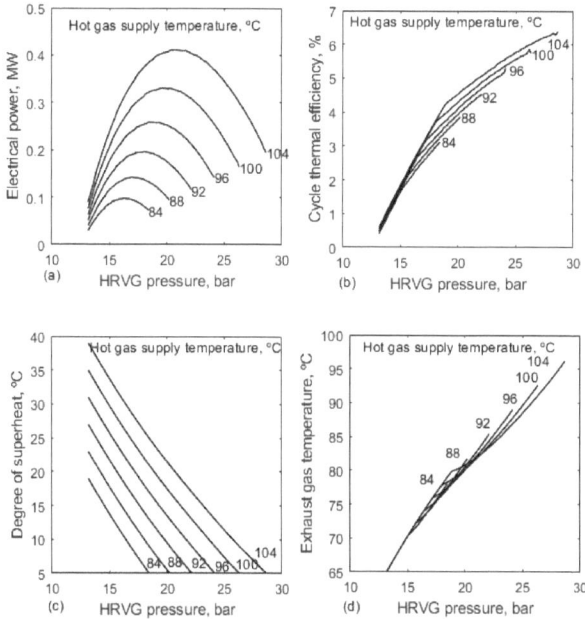

Fig. (12). Influence of source temperature with HRVG pressure on **(a)** net power, **(b)** cycle energy conversion efficiency **(c)** degree of superheat and **(d)** exhaust gas temperature with R134a fluid.

Fig. (**13**) shows the optimum HRVG pressure with R245fa fluid. With the constraints in sink and critical point, the source temperature is changed from 100 °C to 150 °C.

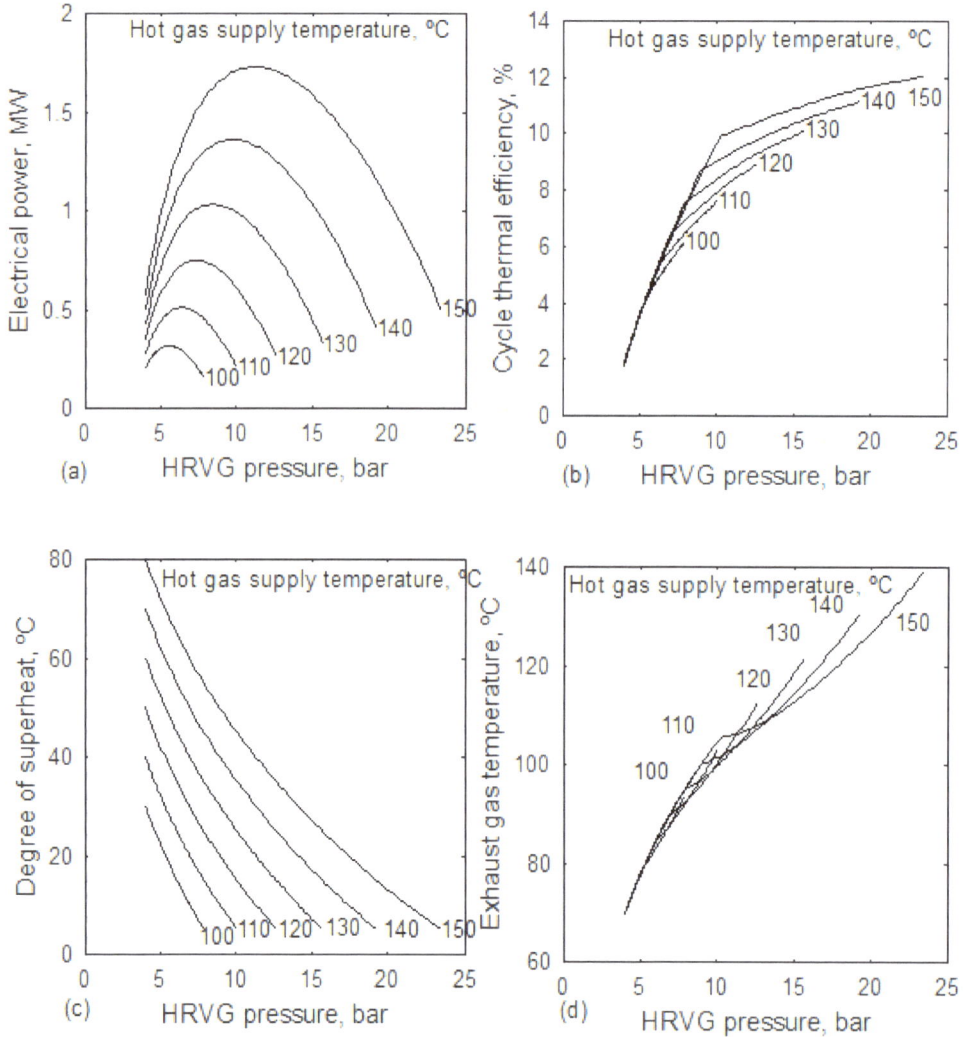

Fig. (13). Influence of source temperature with HRVG pressure on (**a**) net power generation, (**b**) cycle energy conversion efficiency (**c**) degree of superheat and (**d**) exhaust gas temperature with R245fa fluid.

Fig. (**14**) results the optimum HRVG pressure for maximum power as a function of temperature and pressure at a fixed condenser state Figs. (**14a** - **14d**) shows respectively thermal efficiency, DSH and exhaust gas temperature.

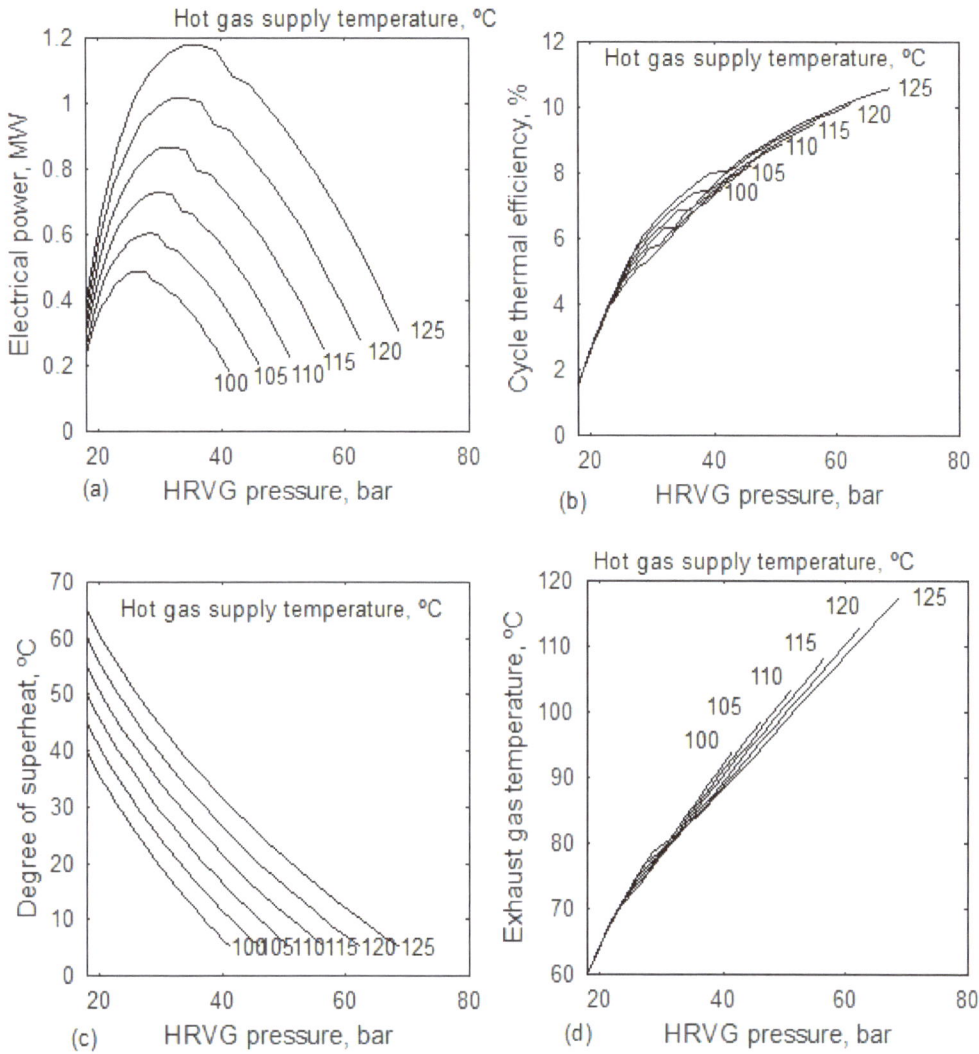

Fig. (14). Influence of source temperature with HRVG pressure on **(a)** net power, **(b)** cycle energy conversion efficiency **(c)** degree of superheat and **(d)** exhaust gas temperature with R717 fluid.

R407C is a zeotropic mixture where the HRVG vapor temperature is not a constant and it depends on pressure at a fixed composition. The boiling starts from BPT and ends with DPT. Similarly, the condensation starts from DPT and ends with BPT. The optimum HRVG pressures are developed with change in source temperature (Fig. **15**).

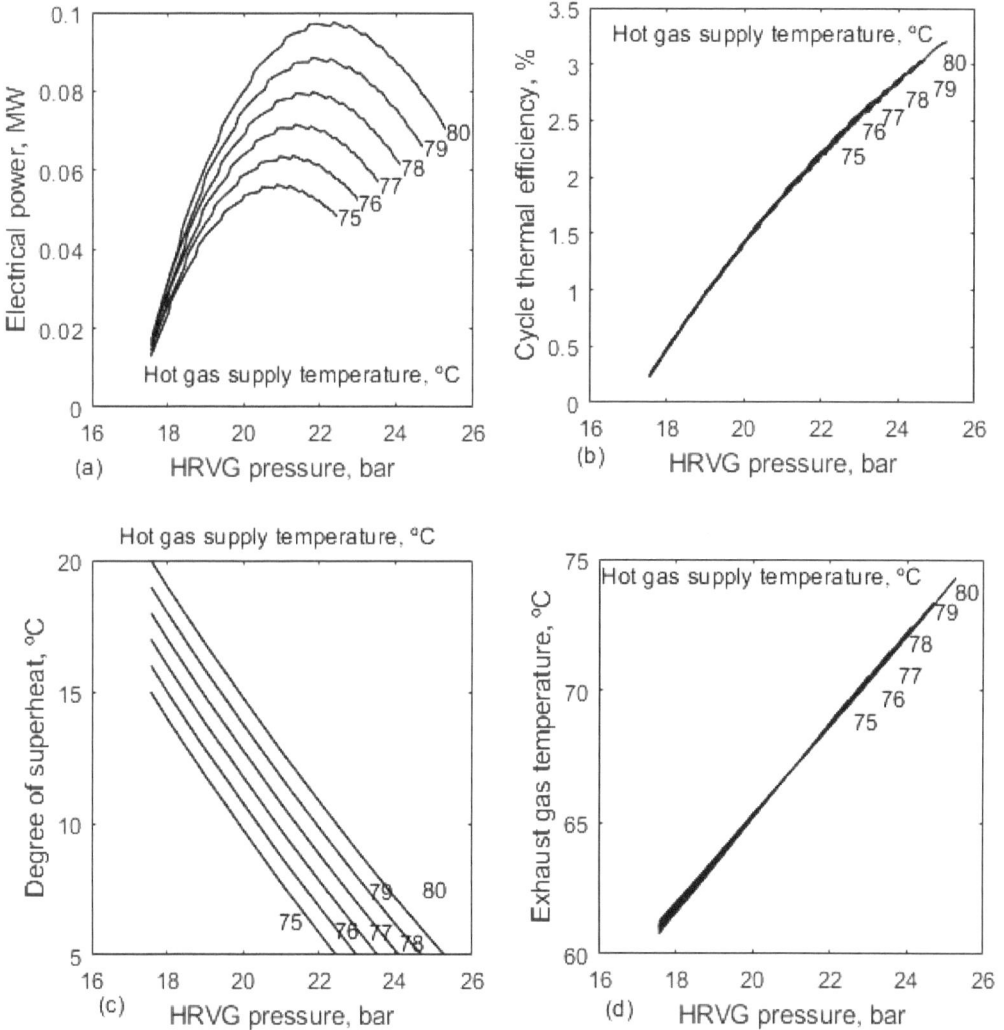

Fig. (15). Influence of source temperature with HRVG pressure on **(a)** net power generation, **(b)** cycle energy conversion efficiency **(c)** degree of superheat and **(d)** exhaust gas temperature with R407C fluid.

The optimum HRVG pressure depends on source temperature and working fluid at a fixed condenser temperature. The data of best optimum pressure at the available source temperature and selected fluid has a significant importance to design a power plant. For every selection, there will be an optimum pressure for maximum power generation. The power generation with heat recovery option needs to focus on maximum power than thermal efficiency. The fuel fired vapor boiler only has to focus on maximum thermal efficiency condition. Therefore, in the present case, the optimum HRVG pressure is selected from the maximum power condition. To formulate the HRVG pressure as a function of source temperature with working fluid, correlations are developed from the best fit method and presented in Table **1**. A non-linear quadratic fit has been developed to find the optimum pressure. Therefore, the optimum pressure can be directly determined from these correlations without repeating the analysis and optimization.

Table 1. Developed coefficients to find the optimum HRVG pressure at variable source temperature and 30 °C of circulating water temperature (condensate temperature is 40 °C).

Working Fluid	Coefficients		
	a_1	a_2	a_3
R123	2.6096	-0.0352	0.0004
R124	3.4378	0.0133	0.0006
R134a	10.9468	-0.0777	0.0017
R245fa	4.8013	-0.0593	0.0007
R717	24.9199	-0.295	0.0031
R407C	35.2449	-0.6218	0.0057

Effect of Sink Temperature.

The optimum boiler pressure with function of source temperature,

$$P_{HRVG\,opt} = a_1 + a_2T + a_3T^2 \tag{27}$$

where T is the hot gas (source) temperature in °C.

The sink temperature of the power plant varies with the seasons. Fig. (**16**) reveals the influence of sink temperature and HRVG pressure on power and thermal efficiency with R123 fluid. As stated earlier, HRVG pressure can be picked between condenser pressure (related to sink temperature) and upper limit of pressure (related to source temperature). The HRVG pressure is plotted against the sink temperature. The sink temperature is changed from 20 °C to 35 °C. Now the lower limit of HRVG pressure is changed with change in sink temperature. The upper limit is same with fixed source temperature. The power characteristics (Fig. **16a**), shows that the optimum HRVG pressure is increasing with increase in sink temperature but with a fall in power. As per the Carnot's theorem, the efficiency of the cycle decreases with increase in the sink temperature (Fig. **16b**). For maximum thermal efficiency, optimum pressure is not resulted due to lack of peak. Finally, the optimum pressure can be chosen for the maximum power from the source and sink temperatures. Similar to R123, the other plots *i.e.* Figs. (**17** -

21) develops the influence of sink temperature on optimum HRVG pressure with the fluids of R124, R134a, R245fa, R717 and R407C. For every case the higher side of the source temperature is fixed. For example, for the R123 sink characteristics, the source temperature is fixed at 180 °C. The hot gas supply temperature is 120 °C, 100 °C, 150 °C, 125 °C and 78 °C respectively maintained with R124, R134a, R245fa, R717 and R407C fluids to study the optimum HRVG pressure with a change in sink temperature.

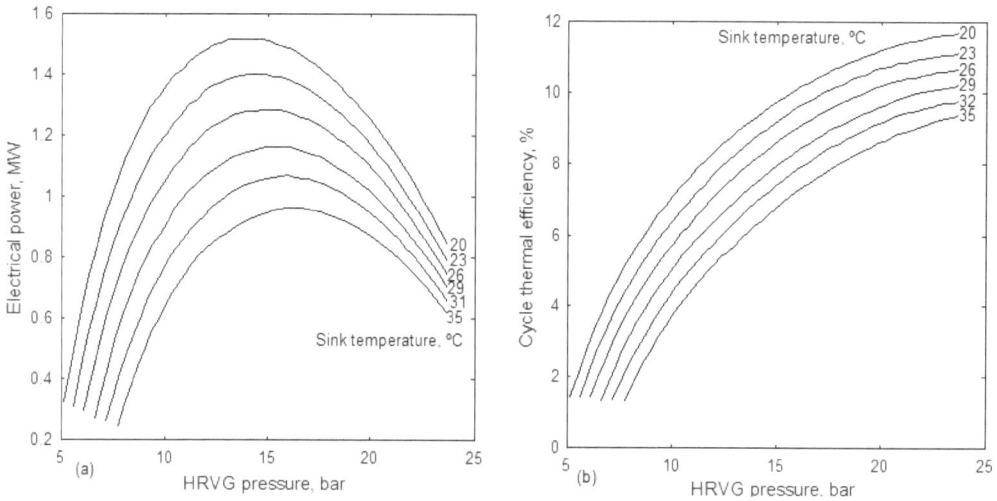

Fig. (16). Influence of sink temperature with HRVG pressure on **(a)** net power generation and **(b)** cycle energy conversion efficiency with R123.

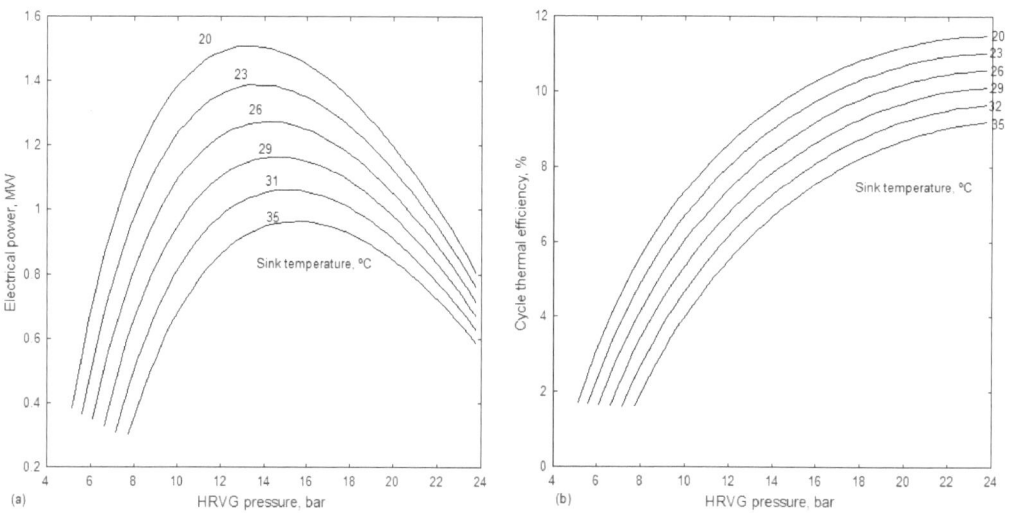

Fig. (17). Influence of sink temperature with HRVG pressure on **(a)** net power generation and **(b)** cycle energy conversion efficiency with R124.

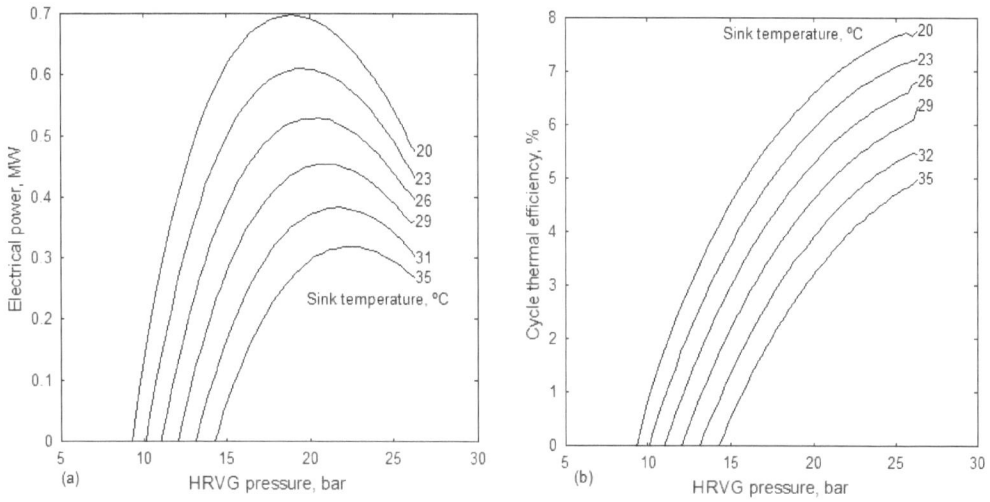

Fig. (18). Influence of sink temperature with HRVG pressure on **(a)** net power generation and **(b)** cycle energy conversion efficiency with R134a.

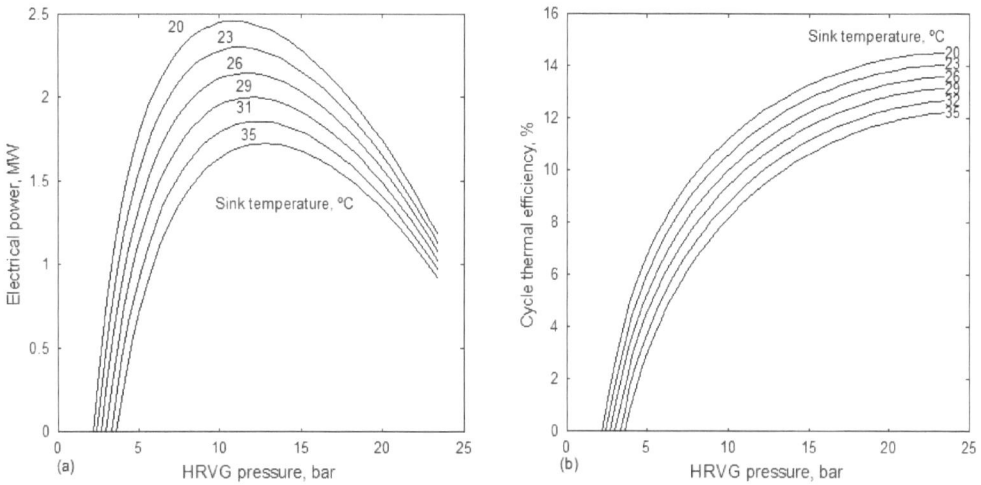

Fig. (19). Influence of sink temperature with HRVG pressure on **(a)** net power generation and **(b)** cycle energy conversion efficiency with R245fa.

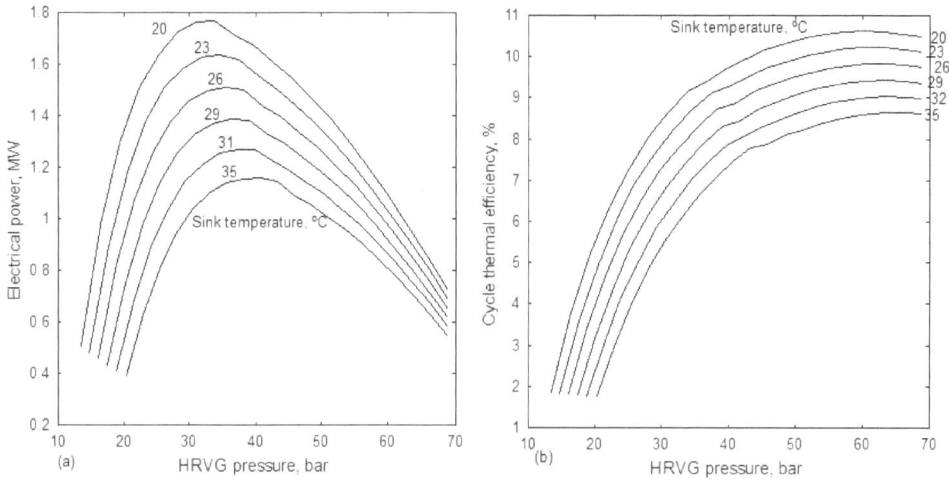

Fig. (20). Influence of sink temperature with HRVG pressure on **(a)** net power generation and **(b)** cycle energy conversion efficiency with R717.

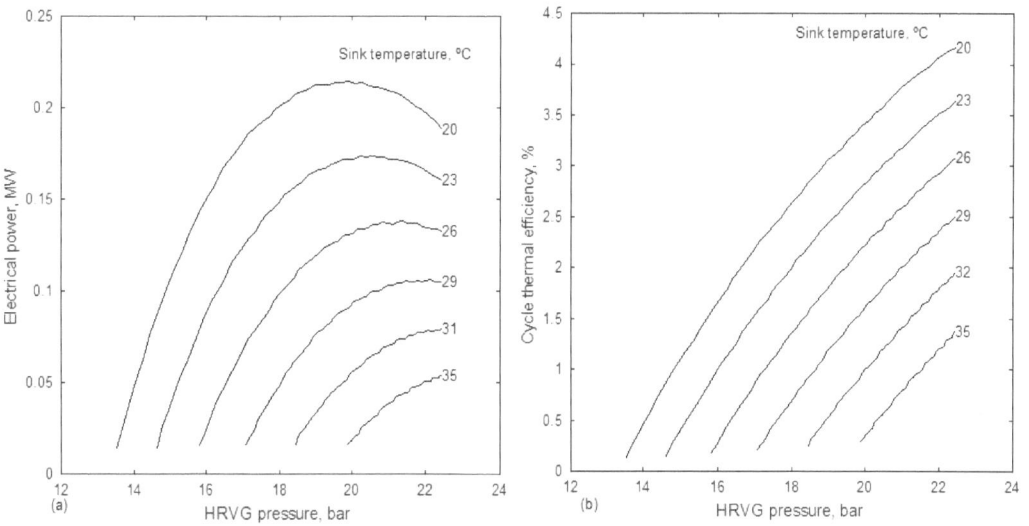

Fig. (21). Influence of sink temperature with HRVG pressure on **(a)** net power generation and **(b)** cycle energy conversion efficiency with R407C.

Property Charts and Tables at the Optimized Conditions

Figs. (**22 - 27**) shows the **(a)** pressure-specific enthalpy and **(b)** temperature-entropy plots for the regenerative ORC with the working fluids of R123, R124, R134a, R245fa, R717 and R407C. The earlier sections detailed about the developing the optimum HRVG pressure as a function of source temperature and fluid with the equation fit. The property charts and tables are prepared at the optimum pressure and with the turbine inlet temperature closed to critical temperature. The optimum conditions are prepared by changing the temperature and pressure below the critical temperature and critical pressure respectively.

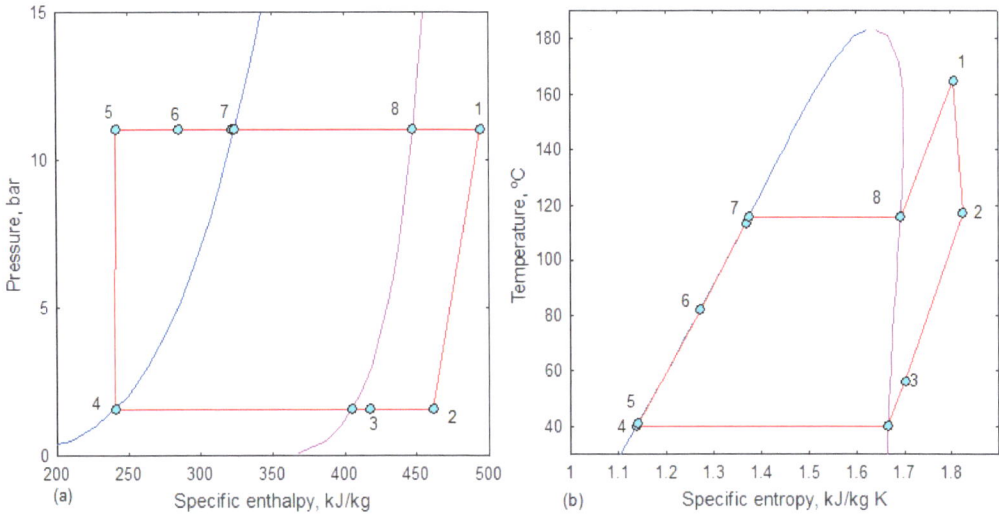

Fig. (22). Temperature-entropy diagram of ORC with regenerator with the fluid of R123.

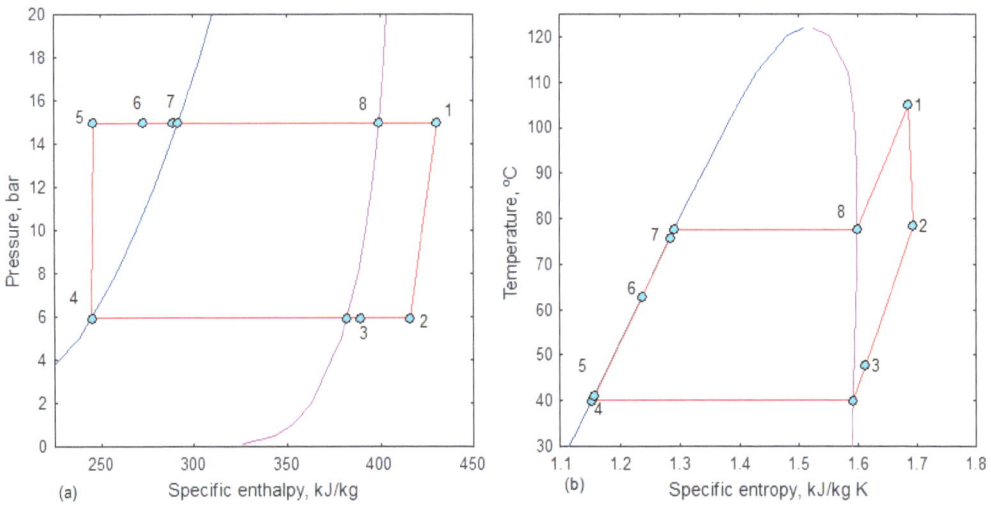

Fig. (23). Temperature-entropy diagram of ORC with regenerator with the fluid of R124.

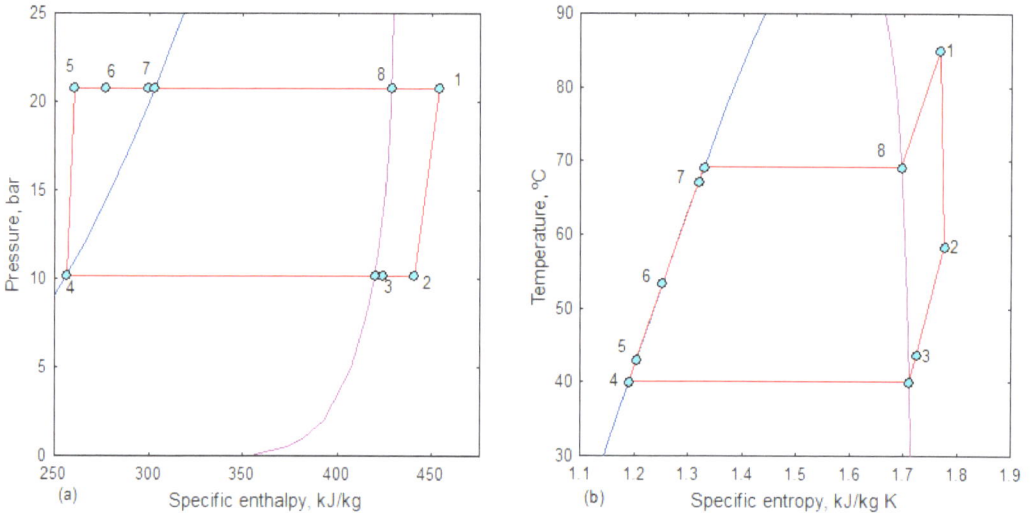

Fig. (24). Temperature-entropy diagram of ORC with regenerator with the fluid of R134a.

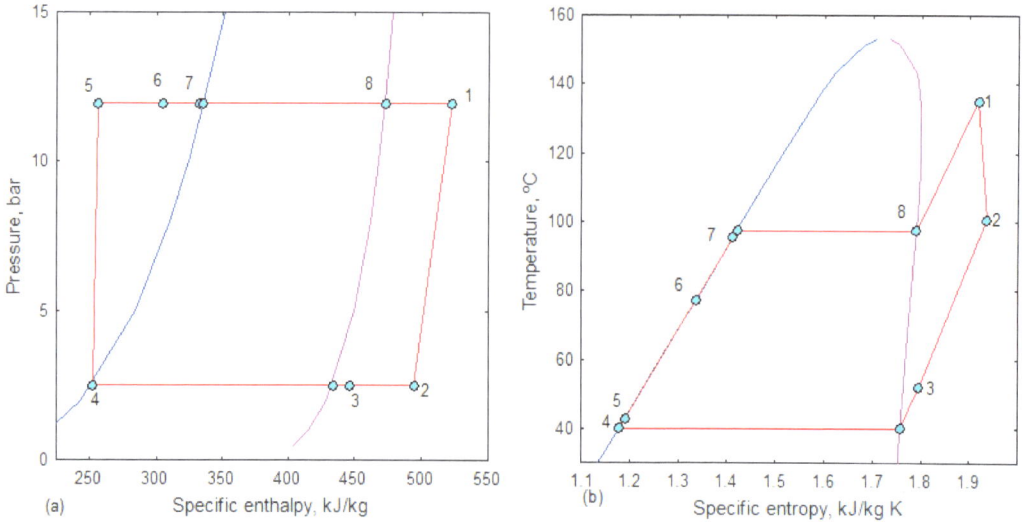

Fig. (25). Temperature-entropy diagram of ORC with regenerator with the fluid of R245fa

The data of pressure, temperature, flow rate, specific enthalpy and specific entropy is tabulated in tables from Table **2** to **7** respectively for the working fluids of R123, R124, R134a, R245fa, R717 and R407C. The slope lines of phase change for R407C can be observed in Fig. (**27b**) for condensation and evaporation.

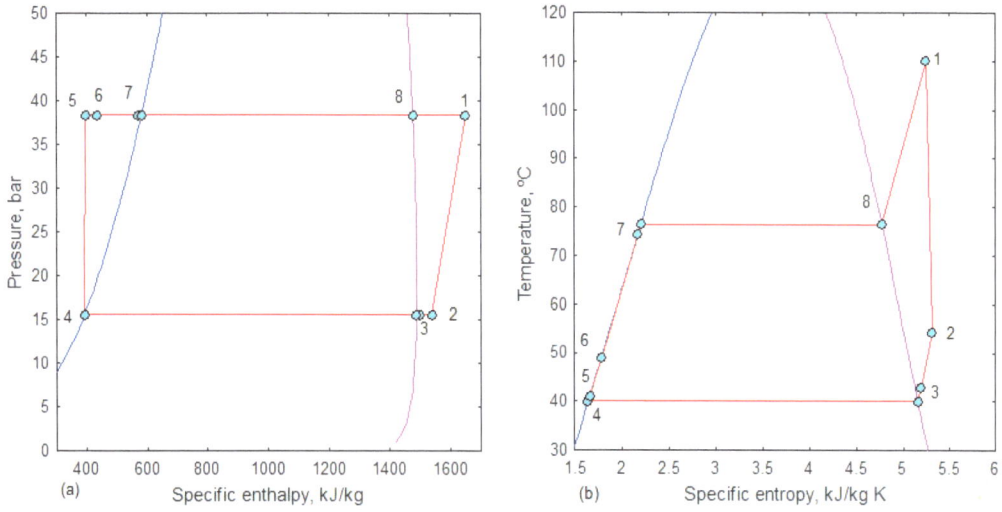

Fig. (26). Temperature-entropy diagram of ORC with regenerator with the fluid of R717.

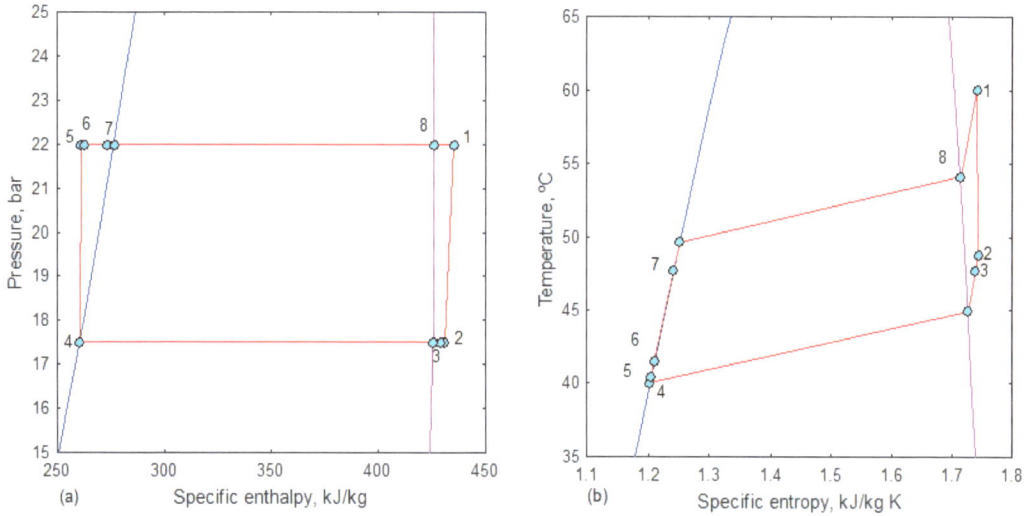

Fig. (27). Temperature-entropy diagram of ORC with regenerator with the fluid of R407C.

Table 2. ORC material balance results with R123 fluid at hot gas supply of 180 °C.

State	P, bar	t, °C	m, kg/s	h, kJ/kg	s, kJ/kg K
1.	9.23	165.00	123.02	495.81	1.82
2.	1.54	122.89	123.02	465.55	1.84
3.	1.54	52.39	123.02	414.51	1.69
4.	1.54	40.00	123.02	240.59	1.14

(Table 2) cont.....

State	P, bar	t, °C	m, kg/s	h, kJ/kg	s, kJ/kg K
5.	9.23	41.00	123.02	241.31	1.14
6.	9.23	87.76	123.02	291.90	1.29
7.	9.23	105.24	123.02	311.82	1.34
8.	9.23	107.24	123.02	443.47	1.69
9.	1.01	25.00	29.72	0.00	0.00
10.	1.01	35.00	339.21	4.09	0.01
11.	1.01	900.00	368.90	1006.40	1.54
12.	1.01	180.00	368.90	162.56	0.44
13.	1.01	163.67	368.90	145.10	0.40
14.	1.01	122.24	368.90	101.20	0.29
15.	1.01	115.95	368.90	94.56	0.28
16.	1.01	25.00	639.83	0.00	0.00
17.	1.01	33.00	639.83	33.44	0.11

Table 3. ORC with regenerator material balance Sankey and Grassmann Diagrams results with R124 fluid at hot gas supply of 120 °C.

State	P, bar	t, °C	m, kg/s	h, kJ/kg	s, kJ/kg K
1.	13.67	105.00	81.96	432.44	1.69
2.	5.93	81.72	81.96	419.04	1.70
3.	5.93	42.44	81.96	384.56	1.60
4.	5.93	40.00	81.96	245.07	1.15
5.	13.67	41.00	81.96	245.86	1.16
6.	13.67	69.22	81.96	280.77	1.26
7.	13.67	71.54	81.96	283.81	1.27
8.	13.67	73.54	81.96	398.13	1.60
9.	1.01	25.00	29.72	0.00	0.00
10.	1.01	35.00	339.21	4.09	0.01
11.	1.01	900.00	368.90	1006.40	1.54
12.	1.01	120.00	368.90	98.84	0.29
13.	1.01	88.56	368.90	65.82	0.20
14.	1.01	88.54	368.90	65.82	0.20
15.	1.01	87.91	368.90	65.14	0.20
16.	1.01	25.00	341.87	0.00	0.00
17.	1.01	33.00	341.87	33.44	0.11

Table 4. ORC with regenerator material balance results with R134a fluid at hot gas supply of 100 °C.

State	P, bar	t, °C	m, kg/s	h, kJ/kg	s, kJ/kg K
1.	20.18	85.00	41.93	454.87	1.77
2.	10.17	59.81	41.93	442.12	1.78
3.	10.17	47.63	41.93	428.16	1.74
4.	10.17	40.00	41.93	256.41	1.19
5.	20.18	42.98	41.93	260.41	1.20
6.	20.18	51.53	41.93	273.93	1.24
7.	20.18	65.87	41.93	297.22	1.31
8.	20.18	67.87	41.93	428.34	1.70
9.	1.01	25.00	29.72	0.00	0.00
10.	1.01	35.00	339.21	4.09	0.01
11.	1.01	900.00	368.90	1006.40	1.54
12.	1.01	100.00	368.90	77.82	0.23
13.	1.01	82.89	368.90	59.90	0.18
14.	1.01	82.87	368.90	59.90	0.18
15.	1.01	80.35	368.90	57.25	0.18
16.	1.01	25.00	215.37	0.00	0.00
17.	1.01	33.00	215.37	33.44	0.11

Table 5. ORC with regenerator material balance results with R245fa fluid at hot gas supply of 150 °C.

State	P, bar	t, °C	m, kg/s	h, kJ/kg	s, kJ/kg K
1.	11.66	135.00	78.34	523.23	1.92
2.	2.50	101.38	78.34	495.77	1.94
3.	2.50	50.50	78.34	444.32	1.79
4.	2.50	40.00	78.34	252.57	1.18
5.	11.66	42.73	78.34	256.20	1.19
6.	11.66	78.69	78.34	307.23	1.34
7.	11.66	94.36	78.34	330.98	1.41
8.	11.66	96.36	78.34	472.11	1.79
9.	1.01	25.00	29.72	0.00	0.00
10.	1.01	35.00	339.21	4.09	0.01
11.	1.01	900.00	368.90	1006.40	1.54
12.	1.01	150.00	368.90	130.58	0.37
13.	1.01	111.39	368.90	89.75	0.26

(Table 5) cont.....

State	P, bar	t, °C	m, kg/s	h, kJ/kg	s, kJ/kg K
14.	1.01	111.36	368.90	89.75	0.26
15.	1.01	106.59	368.90	84.71	0.25
16.	1.01	25.00	449.24	0.00	0.00
17.	1.01	33.00	449.24	33.44	0.11

Table 6. ORC with regenerator material balance results with R717 fluid at hot gas supply of 125 °C.

State	P, bar	t, °C	m, kg/s	h, kJ/kg	s, kJ/kg K
1.	36.48	110.00	12.60	1657.64	5.28
2.	15.55	57.87	12.60	1552.62	5.35
3.	15.55	50.76	12.60	1527.68	5.27
4.	15.55	40.00	12.60	390.64	1.64
5.	36.48	41.00	12.60	395.46	1.66
6.	36.48	45.89	12.60	420.13	1.74
7.	36.48	72.22	12.60	557.15	2.14
8.	36.48	74.22	12.60	1480.38	4.80
9.	1.01	25.00	29.72	0.00	0.00
10.	1.01	35.00	339.21	4.09	0.01
11.	1.01	900.00	368.90	1006.40	1.54
12.	1.01	125.00	368.90	104.11	0.30
13.	1.01	89.24	368.90	66.53	0.20
14.	1.01	89.22	368.90	66.53	0.20
15.	1.01	84.76	368.90	61.85	0.19
16.	1.01	25.00	428.38	0.00	0.00
17.	1.01	33.00	428.38	33.44	0.11

The component's irreversibilities are presented in percentage of exergy of hot gas. These exergy results are developed at the optimized HRVG pressure with the turbine inlet temperature nearer to critical temperature. The (a) energy balance results and (b) exergy balance results are shown from Figs. (28 - 33) with working fluids listed earlier. The Sankey charts and Grassmann diagrams are showing the first law output and second law output respectively. On an average, the energy of hot gas supply is lost 60% with exhaust and 35% at condenser. Nearly 2-6% is the output from the plant's energy supply. So the energy balance reveals that the majority of energy lost is through the heat rejection by exhaust gas and hot water from water condenser. The balanced energy is resulted as power generation. Grassmann pictures show the exergy losses in percentage loss of exergy of hot gas. The depicted exergy losses are HRVG, turbine, regenerator, condenser, pump, exhaust and hot water. The exergy output (work) is obtained from the exergy balance applied to plant. Similar to first law results, the second law losses are also more with exhaust (25% to 40%). After the irreversibility of exhaust, HRVG causes major losses. The balance components offer nominal exergy losses.

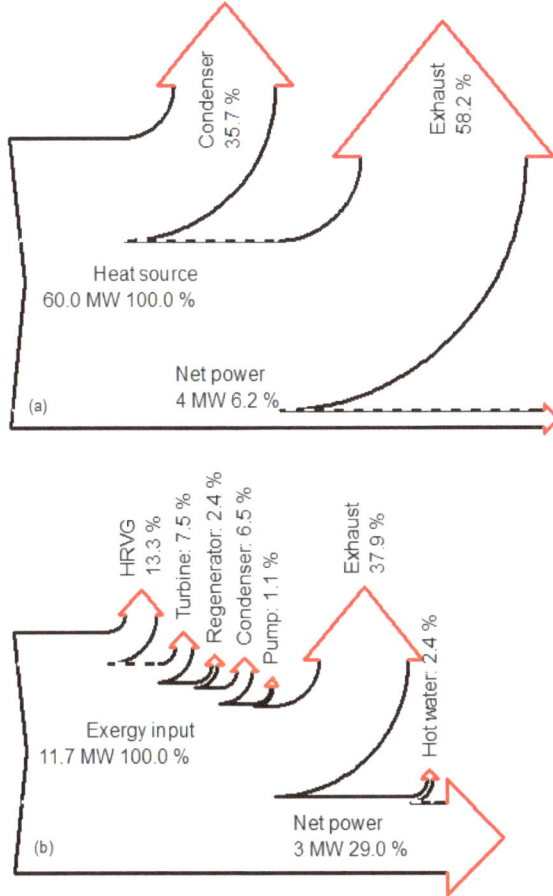

Fig. (28). (a) Sankey diagram and **(b)** Grassmann diagram of ORC with R123.

Table 7. ORC with regenerator material balance results with R407C zeotrope at hot gas supply of 78 °C.

State	P, bar	t, °C	m, kg/s	h, kJ/kg	s, kJ/kg K
1.	21.42	63.00	22.41	441.67	1.76
2.	17.50	53.54	22.41	437.36	1.76
3.	17.50	50.95	22.41	433.78	1.75
4.	17.50	40.00	22.41	260.34	1.20

(Table 7) cont.....

State	P, bar	t, °C	m, kg/s	h, kJ/kg	s, kJ/kg K
5.	21.42	40.37	22.41	260.83	1.20
6.	21.42	42.61	22.41	264.57	1.22
7.	21.42	46.52	22.41	271.16	1.24
8.	21.42	53.05	22.41	426.26	1.71
9.	1.01	25.00	29.72	0.00	0.00
10.	1.01	35.00	339.21	4.09	0.01
11.	1.01	900.00	368.90	1006.40	1.54
12.	1.01	78.00	368.90	54.82	0.17
13.	1.01	68.07	368.90	44.47	0.14
14.	1.01	68.05	368.90	44.47	0.14
15.	1.01	67.69	368.90	44.07	0.14
16.	1.01	25.00	116.22	0.00	0.00
17.	1.01	33.00	116.22	33.44	0.11

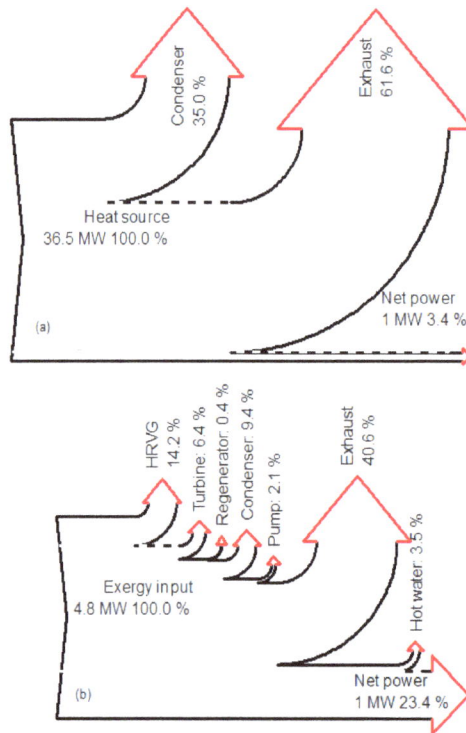

Fig. (29). (a) Sankey diagram and (b) Grassmann diagram of ORC with R124.

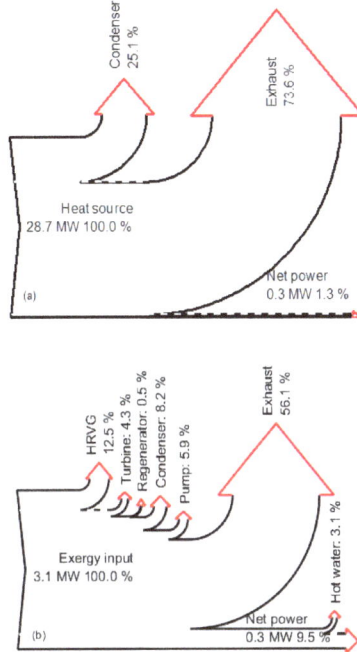

Fig. (30). (a) Sankey diagram and **(b)** Grassmann diagram of ORC with R134a.

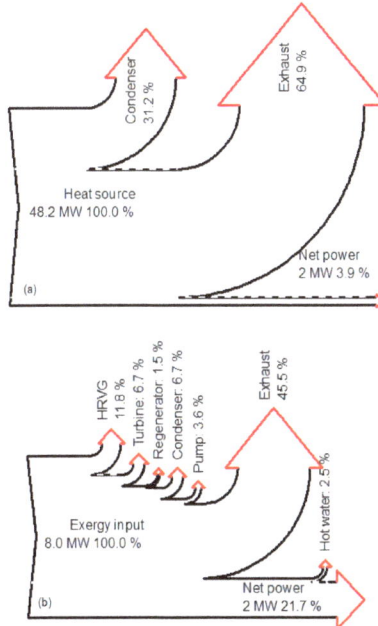

Fig. (31). (a) Sankey diagram and **(b)** Grassmann diagram of ORC with R245fa.

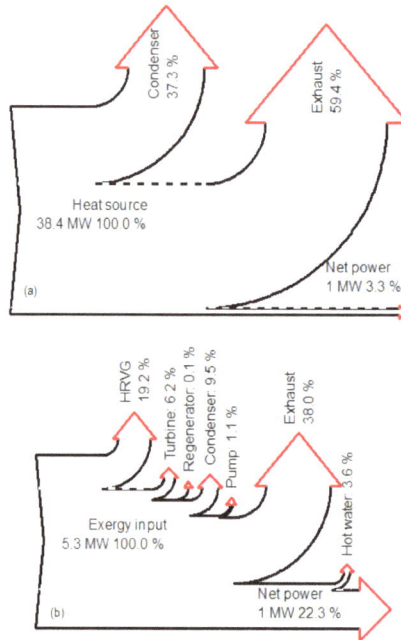

Fig. (32). (a) Sankey diagram and **(b)** Grassmann diagram of ORC with R717.

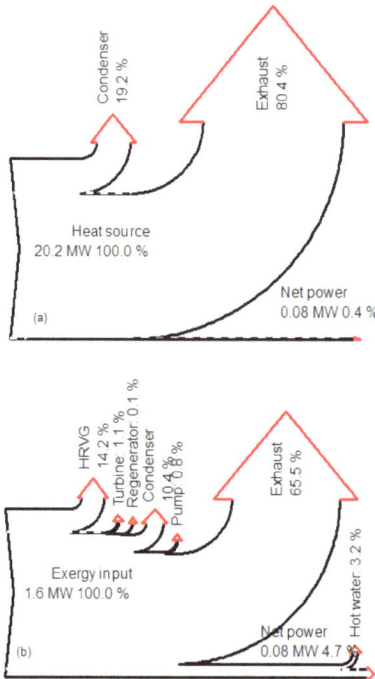

Fig. (33). (a) Sankey diagram and **(b)** Grassmann diagram of ORC with R407C.

Performance Comparison of ORC with Fluids

Fig. (**34**) compares the performance (power and efficiency) changes with change in working fluid along with temperature. The six fluids are studied under the variable source temperature to recommend the suitable fluid to the source temperature. Out of six fluids, one fluid is the zeotrope mixture. It gives the advantage of variable temperature during the phase change. The suitable optimum HRVG pressure is chosen with the working fluid from the source and sink temperatures as discussed in the earlier section.

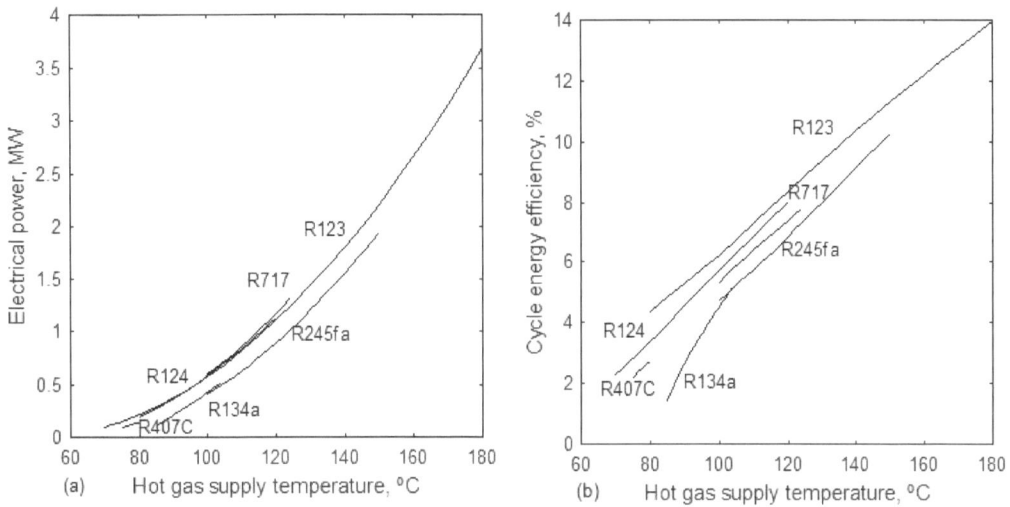

Fig. (34). Performance comparison of ORC with working fluids.

The fluids R123, R124 and R717 generate nearly same amount of power under the specified working conditions but with different temperature ranges (Fig. **34a**). The adopted temperature range in R123 is more compared to the other fluids. R407C and R134a works only at lower source temperature. Compared to R134a, R407C can be used still low temperature. But they are not able to dominate the power and efficiency levels of other fluids *i.e.* R123, R124 and R717. These three fluids are generating nearly equal power generation. On basis of thermal efficiency, R124 proves well compared to R123 and R717. R245fa is not able to cross the performance levels of R123, R124 and R717. The ORC with R407C is not competitive for thermal efficiency as it is suitable at lowest temperature. Finally, R123 is recommended to generate more power with wide range of source temperature.

Table **8** compares the specifications of ORC with regenerator at the exit of turbine. The critical temperature and critical pressure of the selected fluids are listed to compare the sub-critical temperature of the fluid. R123 has high critical temperature and suitable for large temperature variation in source. The optimum HRVG pressure is presented with fluid, source temperature and sink temperature. A maximum optimum pressure is resulted with R717. A minimum boiler pressure is resulted with R123. The condenser pressure is determined from the sink temperature. A minimum condenser pressure also resulted with R123. The maximum condenser pressure is with R407C and R717. The specific power is the power per unit mass of working fluid. The mass of working fluid is taken at the inlet of turbine. The maximum specific power is resulted with R717. R123 results a maximum thermal efficiency and exergy efficiency with high temperature. R407C and R134a are suitable to low temperature and these efficiencies are low. To justify the efficiency irrespective of the source temperature, relative cycle efficiency has significant weightage which is defined as the ratio of thermal efficiency to the Carnot efficiency. The highest relative efficiency is resulted with R407C and R123.

Table 8. Comparison of ORC specifications with working fluids at optimized HRVG pressure and below the critical temperature.

Working Fluid	R123	R124	R134a	R245fa	R717	R407C
Critical temperature, °C	183.68	122.47	101.08	154.10	132.40	86.03
Critical pressure, bar	36.68	36.34	40.60	36.40	113.40	46.30
Hot fluid inlet temperature, °C	180.00	120.00	100.00	150.00	125.00	78.00
Turbine inlet temperature, °C	165.00	105.00	85.00	135.00	110.00	63.00
Optimized HRVG pressure, Bar	9.23	13.67	20.17	11.55	36.48	21.42
Condenser pressure, Bar	1.55	5.93	10.17	2.50	15.55	17.50
Specific power, kW/kg	28.33	12.06	8.17	22.67	87.12	3.64
Plant power, MW	3.41	0.96	0.33	1.72	1.18	0.08
Cycle thermal efficiency, %	13.59	7.78	4.33	10.21	8.03	2.00
Relative cycle efficiency, %	52.90	46.55	41.50	49.31	49.01	55.83
Plant exergy efficiency, %	29.02	20.00	10.56	21.65	22.25	4.92

ORGANIC RANKINE CYCLE WITH SOURCE TEMPERATURE ABOVE CRITICAL TEMPERATURE

The same configuration shown in Fig. (**3**) is used for ORC at the source temperature above the critical temperature. Similar to the earlier sections, optimum HRVG pressure has been evaluated at the source temperature but above the critical temperature of the fluid. At this temperature, the effect of sink is analyzed to observe any deviations in the pressure optimization.

Optimum Boiler Pressure with Source Temperature

Fig. (**35a**) gave some interesting results to choose the HRVG pressure at known source temperature to gain the advantage of high power. The optimum HRVG pressure is increasing with increase in source temperature to result maximum power. Maximum thermal efficiency condition is demanding higher HRVG pressure greater than optimum pressure (Fig. **35b**). Fig. (**35c**) is prepared to study the variations in DSH resulted with changes in temperature and pressure. The DSH is increasing with increase in source temperature and decreasing with rise in HRVG pressure. The exhaust gas temperature is increasing with increase in temperature and pressure (Fig. **35d**).

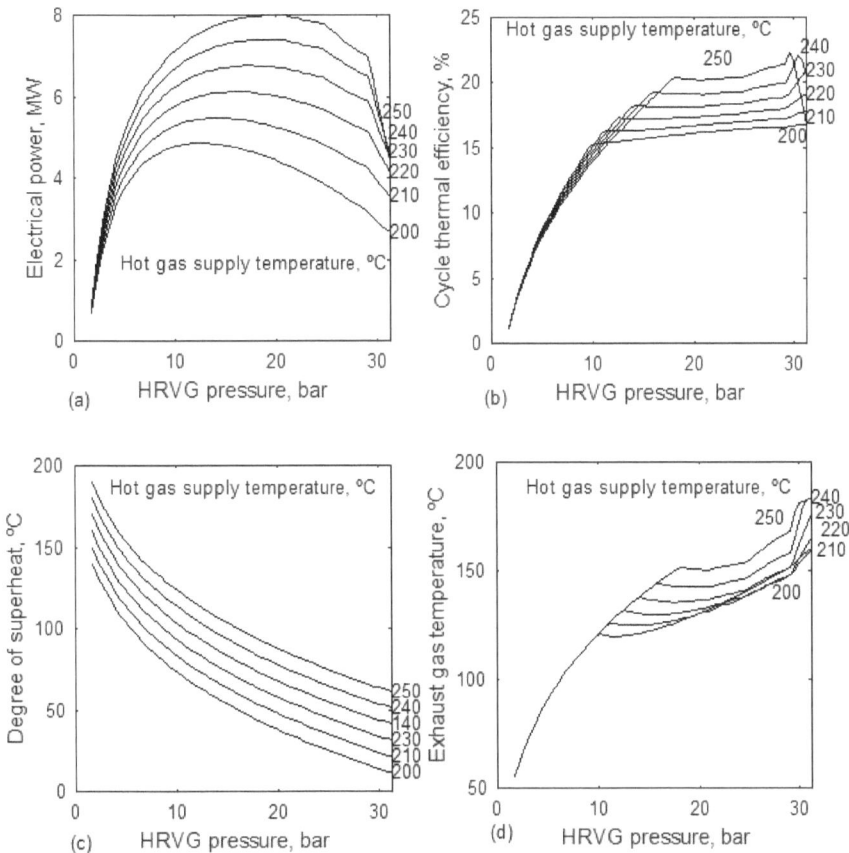

Fig. (35). Influence of source temperature with HRVG pressure on (**a**) net power generation, (**b**) cycle energy conversion efficiency (**c**) degree of superheat and (**d**) exhaust gas temperature with R123 fluid.

Fig. (**36**) shows the role of source temperature and HRVG pressure on ORC output (Fig. **36a**) and its thermal efficiency (Fig. **36b**) for the working fluid of R124. The changes in DSH with temperature and pressure also included to analyze the role of super heater on performance (Fig. **36c**). As per the earlier results, the optimum boiler pressure is increasing with increase in temperature. DSH decreases with increase in pressure due to increase in saturation temperature at a fixed superheated temperature. As per the properties of R124, the source temperature is varied from 140 °C to 185 °C. The corresponding pressure is changed from 7 bar to 30 bar.

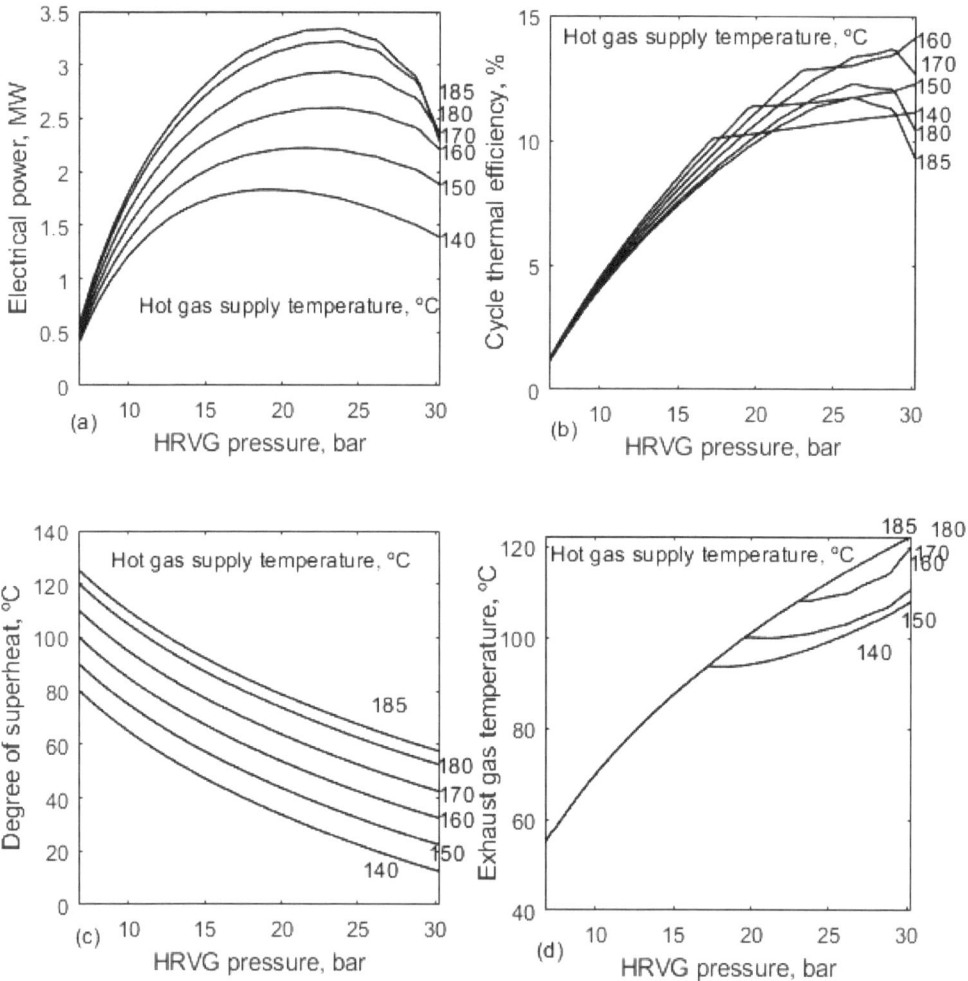

Fig. (36). Influence of source temperature with HRVG pressure on (**a**) net power generation, (**b**) cycle energy conversion efficiency (**c**) degree of superheat and (**d**) exhaust gas temperature with R124 fluid.

Fig. (**37**) indicates the (**a**) plant power generation, (**b**) thermal efficiency (**c**) degree of superheat and exhaust gas temperature with change in source temperature and boiler pressure for R134a.

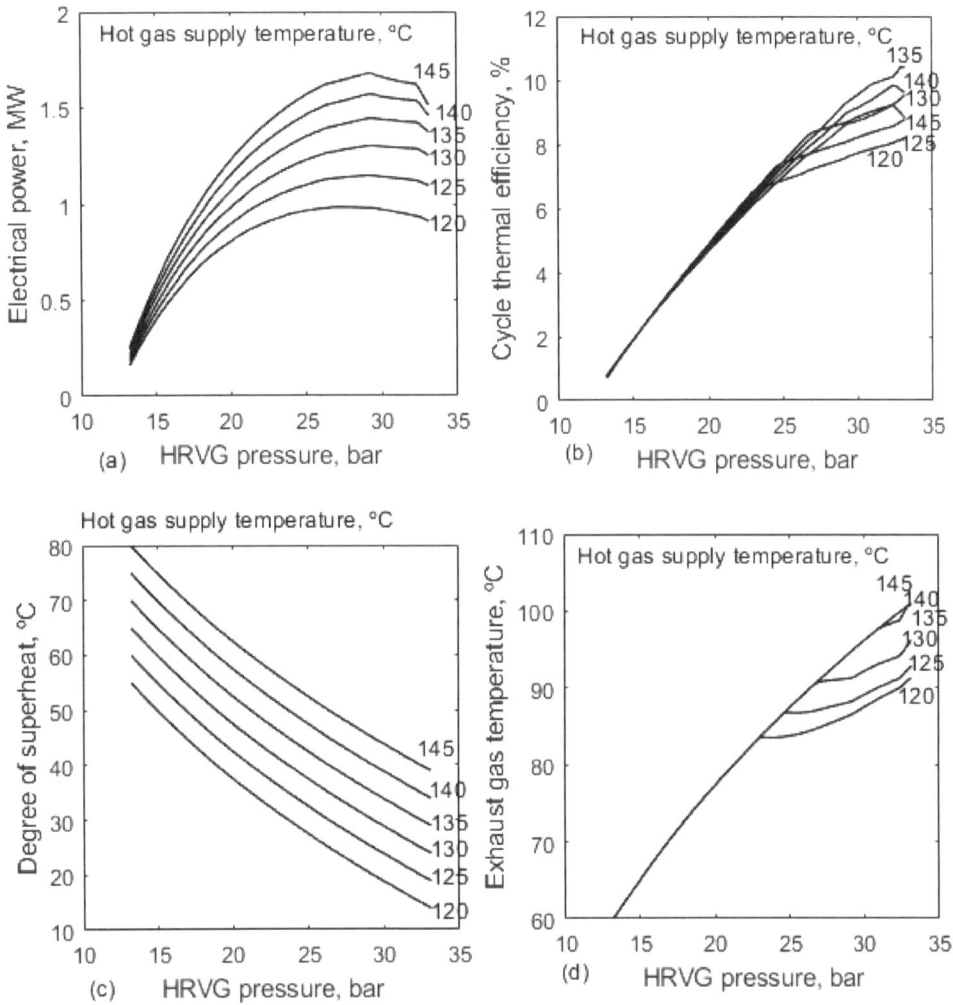

Fig. (37). Influence of source temperature with HRVG pressure on (**a**) net power, (**b**) cycle energy conversion efficiency (**c**) degree of superheat and (**d**) exhaust gas temperature with R134a fluid.

Fig. (**38**) shows the optimum HRVG pressure of R245fa. The source temperature is changed from 170 °C to 220 °C.

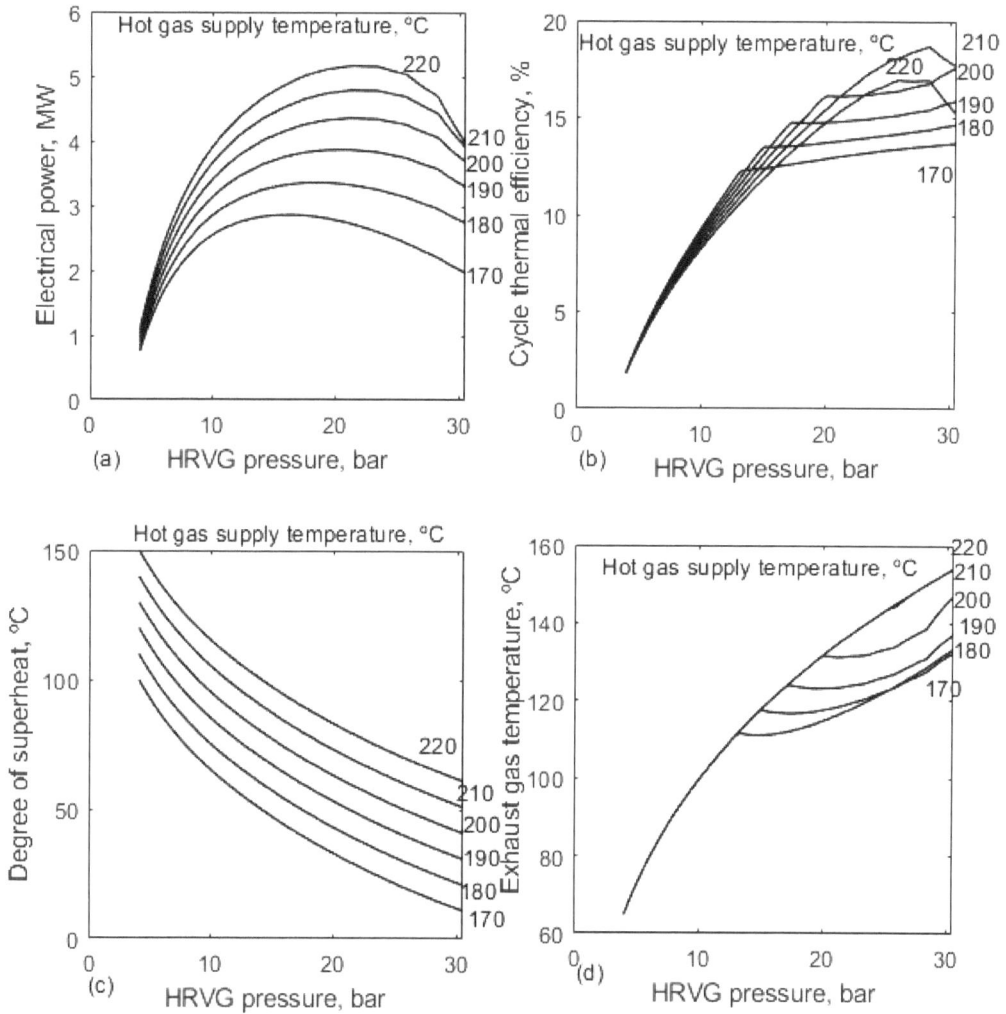

Fig. (38). Influence of source temperature with HRVG pressure on **(a)** net power generation, **(b)** cycle energy conversion efficiency **(c)** degree of superheat and **(d)** exhaust gas temperature with R245fa fluid.

Fig. (**39**) results the optimum HRVG pressure to generate maximum power as a function of temperature and pressure at a fixed condenser state (Fig. **39a**). Figs. (**39b** - **d**) shows respectively thermal efficiency, DSH and exhaust gas temperature.

The optimum HRVG pressures are developed with change in source temperature with R407C fluid (Fig. **40**).

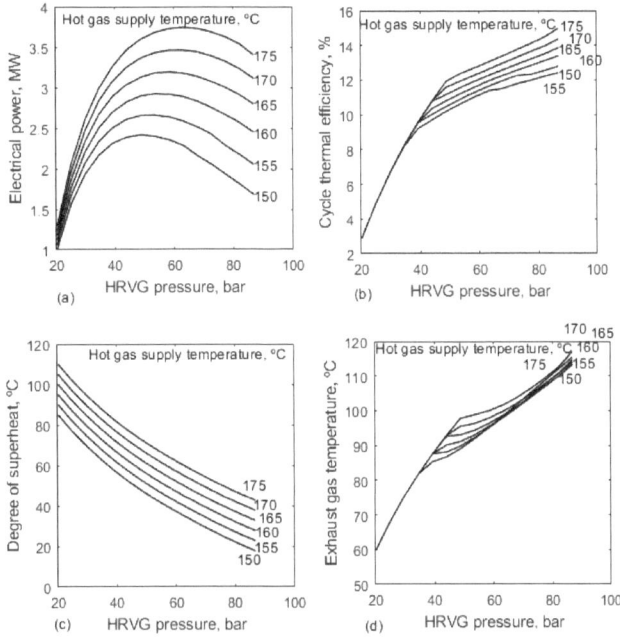

Fig. (39). Influence of source temperature with HRVG pressure on **(a)** net power generation, **(b)** cycle energy conversion efficiency **(c)** degree of superheat and **(d)** exhaust gas temperature with R717 fluid.

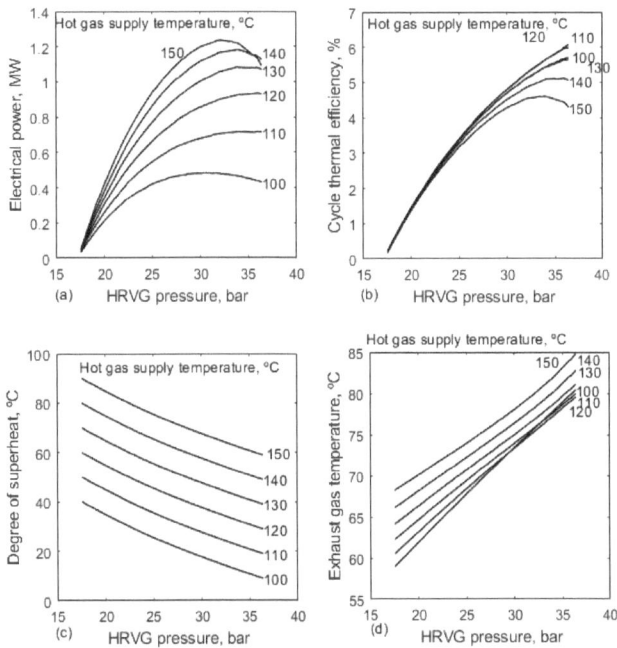

Fig. (40). Influence of source temperature with HRVG pressure on **(a)** net power, **(b)** cycle energy conversion efficiency **(c)** degree of superheat and **(d)** exhaust gas temperature with R407C fluid.

Similar to the earlier section, the optimum pressure of HRVG has been evaluated from the curve fitting equation. The coefficients of non-linear equation have been listed in Table **9** for the six working fluid.

Table 9. Developed coefficients to find the optimum HRVG pressure at source temperature above the critical temperature of fluid and 30 °C of circulating water temperature (condensate temperature is 40 °C).

Working Fluid	Coefficients		
	a_1	a_2	a_3
R123	-25.7219	0.2197	-0.0001
R124	-102.499	1.4549	-0.0042
R134a	-127.426	2.3111	-0.0085
R245fa	-141.858	1.5706	-0.0038
R717	-129.959	1.7036	-0.0034
R407C	-67.9813	1.6551	-0.0066

Effect of Sink Temperature

The sink temperature is changed with the source temperature above the critical temperature. Fig. (**41**) shows the performance of ORC-R123 with change in sink temperature on power and efficiency. The HRVG pressure is changed from condenser pressure to close to a critical pressure. In the results, the HRVG pressure has been analyzed to study its changes with the sink temperature. The lower limit of pressure range is increasing with increase in sink temperature. The sink temperature is increased from 20 °C to 35 °C. Since the source temperature is fixed, the final pressure in the analysis is same to all the sink temperatures. The results Fig. (**41a**), shows that a minor increment in optimum pressure with increase in sink temperature for maximum power. The efficiency of the cycle decreasing with increase in sink temperature as per the law (Fig. **41b**). For maximum thermal efficiency, optimum pressure is not resulted due to lack of maximum efficiency. Finally, the optimum pressure can be chosen for the maximum power production from the source and sink temperatures. The analysis of sink on performance is extended with the other fluids *viz.* R124, R134a, R245fa, R717 and R407C (Figs. **42 - 46**). The source temperature is fixed above the critical temperature but the pressure is limited up to critical pressure. For example, for the R123 sink characteristics, the source temperature is fixed at 250 °C. The hot gas supply temperature is 180 °C, 145 °C, 220 °C, 175 °C and 120 °C respectively with R124, R134a, R245fa, R717 and R407C to study the optimum pressure with sink temperature.

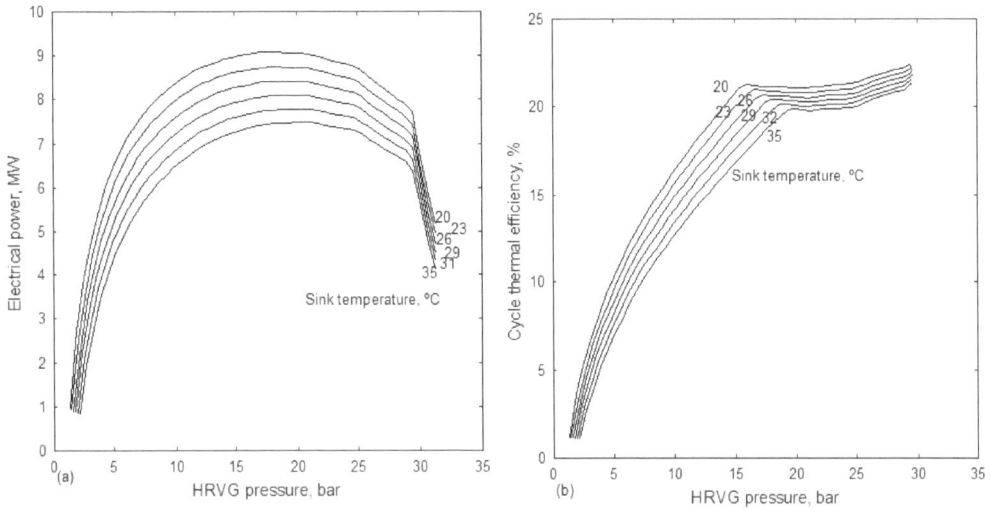

Fig. (41). Influence of sink temperature with HRVG pressure on **(a)** net power generation and **(b)** cycle energy conversion efficiency with R123.

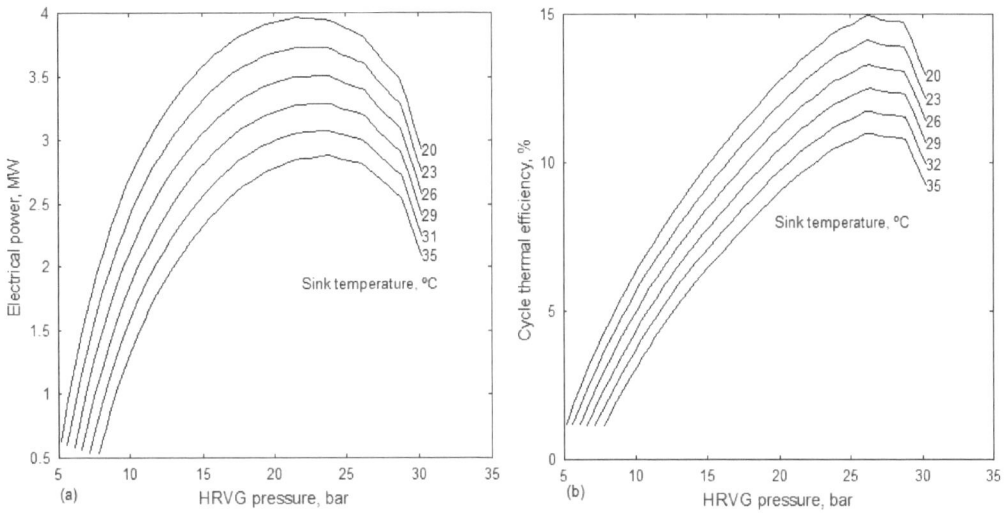

Fig. (42). Influence of sink temperature with HRVG pressure on **(a)** net power generation and **(b)** cycle energy conversion efficiency with R124.

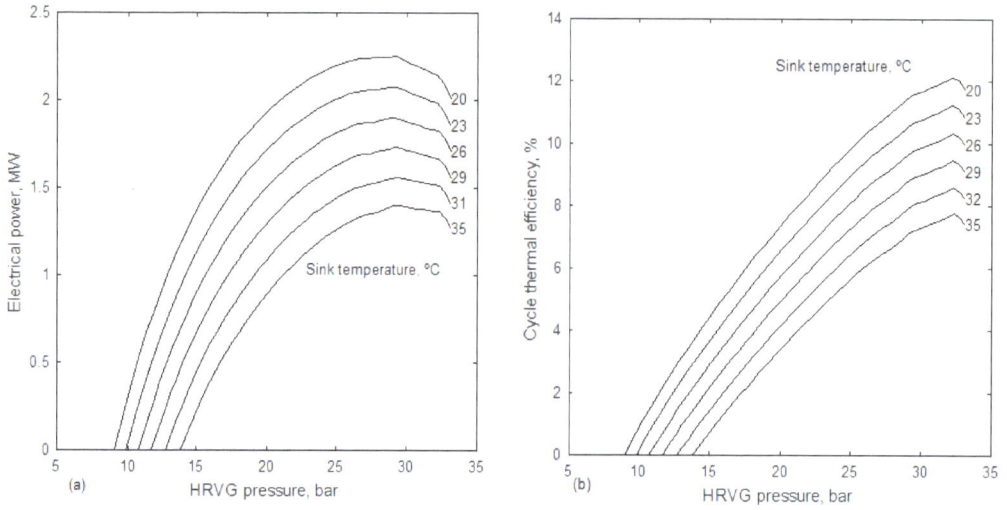

Fig. (43). Influence of sink temperature with HRVG pressure on **(a)** net power generation and **(b)** cycle thermal efficiency with R134a.

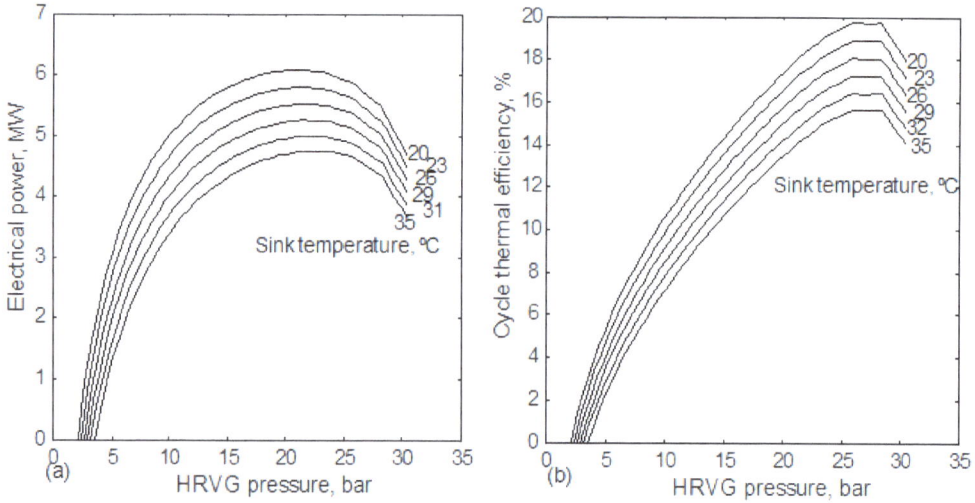

Fig. (44). Influence of sink temperature with HRVG pressure on **(a)** net power generation and **(b)** cycle thermal efficiency with R245fa.

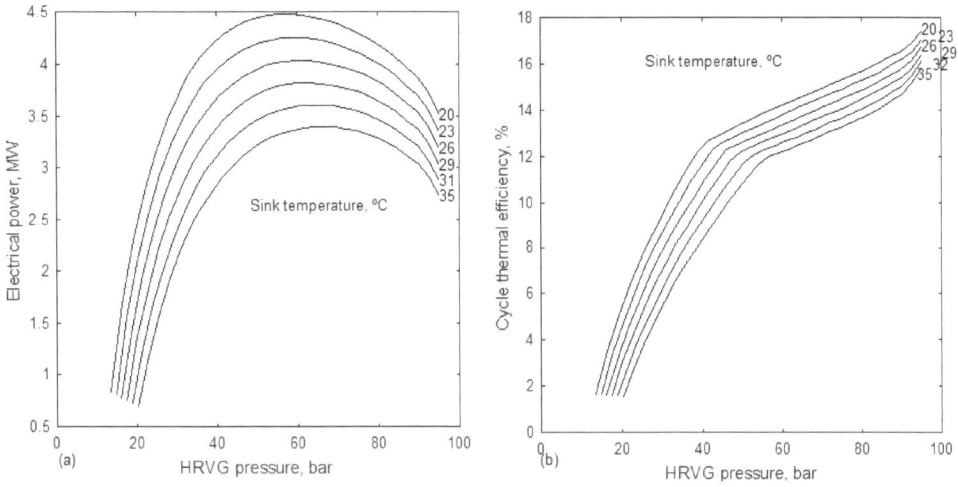

Fig. (45). Influence of sink temperature with HRVG pressure on **(a)** net power generation and **(b)** cycle thermal efficiency with R717.

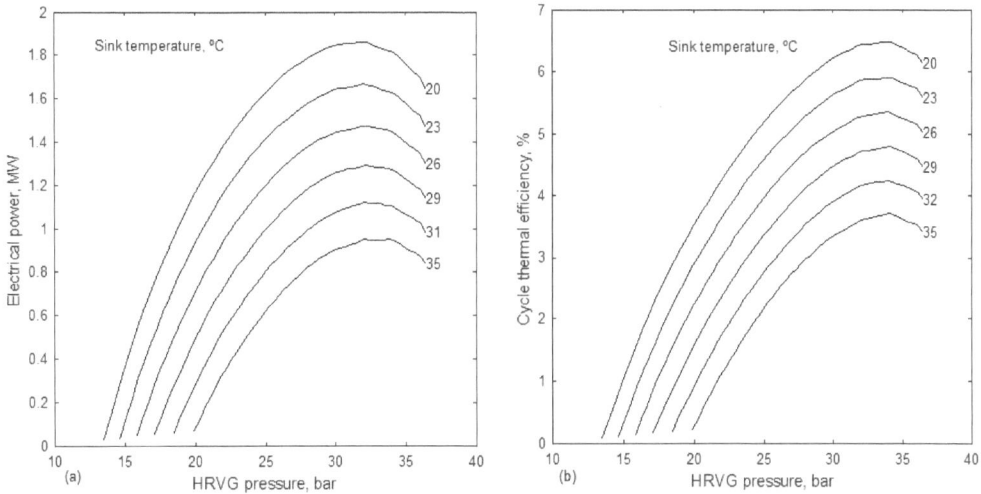

Fig. (46). Influence of sink temperature with HRVG pressure on **(a)** net power generation and **(b)** cycle thermal efficiency with R407C.

Property Charts and Tables at the Optimized Conditions

Figs. (**47 - 52**) shows the **(a)** pressure-specific enthalpy and **(b)** temperature-entropy plots for the regenerative ORC with the working fluids of R123, R124, R134a, R245fa, R717 and R407C. The property charts and tables are prepared at the optimum pressure and with the turbine inlet temperature above the critical temperature.

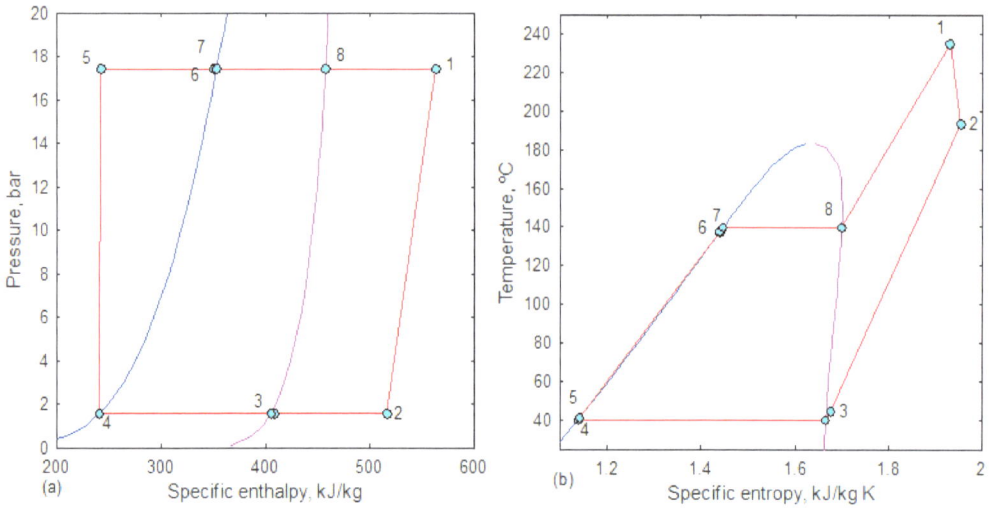

Fig. (47). Temperature-entropy diagram of ORC with regenerator with the fluid of R123.

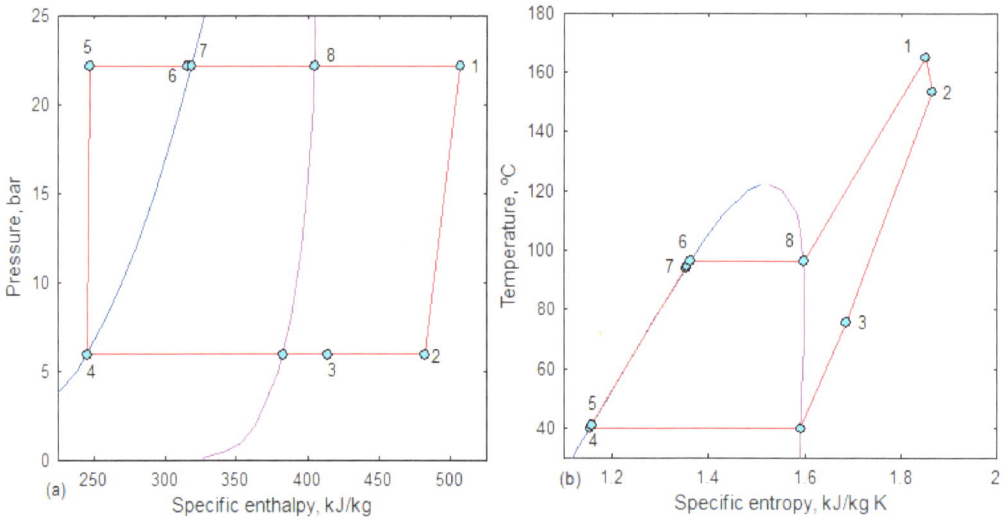

Fig. (48). Temperature-entropy diagram of ORC with regenerator with the fluid of R124.

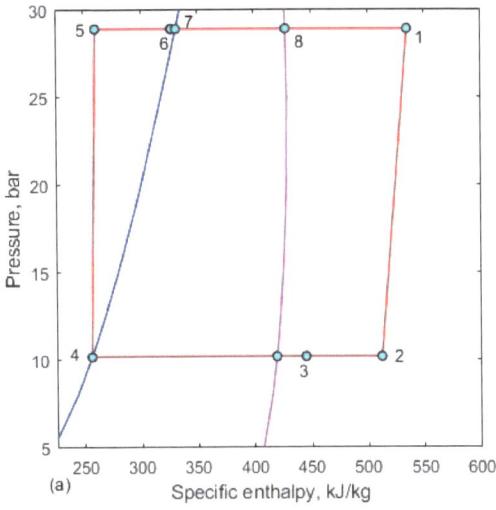

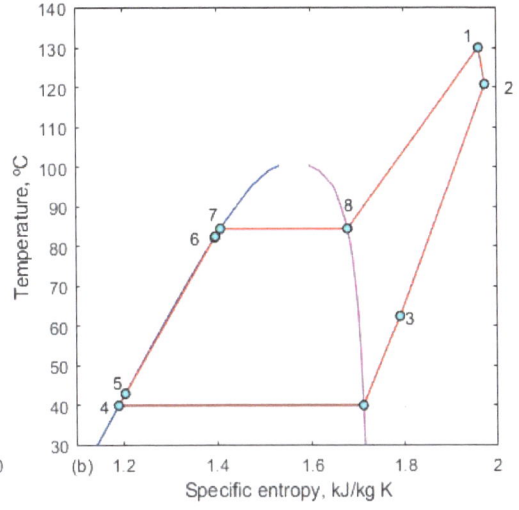

Fig. (49). Temperature-entropy diagram of ORC with regenerator with the fluid of R134a.

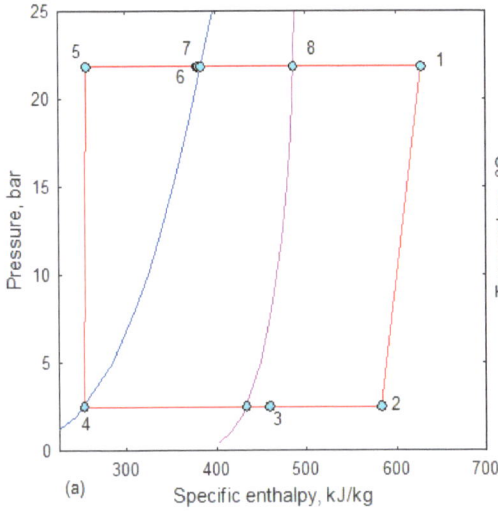

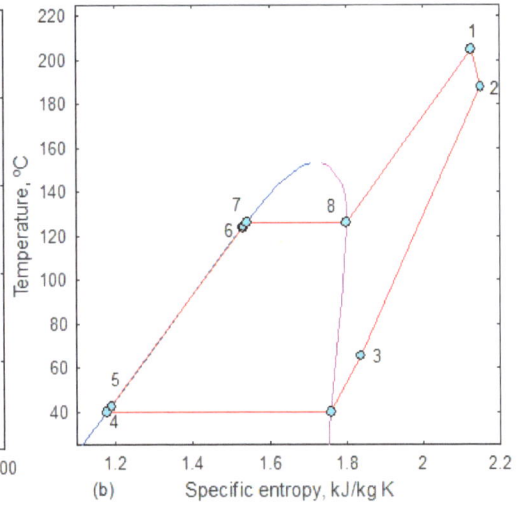

Fig. (50). Temperature-entropy diagram of ORC with regenerator with the fluid of R245fa.

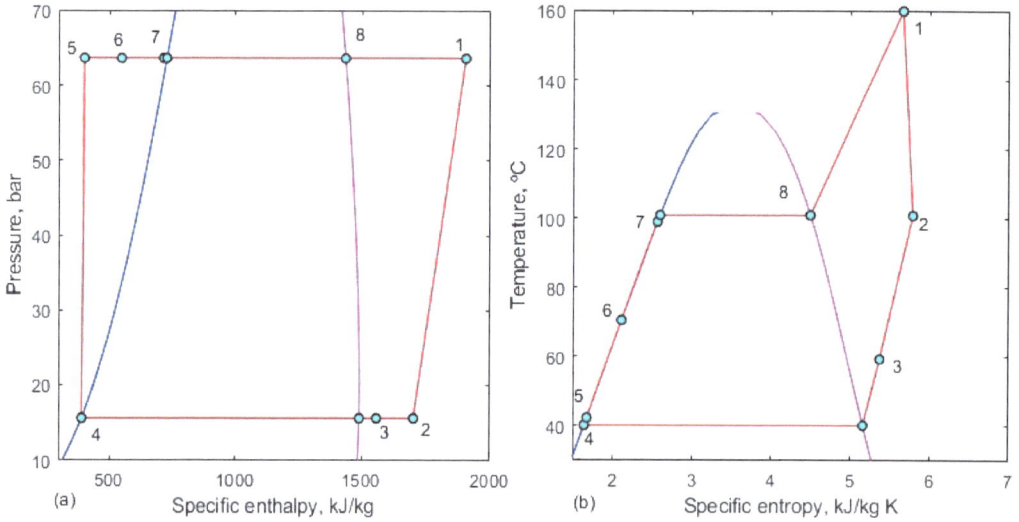

Fig. (51). Temperature-entropy diagram of ORC with regenerator with the fluid of R717.

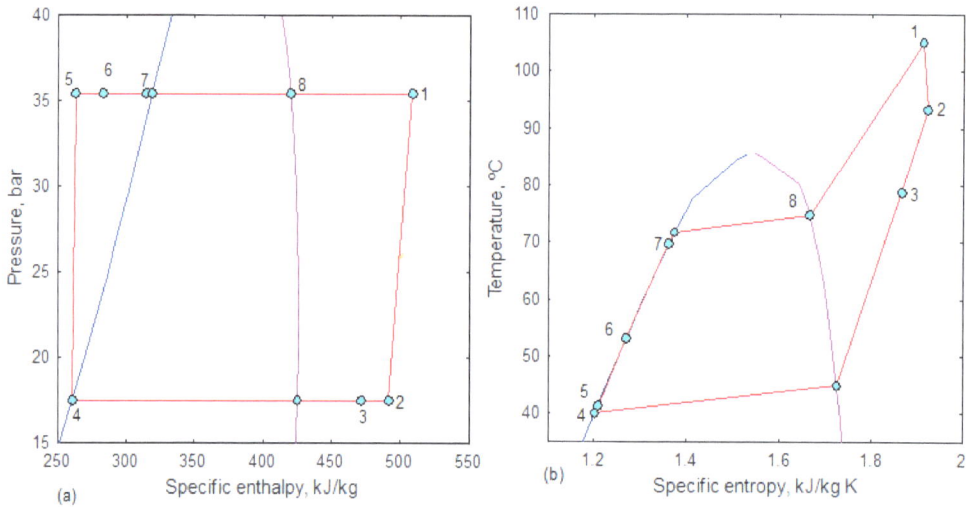

Fig. (52). Temperature-entropy diagram of ORC with regenerator with the fluid of R407C.

The properties of pressure, temperature, flow rate, specific enthalpy and specific entropy are tabulated from Tables (**10** to **15**) respectively for the working fluids of R123, R124, R134a, R245fa, R717 and R407C.

Table 10. ORC material balance results with R123 fluid at hot gas supply of 250 °C.

State	P, bar	t, °C	m, kg/s	h, kJ/kg	s, kJ/kg K
1.	22.95	235.00	169.54	570.57	1.93
2.	1.54	196.71	169.54	519.00	1.96
3.	1.54	58.64	169.54	419.04	1.71
4.	1.54	40.00	169.54	240.59	1.14
5.	22.95	41.45	169.54	242.59	1.14
6.	22.95	130.68	169.54	342.11	1.42
7.	22.95	153.02	169.54	371.34	1.49
8.	22.95	155.02	169.54	462.03	1.70
9.	1.01	25.00	29.72	0.00	0.00
10.	1.01	35.00	339.21	4.09	0.01
11.	1.01	900.00	368.90	1006.40	1.54
12.	1.01	250.00	368.90	238.12	0.59
13.	1.01	165.05	368.90	146.56	0.40
14.	1.01	165.02	368.90	146.56	0.40
15.	1.01	152.43	368.90	133.13	0.37
16.	1.01	25.00	904.69	0.00	0.00
17.	1.01	33.00	904.69	33.44	0.11

Table 11. ORC with regenerator material balance results with R124 fluid at hot gas supply of 180 °C.

State	P, bar	t, °C	m, kg/s	h, kJ/kg	s, kJ/kg K
1.	23.30	165.00	146.37	508.83	1.85
2.	5.93	155.14	146.37	483.51	1.87
3.	5.93	73.13	146.37	411.50	1.68
4.	5.93	40.00	146.37	245.07	1.15
5.	23.30	41.09	146.37	246.85	1.16
6.	23.30	96.84	146.37	318.73	1.36
7.	23.30	96.92	146.37	318.86	1.36
8.	23.30	98.92	146.37	405.06	1.60
9.	1.01	25.00	29.72	0.00	0.00
10.	1.01	35.00	339.21	4.09	0.01
11.	1.01	900.00	368.90	1006.40	1.54
12.	1.01	180.00	368.90	162.56	0.44
13.	1.01	108.94	368.90	87.18	0.26

(Table 11) cont.....

State	P, bar	t, °C	m, kg/s	h, kJ/kg	s, kJ/kg K
14.	1.01	108.92	368.90	87.18	0.26
15.	1.01	108.89	368.90	87.13	0.26
16.	1.01	25.00	728.49	0.00	0.00
17.	1.01	33.00	728.49	33.44	0.11

Table 12. ORC with regenerator material balance results with R134a fluid at hot gas supply of 145 °C.

State	P, bar	t, °C	m, kg/s	h, kJ/kg	s, kJ/kg K
1.	28.97	130.00	94.42	535.19	1.96
2.	10.17	120.76	94.42	511.90	1.97
3.	10.17	62.34	94.42	445.01	1.79
4.	10.17	40.00	94.42	256.41	1.19
5.	28.97	42.98	94.42	260.41	1.20
6.	28.97	82.33	94.42	326.89	1.40
7.	28.97	82.51	94.42	327.32	1.40
8.	28.97	84.51	94.42	427.86	1.68
9.	1.01	25.00	29.72	0.00	0.00
10.	1.01	35.00	339.21	4.09	0.01
11.	1.01	900.00	368.90	1006.40	1.54
12.	1.01	145.00	368.90	125.27	0.35
13.	1.01	94.53	368.90	72.07	0.22
14.	1.01	94.51	368.90	72.07	0.22
15.	1.01	94.43	368.90	71.96	0.22
16.	1.01	25.00	532.53	0.00	0.00
17.	1.01	33.00	532.53	33.44	0.11

Table 13. ORC with regenerator material balance results with R245fa fluid at hot gas supply of 220 °C.

State	P, bar	t, °C	m, kg/s	h, kJ/kg	s, kJ/kg K
1.	19.75	205.00	138.22	624.59	2.12
2.	2.50	185.81	138.22	581.13	2.14
3.	2.50	72.01	138.22	466.08	1.86
4.	2.50	40.00	138.22	252.57	1.18
5.	19.75	42.73	138.22	256.20	1.19
6.	19.75	118.98	138.22	370.81	1.51
7.	19.75	119.19	138.22	371.25	1.51

(Table 13) cont.....

State	P, bar	t, °C	m, kg/s	h, kJ/kg	s, kJ/kg K
8.	19.75	121.19	138.22	484.83	1.80
9.	1.01	25.00	29.72	0.00	0.00
10.	1.01	35.00	339.21	4.09	0.01
11.	1.01	900.00	368.90	1006.40	1.54
12.	1.01	220.00	368.90	205.57	0.53
13.	1.01	131.22	368.90	110.65	0.32
14.	1.01	131.19	368.90	110.65	0.32
15.	1.01	131.06	368.90	110.49	0.32
16.	1.01	25.00	882.49	0.00	0.00
17.	1.01	33.00	882.49	33.44	0.11

Table 14. ORC with regenerator material balance results with R717 fluid at hot gas supply of 175 °C.

State	P, bar	t, °C	m, kg/s	h, kJ/kg	s, kJ/kg K
1.	64.05	160.00	20.99	1905.10	5.65
2.	15.55	100.78	20.99	1703.26	5.78
3.	15.55	59.47	20.99	1558.25	5.37
4.	15.55	40.00	20.99	390.64	1.64
5.	64.05	42.16	20.99	401.80	1.68
6.	64.05	70.30	20.99	546.53	2.11
7.	64.05	99.17	20.99	715.73	2.57
8.	64.05	101.17	20.99	1433.10	4.48
9.	1.01	25.00	29.72	0.00	0.00
10.	1.01	35.00	339.21	4.09	0.01
11.	1.01	900.00	368.90	1006.40	1.54
12.	1.01	175.00	368.90	157.21	0.43
13.	1.01	111.19	368.90	89.55	0.26
14.	1.01	111.17	368.90	89.55	0.26
15.	1.01	102.03	368.90	79.92	0.24
16.	1.01	25.00	732.78	0.00	0.00
17.	1.01	33.00	732.78	33.44	0.11

Table 15. ORC with regenerator material balance results with R407C zeotrope at hot gas supply of 120 °C.

State	P, bar	t, °C	m, kg/s	h, kJ/kg	s, kJ/kg K
1.	35.59	105.00	69.76	509.02	1.91
2.	17.50	93.37	69.76	492.34	1.92
3.	17.50	78.83	69.76	472.26	1.87
4.	17.50	40.00	69.76	260.34	1.20
5.	35.59	41.35	69.76	262.60	1.21
6.	35.59	53.11	69.76	282.52	1.27
7.	35.59	69.96	69.76	314.72	1.36
8.	35.59	74.98	69.76	419.86	1.67
9.	1.01	25.00	29.72	0.00	0.00
10.	1.01	35.00	339.21	4.09	0.01
11.	1.01	900.00	368.90	1006.40	1.54
12.	1.01	120.00	368.90	98.84	0.29
13.	1.01	85.00	368.90	62.10	0.19
14.	1.01	84.98	368.90	62.10	0.19
15.	1.01	79.16	368.90	56.01	0.17
16.	1.01	25.00	442.08	0.00	0.00
17.	1.01	33.00	442.08	33.44	0.11

Sankey and Grassmann Diagrams

The exergy losses of plant components are expressed as percentage exergy loss of exergy of hot gas at the plant inlet. The energy analysis and exergy analysis are expressed in Sankey diagram and Grassmann diagram respectively. These results are developed at the optimum HRVG pressure with the source temperature above the critical temperature of working fluid. The **(a)** energy balance results and **(b)** exergy balance results are shown from Figs. (**53** - **58**) with working fluids. The energy analysis shows that the major energy loss is with the exhaust gas. As per the second law analysis, Grassmann diagram depicts exergy losses of components as a percentage exergy loss of hot gas. The exergy losses in HRVG, turbine, regenerator, condenser, pump, exhaust and hot water are shown. The energy loss with exhaust is more. After the exhaust, HRVG occupies the major exergy losses.

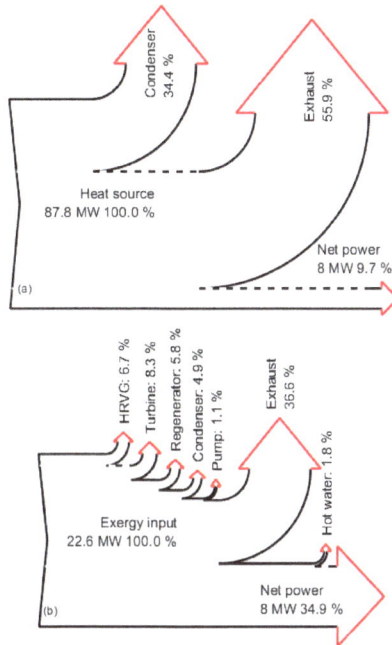

Fig. (53). (a) Sankey diagram and **(b)** Grassmann diagram of ORC with R123.

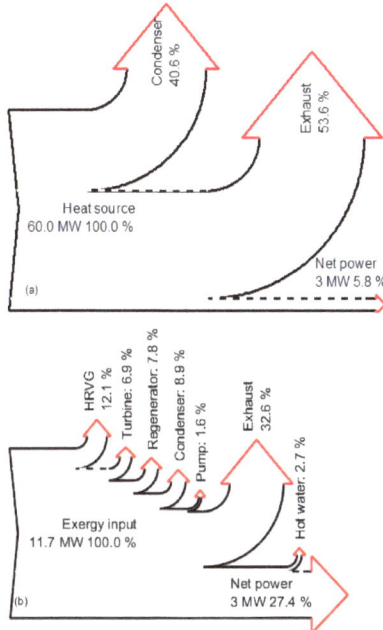

Fig. (54). (a) Sankey diagram and **(b)** Grassmann diagram of ORC with R124.

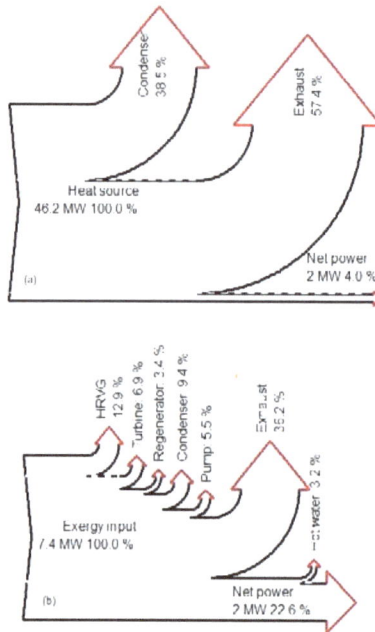

Fig. (55). (a) Sankey diagram and **(b)** Grassmann diagram of ORC with R134a.

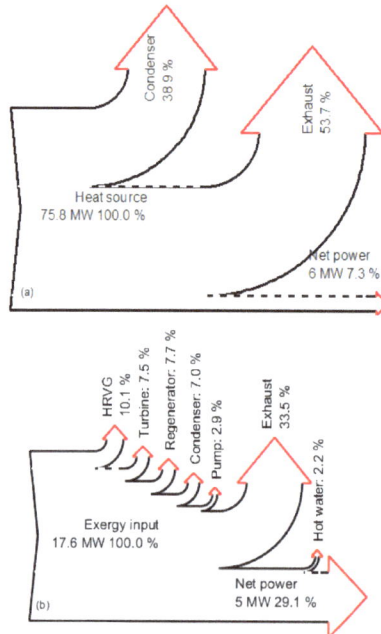

Fig. (56). (a) Sankey diagram and **(b)** Grassmann diagram of ORC with R245fa.

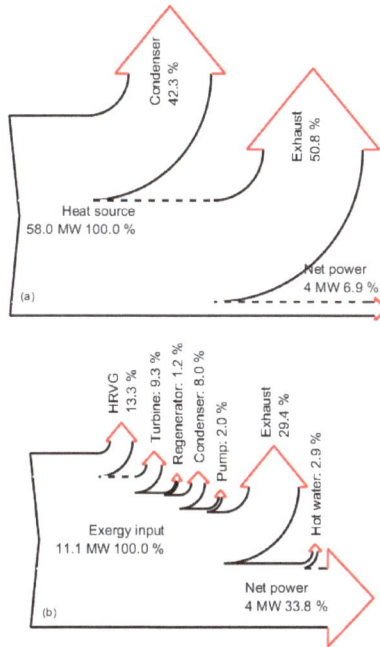

Fig. (57). (a) Sankey diagram and **(b)** Grassmann diagram of ORC with R717.

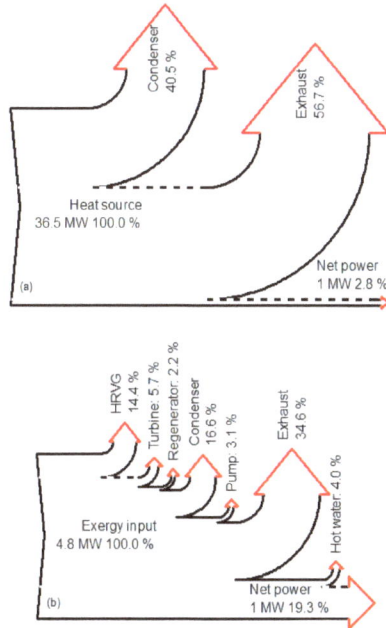

Fig. (58). (a) Sankey diagram and **(b)** Grassmann diagram of ORC with R407C.

Performance Comparison of ORC with Fluids

Figs. (**59a - b**) shows the power and thermal efficiency of the ORC with the working fluids and source temperature. R123 generating highest power compared to the other fluids due to its higher critical temperature. After R123, R245fa is good for higher power production. Since R134a and R407C are operating at lower source temperature, the output and efficiency are low compared to others. The thermal efficiency of R123 and R245fa also high with the high temperature and pressure.

Table **16** summarizes the specifications of ORC with six working fluids. The source temperature is above the critical temperature of the fluid with the pressure below the critical pressure as the current cycles are sub-critical cycles. Since the critical temperature of R123 is greater than the other fluids, its turbine inlet temperature also higher than the other cycles. Based on the parametric optimization, the highest optimum HRVG pressure is with R717 fluid. The lowest HRVG pressure is in ORC-R245fa. But the lowest condenser pressure is in R123 fluid. Since the optimum pressure is high with R717, its specific power production is maximum. But the total power from the plant is high with R123. The first law efficiency (energy) and the second law efficiency (exergy) are maximum with R123 fluid as its turbine inlet temperature is high compared to the others. A high relative cycle thermal efficiency is resulted with 407C and the lowest relative efficiency is with R134a.

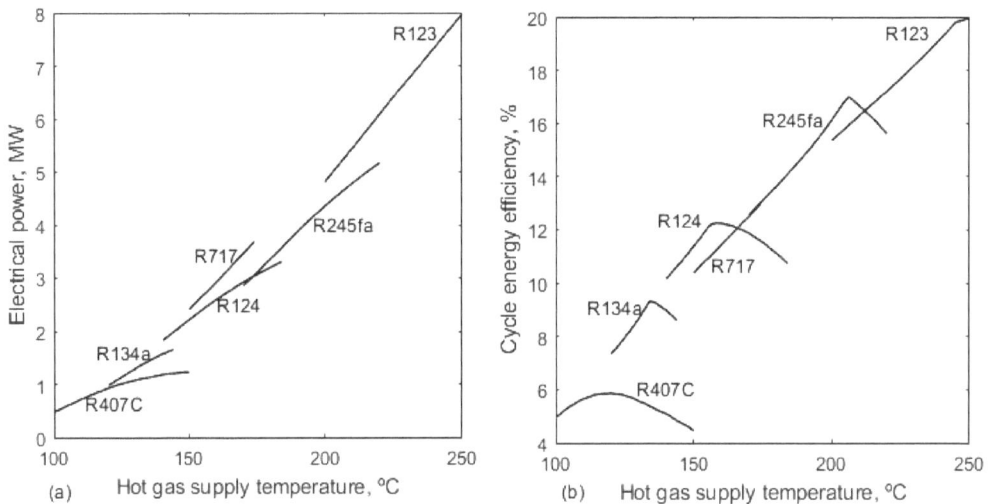

Fig. (59). Performance comparison of ORC with working fluids.

Table 16. Comparison of ORC specifications with working fluids at optimized HRVG pressure and above the critical temperature.

Working Fluid	R123	R124	R134a	R245fa	R717	R407C
Hot fluid inlet temperature, °C	250.00	180.00	145.00	220.00	175.00	120.00
Turbine inlet temperature, °C	235.00	165.00	130.00	205.00	160.00	105.00
Optimized HRVG pressure, bar	22.95	23.30	28.97	19.75	64.04	35.59
Condenser pressure, bar	1.55	5.93	10.17	2.50	15.55	17.50
Specific power, kW/kg	47.48	21.96	17.70	37.11	178.26	13.27
Plant power, MW	7.87	3.21	1.67	5.13	3.74	0.93
Cycle thermal efficiency, %	20.32	11.55	8.52	14.62	13.12	5.89
Relative cycle efficiency, %	58.56	55.11	50.74	52.58	56.53	85.11
Plant exergy efficiency, %	34.87	27.37	22.60	29.12	33.77	19.27

SUMMARY

The performance of ORC power plants is studied with turbine inlet temperature below and above the critical temperature. Six working fluids are used in the ORC *viz.* R123, R124, R134a, R245fa, R717 and R407C. The optimum HRVG pressure at the various source temperatures are presented after maximizing the power and thermal efficiencies. The exergetic losses in cycle components are compared with the fluids. The influence of sink temperature is tested with the association of HRVG pressure. The source temperature with R407C and R134a is low hence the power and thermal efficiency are less than the others. R123 is recommended for maximum power and thermal efficiency with large range in temperature. Naturally thermal efficiency increases with increase in source temperature. Therefore, to justify the thermal power cycle's potential, relative efficiencies are compared without constraint of source temperature. A higher relative efficiency is resulted with R407C. A higher specific power is generated with R717 fluid.

CHAPTER 5

Organic Flash Cycle

Abstract: In organic flash cycle (OFC), a small amount of liquid is flashed to a low pressure at the exit of the economizer. The resulted vapour from the flasher is used in turbine for power augmentation. It increases the economizer's load hence a loss in thermal efficiency. OFC has been recommended to power plant operated by waste heat recovery as the focus is on maximum power. The modeling, simulation, analysis and optimization of OFC process conditions are presented to highlight its merits compared to organic Rankine cycle (ORC). The performance variations are analyzed with the source temperature below and above the critical temperature of the working fluid. The studied working fluids in OFC are R123, R124, R134a, R245fa, R717 and R407C.

Keywords: Optimum pressure, Organic flash cycle, Power augmentation, Waste heat recovery.

INTRODUCTION

Organic flash cycle (OFC) is a modified cycle of organic Rankine cycle (ORC) with added flashing unit(s). OFC may have a single flasher or multi flashers. The main goal of the OFC is power augmentation without bothering about the penalty on efficiency. OFC demands more heat addition to the cycle at the same source temperature and fluid compared to ORC. A partial liquid is bled from an economizer in a heat recovery vapour generator (HRVG) and flashed to liquid and vapour. The vapour is separated after flashing and supplied to turbine at the appropriate pressure. The balanced liquid from the separator is recycled to the boiler. The extra vapour supplied to the turbine generated more power and meets the objective of heat recovery. The excess heat addition to economizer with flash option decreases the efficiency. Therefore, OFC is recommended only to waste heat recovery units. A power plant with fuel burning in a furnace demand maximum thermal efficiency to save the fuel. In this chapter, more focus has been given to the double flash power plant due to its potential over the single flash plant. The location of low pressure flasher (LPF) and high pressure flasher (HPF) in a double flash plant has been optimized. The source and sink conditions are also examined to maximize the power.

Tangellapalli Srinivas

ORGANIC FLASH CYCLE WITH SOURCE TEMPERATURE BELOW THE CRITICAL TEMPERATURE

As per the available heat source, the turbine inlet temperature may fall below or above the critical temperature of the fluid. The performance characteristics differs in these regions of temperature. The flash cycle in the current study is different compared to the flash cycle used in a geothermal power plant. The schematic flow diagram of conventional flash cycle is shown in Fig. (1) to understand the difference between the current plant and conventional plant. In conventional plant, the complete liquid from the source (1) is flashed into liquid-vapor mixture (2) without subjecting evaporation and superheating. The separated vapor (4) after flashing is expanded in turbine. After expansion, the vapor (5) from the turbine and liquid from the separator (3) are mixed and condensed (6-7) in a closed cycle. The condensed fluid (7) is pumped to well for recirculation.

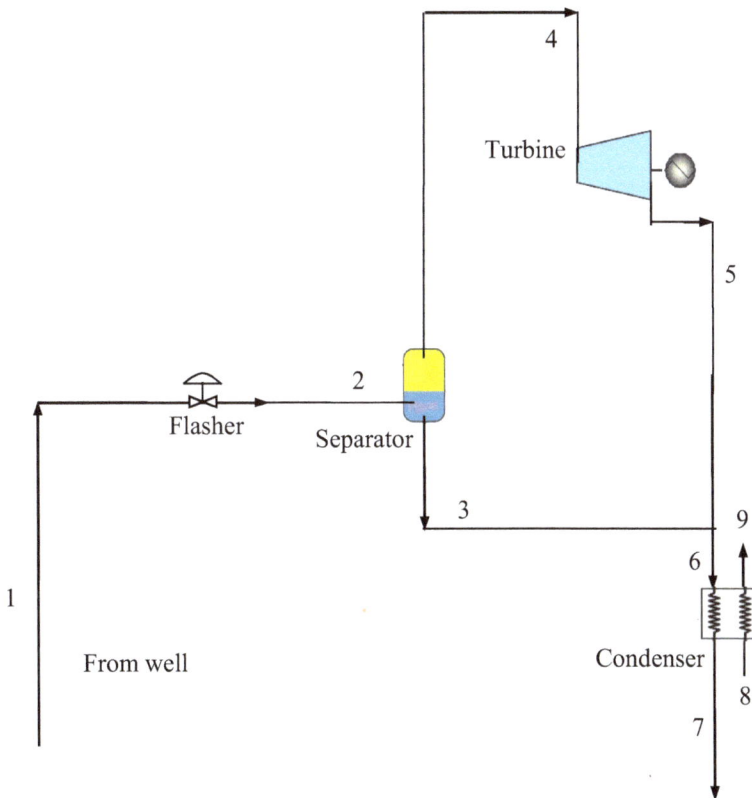

Fig. (1). Flash cycle in a geothermal power plant.

Fig. (2) shows the schematic of a single flash cycle suitable for heat recovery and similar energy sources. The vapour temperature at the exit of the turbine allows a regenerator for internal heat recovery and improves the thermal efficiency. So a regenerator is located in OFC similar to ORC. The pumped fluid (14) is flashed in the flasher from high pressure to a low pressure (15) to generate a liquid-vapor mixture. This mixture is separated into liquid (16) and vapor (17) at the exit of the flasher. The vapour is expanded in the turbine (1-4), and the heat has been recovered in the regenerator (4-5). After the internal heat recovery, the vapour is condensed (5-6) to saturated liquid condition (6). The liquid is pumped (6-7) to flasher's pressure (intermediate pressure). The mixed liquid (9) is again pumped (9-10) from the flash pressure to boiler pressure. The temperature of the liquid is increased to the saturation temperature in the economizer (10-11). The fluid is heated from condenser temperature to saturation temperature. From this saturated state, the liquid is divided into flasher (14) and HRVG (12).

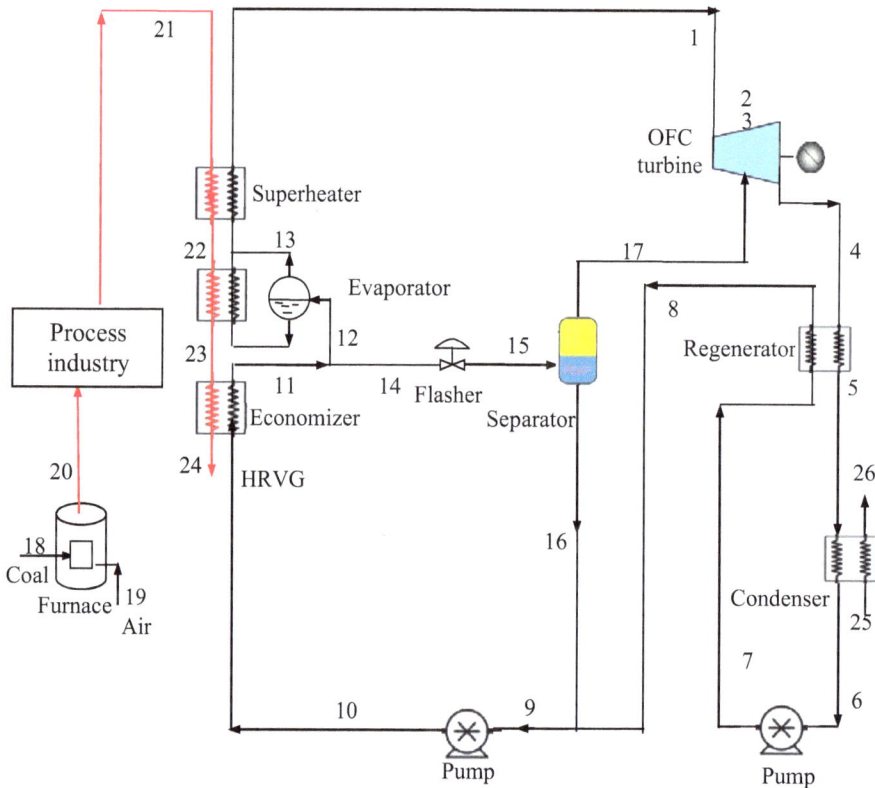

Fig. (2). Organic flash cycle with single flasher.

The liquid in the flasher expands from high pressure to flash pressure through a valve (14-15). It is separated into liquid and vapor. The vapor (17) is supplied to turbine and the balanced liquid (16) is mixed with the condensate. Even through the vapour quantity is low, it enhances the power from the plant. The addition of flasher is compared under the similar state at turbine inlet. The economizer load increases with additional working fluid. The additional fluid is equal to the fluid used in flasher.

The temperature-specific entropy plot of the single flash cycle is shown in Fig. (**3**). The flashing process is shown as a vertical line (14-15) from high pressure to intermediate pressure with a small rise in entropy.

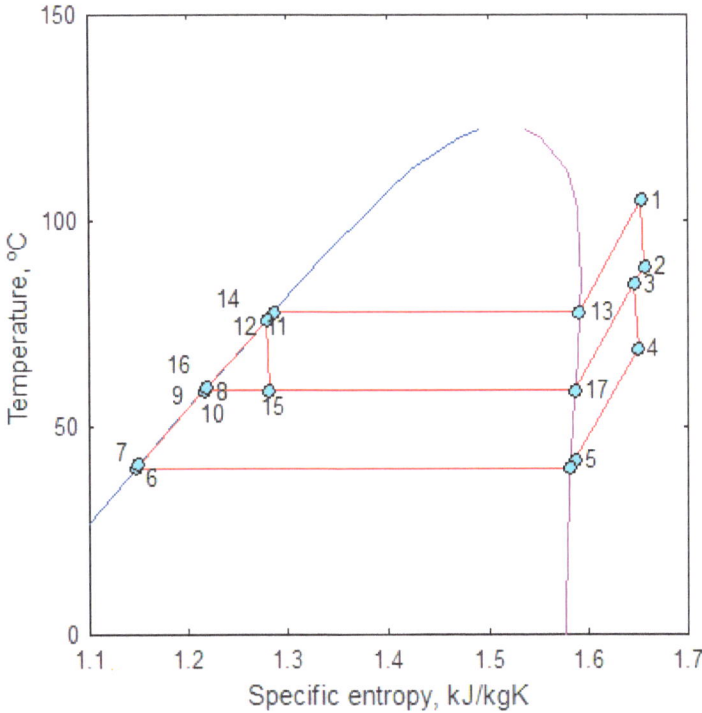

Fig. (3). Temperature-specific entropy diagram of OFC with single flasher.

Fig. (**4**) is a schematic flow diagram of an OFC with two flashers. Two flashers are preferred in this study with more power augmentation compared to single flasher. The two flashers in OFC are high pressure flasher (HPF) and low pressure flasher (LPF). The OFC is operating on four pressures *viz.* HRVG's pressure, HPF pressure, LPF pressure and condenser pressure. The high pressure (HP) and low pressure (LP) are respectively HRVG pressure and condenser pressure. The hot gas temperature is decreased in HRVG (26-29) with a rise in liquid temperature to superheated temperature (1). A part of saturated and pressurized hot liquid is collected from HRVG (14) and supplied to HPF unit.

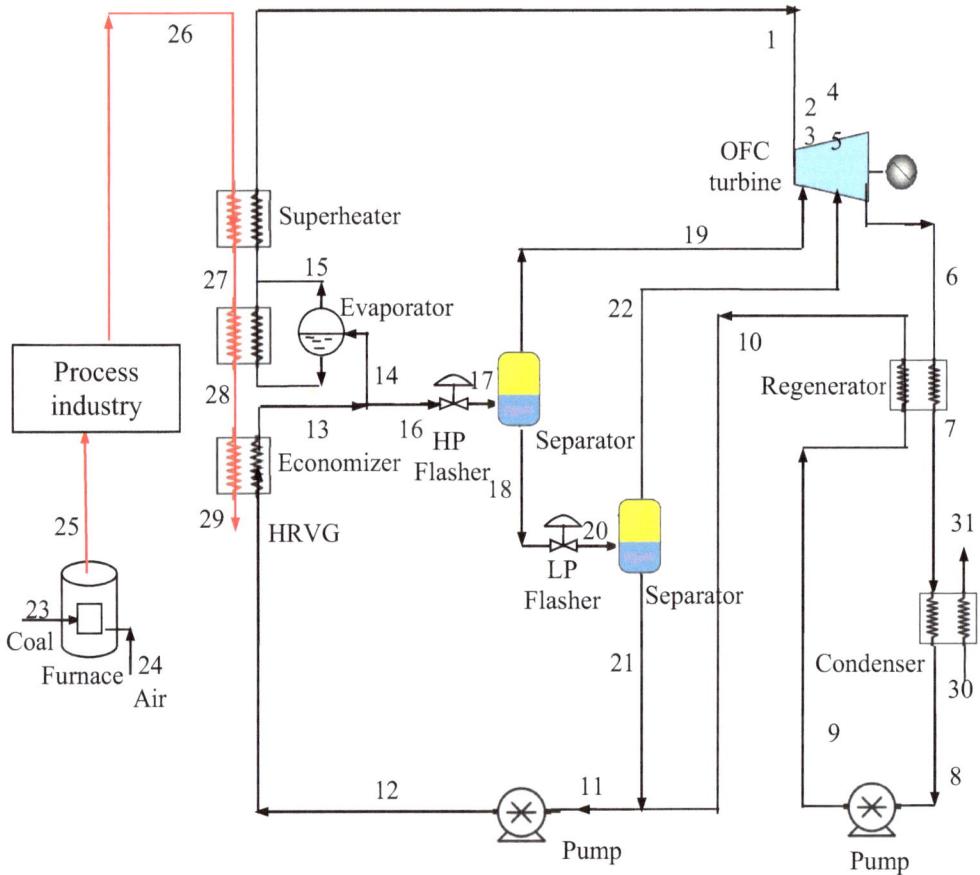

Fig. (4). Organic flash cycle with two flashers.

In HPF, the liquid (16) is flashed into wet mixture (17). The wet mass consists of liquid (18) and saturated vapor (19) at the exit of flasher. The vapor (19) is supplied to the turbine. The liquid (18) is connected to second flasher (LPF). The liquid from the HPF (18) is flashed into liquid-vapor mixture (20) with a separation of liquid (21) and saturated vapor (22). The turbine is receiving two extra streams of fluid, one from HPF and the other from the LPF. OFC increases the power with an increase in load in the economizer and pump. The vapor (6) from the turbine is fully condensed into liquid (8) with circulating water (30-31) in condenser after regenerator (6-7). The condensate is pumped to HRVG *via* regenerator to repeat the cycle.

Fig. (**5**) is the temperature-entropy diagram of OFC with two flashers. In OFC, the flashed vapour is mixed with the superheated vapour to state '3' and '5' respectively at HPF and LPF pressures. In Kalina cycle (KC), the fluid in turbine is less than the HRVG due to separation of liquid, which is detailed in the next chapter. The enthalpy of saturated liquid before flashing is equal to the total enthalpy of liquid and vapor mixture after the flashing. The dryness fraction of vapor after flashing can be determined from the enthalpy balance. In OFC turbine, the saturated vapour is mixed with the superheated vapour. Therefore, the dryness fraction of vapour in OFC is slightly less than the ORC.

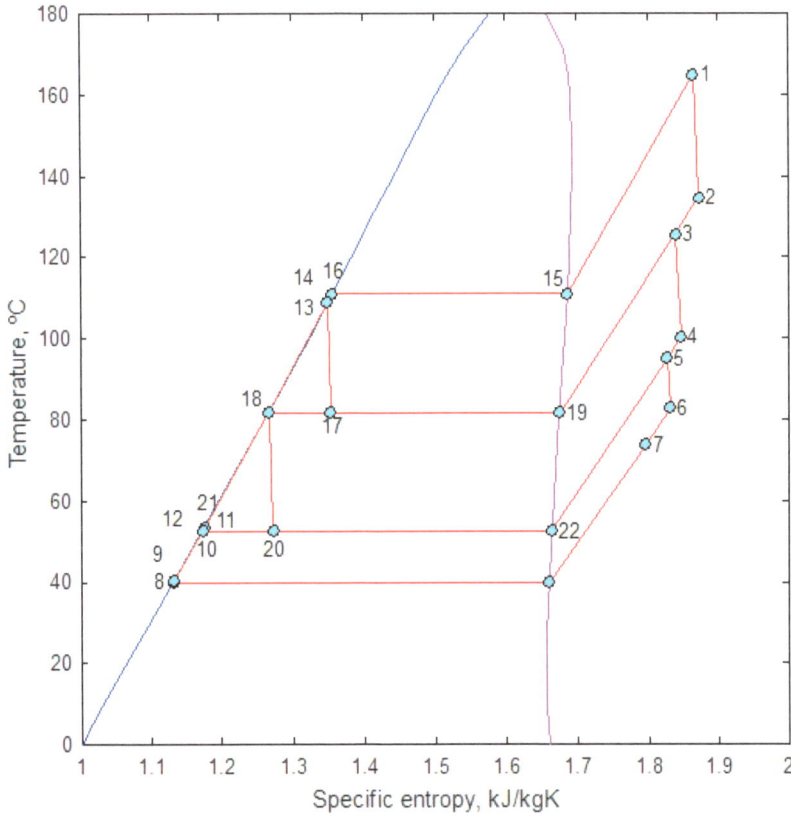

Fig. (5). (a) Pressure-specific volume and **(b)** temperature-specific entropy diagram of OFC with two flashers.

Fig. **(6)** compares the performance of OFC with single flash and double flash integration. The power augmentation and efficiency loss are increasing with increase in source temperature from single flash to double flash. At temperature, single flash efficiency is greater than the double flash efficiency.

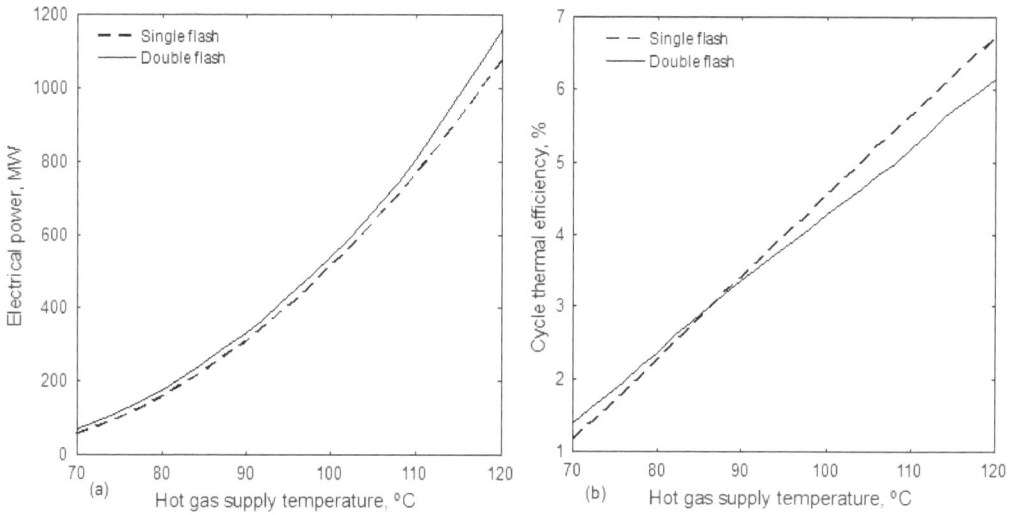

Fig. (6). Comparison of **(a)** power and **(b)** thermal efficiency from a single flash OFC and double flash OFC.

Evaluation of OFC

Following are the assumptions and followed by formulation used in the evaluation of the double flash OFC power plant using waste heat of a thermal industry. The hot gas from a typical cement factory is 1000000 Nm³/h. HRVG pressure has been selected from the optimized model at the source temperature. Terminal temperature difference (TTD) in superheater is 15 K. Pinch point in the evaporator is 10 K. Approach point (AP) at the inlet of evaporator is 2 K. Isentropic efficiencies of pump and turbine respectively are 75% and 80%. Mechanical efficiency of generator, pump and turbine is 98%. In condenser the circulating water inlet temperature is 30 °C with a temperature rise of 8 °C.

The liquid exit temperature from regenerator is considered equal to the second flasher temperature *i.e.* $T_{10} = T_{21}$.

The energy balance equation at regenerator gives the vapor exit temperature, T_7.

The energy balance in HRVG at evaporator and superheater results the vapor from HRVG.

$$m_{14} = \frac{m_{26}(h_{26} - h_{28})}{h_1 - h_{14}} \tag{1}$$

The gas enthalpy at temperature T,

$$h_{gas,T} = b_1 h_{CO_2,T} + b_2 h_{H_2O,T} + b_3 h_{O_2,T} + b_4 h_{N_2,T} \tag{2}$$

The fluid used for flashing from the economizer is 25% of the fluid in the economizer. The temperature ratio (Eq.8 and Eq.9) in high pressure flasher and low pressure flasher is 0.5.

Flashing is a process of expansion of pressurized saturated liquid into liquid and vapor mixture. The enthalpy of saturated liquid before flashing is equal to the total enthalpy of liquid and vapor mixture after the flashing.

$$h_{16} = h_{17} = h_{18} + d_{17}(h_{19} - h_{18}) \tag{3}$$

$$h_{18} = h_{20} = h_{21} + d_{20}(h_{22} - h_{21}) \tag{4}$$

The dryness fractions of steam after the expansion in the flashers (Fig. **4**) at state 17 and state 20 are determined from Eq.3 and Eq.4 respectively.

Let flash mass ratio (FMR) is mass ratio of partial amount of fluid flashed to total fluid in economizer.

$$FMR = \frac{m_{16}}{m_{13}} = \frac{m_{16}}{m_1 + m_{16}} \tag{5}$$

after the simplification,

$$m_{16} = \frac{FMR\, m_1}{(1 - FMR)} \tag{6}$$

From the energy balance equation in HRVG' economizer,

$$T_{29} = T_{28} - \frac{m_{13}(h_{13} - h_{12})}{m_{28}(h_{29} - h_{28})} \tag{7}$$

To find the optimum temperature or pressure of HP flasher and LP flasher, let θ_{HPF} and θ_{LPF} are the temperature difference ratios in HP flasher and LP flasher respectively as follows:

$$\theta_{HPF} = \frac{T_{HPF} - T_{LPF}}{T_{HRVG} - T_{LPF}} \tag{8}$$

$$\theta_{LPF} = \frac{T_{LPF} - T_{cond}}{T_{HPF} - T_{cond}} \tag{9}$$

T_{HRVG} and T_{cond} are respectively boiler and condenser saturation temperatures. With the initial assumptions of θ_{HPF} and θ_{LPF} (to be optimized), the saturation temperatures for high pressure flashing (HPF) and low pressure flashing (LPF) can be determined with the simplification of above two equations.

Eliminating T_{LPF} in θ_{HPF}

$$\theta_{HPF} = \frac{T_{HPF} - \left(T_{cond} + \theta_{LPF}(T_{HPF} - T_{cond})\right)}{T_{HRVG} - \left(T_{cond} + \theta_{LPF}(T_{HPF} - T_{cond})\right)}$$

$$= \frac{T_{HPF} - T_{cond} - \theta_{LPF}T_{HPF} + \theta_{LPF}T_{cond}}{T_{HRVG} - T_{cond} - \theta_{LPF}T_{HPF} + \theta_{LPF}T_{cond}}$$

$$(1 - \theta_{LPF} + \theta_{HPF}\theta_{LPF})T_{HPF} = \theta_{HPF}T_{HRVG} + (-\theta_{HPF} + \theta_{HPF}\theta_{LPF} - \theta_{LPF} + 1)T_{cond} \tag{10}$$

$$T_{HPF} = \frac{T_{HRVG}\theta_{HPF} + T_{cond}(1 - \theta_{HPF} - \theta_{LPF} + \theta_{HPF}\theta_{LPF})}{\left(1 - \theta_{LPF} + \theta_{HPF}\theta_{LPF}\right)}$$

Similarly eliminating T_{HPF} in θ_{LPF}

$$\theta_{LPF} = \frac{T_{LPF} - T_{cond}}{T_{LPF} + \theta_{HPF}(T_{HRVG} - T_{LPF}) - T_{cond}}$$

$$= \frac{T_{LPF} - T_{cond}}{T_{LPF} + \theta_{HPF}T_{HRVG} - \theta_{HPF}T_{LPF} - T_{cond}}$$

$$\theta_{HPF}\theta_{LPF}T_{HRVG} + (1 - \theta_{LPF})T_{cond} = (1 - \theta_{LPF} + \theta_{HPF}\theta_{LPF})T_{LPF} \tag{11}$$

$$T_{LPF} = \frac{T_{HRVG}\theta_{HPF}\theta_{LPF} + T_{cond}(1 - \theta_{LPF})}{\left(1 - \theta_{LPF} + \theta_{HPF}\theta_{LPF}\right)}$$

Eq.10 and 5.11 illustrates the HPF temperature and LPF temperature as a function of boiler temperature, condenser temperature, boiler temperature ratio and condenser temperature ratio. The liquid exit temperature from the regenerator is considered as a saturated liquid at the pressure maintained in LPF. So the liquid temperatures before mixing at the exit of LPF are same. The energy balance in the regenerator results the vapor exit temperature. Fig. (**7**) outlines the method adopted in OFC solutions.

Role of Regenerator on OFC Performance

The role of regenerator has been discussed in the ORC chapter. Similarly, regenerator also plays a significant role in OFC to enhance the efficiency of the plant.

Fig. (**8**) compares the thermal efficiency of OFC with and without regenerator. The regenerator does not influence the vapour generation in HRVG and power production with the pinch point constraint. The regenerator influences the thermal efficiency. The efficiency rise is increasing with increase in source temperature.

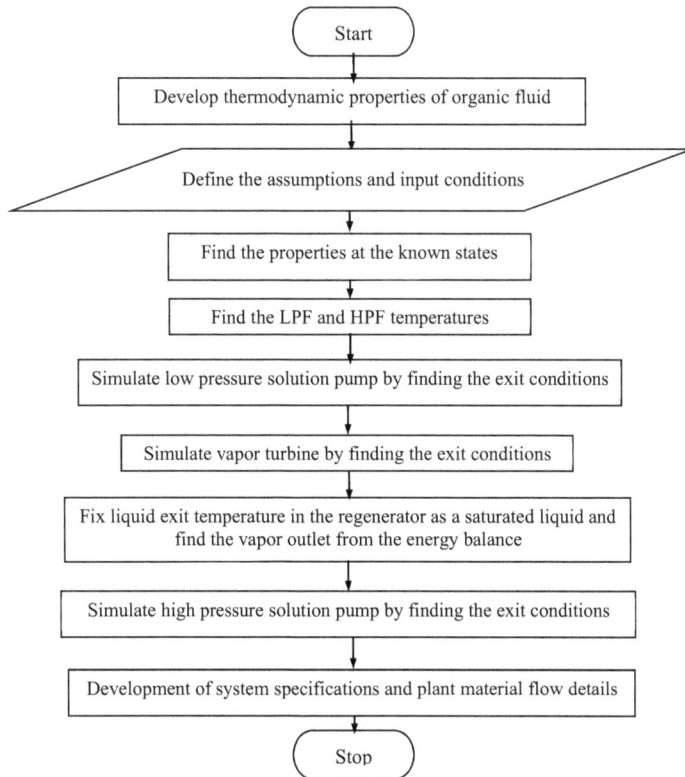

```
                        ┌─────────────┐
                        │    Start    │
                        └─────────────┘
                               │
        ┌──────────────────────────────────────────────┐
        │  Develop thermodynamic properties of organic fluid │
        └──────────────────────────────────────────────┘
                               │
      ╱──────────────────────────────────────────────────╲
          Define the assumptions and input conditions
      ╲──────────────────────────────────────────────────╱
                               │
            ┌──────────────────────────────────────┐
            │  Find the properties at the known states │
            └──────────────────────────────────────┘
                               │
            ┌──────────────────────────────────────┐
            │    Find the LPF and HPF temperatures    │
            └──────────────────────────────────────┘
                               │
      ┌──────────────────────────────────────────────────┐
      │ Simulate low pressure solution pump by finding the exit conditions │
      └──────────────────────────────────────────────────┘
                               │
        ┌────────────────────────────────────────────┐
        │  Simulate vapor turbine by finding the exit conditions │
        └────────────────────────────────────────────┘
                               │
    ┌────────────────────────────────────────────────────┐
    │ Fix liquid exit temperature in the regenerator as a saturated liquid and │
    │      find the vapor outlet from the energy balance      │
    └────────────────────────────────────────────────────┘
                               │
      ┌──────────────────────────────────────────────────┐
      │ Simulate high pressure solution pump by finding the exit conditions │
      └──────────────────────────────────────────────────┘
                               │
      ┌──────────────────────────────────────────────────┐
      │ Development of system specifications and plant material flow details │
      └──────────────────────────────────────────────────┘
                               │
                        ┌─────────────┐
                        │    Stop     │
                        └─────────────┘
```

Fig. (7). Summary of methodology applied in OFC solution.

Optimum Boiler Pressure with Source Temperature

The studied fluids are R123, R124, R134a, R245fa, R717 and R407C. Fig. (**9**) depicts the change in hot gas supply temperature and HRVG pressure on performance, degree of superheat and exhaust gas temperature. Suitable temperature range (80 °C to 180 °) and pressure range (2 bar to 25 bar) are selected to operate the plant under sub-critical condition for R123. The HRVG pressure is increased from condenser pressure to the pressure related to source temperature but below the critical pressure. The condenser pressure is fixed with the sink temperature. Since the source temperature is changing (increasing), proportionately the upper limit of pressure is increasing. The maximum turbine inlet temperature is selected at sub-critical state. The power is maximized at a particular HRVG pressure. The optimum boiler pressure differs with source temperature. It increases with increase in source temperature/turbine inlet temperature. The rise in turbine inlet pressure increases the vapor expansion in turbine and favors the power. The pump input also increases with increase in the boiler pressure. The net power maximizes at an optimum boiler pressure. The efficiency is the function of power and heat supply. The rise in boiler pressure, increases the power and drops the heat supply with a pinch point limit in HRVG. These two factors, favors the thermal efficiency with diminishing rate as shown in Fig. (**9b**). The degree of superheat decreases with increase in pressure and also with source temperature (Fig. **9c**). The exhaust gas temperature increases with increase in pressure and decreases with increase in source temperature (Fig. **9d**).

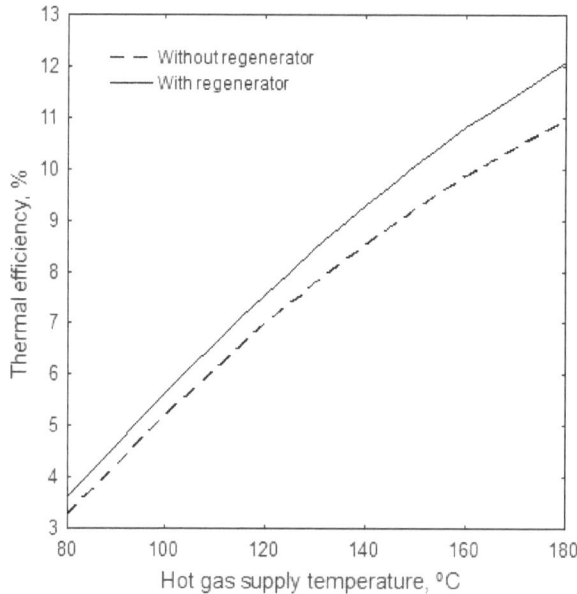

Fig. (8). Use of regenerator in OFC to improve the thermal efficiency.

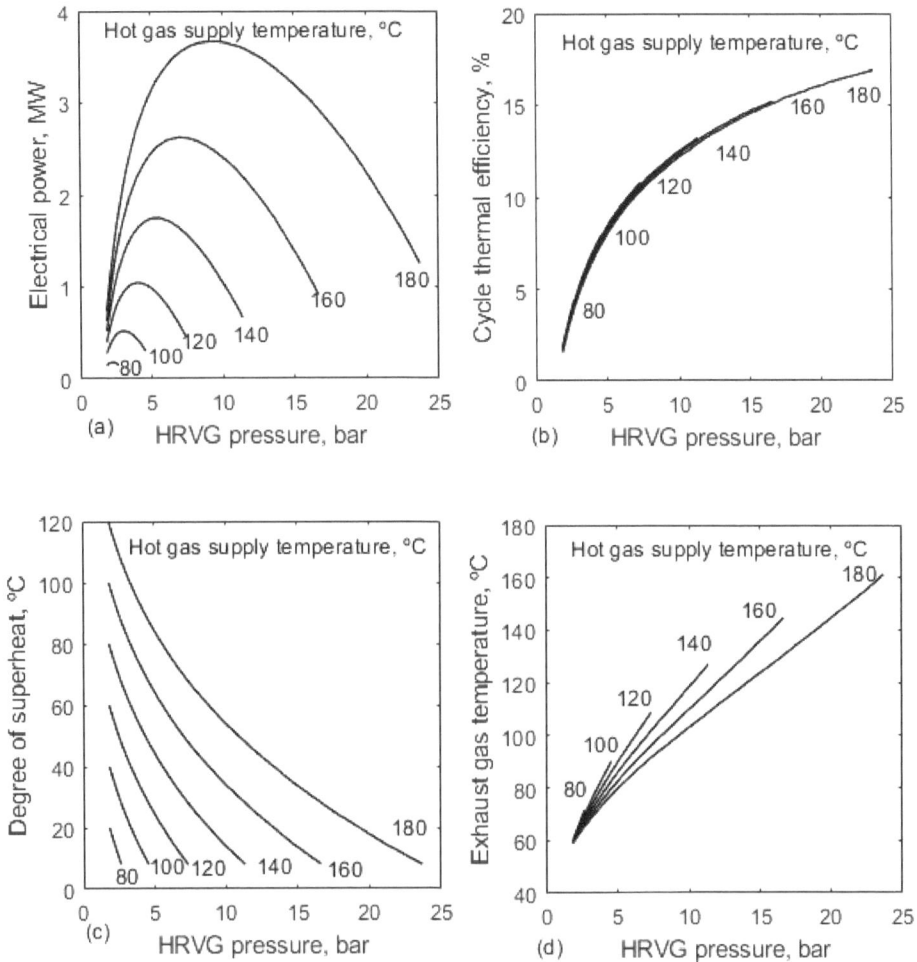

Fig. (9). Influence of source temperature with HRVG pressure on **(a)** net power generation, **(b)** cycle energy conversion efficiency **(c)** degree of superheat and **(d)** exhaust gas temperature with R123 fluid.

Fig. (**10**) is developed to study the performance and process conditions of OFC with R124 fluid. Fig. (**10a**) results can be used to select the best HRVG pressure at a suitable/available source temperature. Similar to earlier plant, the optimum HRVG pressure is increasing with increase in source temperature at maximum power condition. Thermal efficiency follows a continue rise with rise in pressure and temperature. Fig. (**10c**) is prepared to study the variations in DSH resulted with changes in temperature and pressure. The DSH is increasing with increase in source temperature and decreasing with rise in HRVG pressure. The pressure can be determined from saturation temperature.

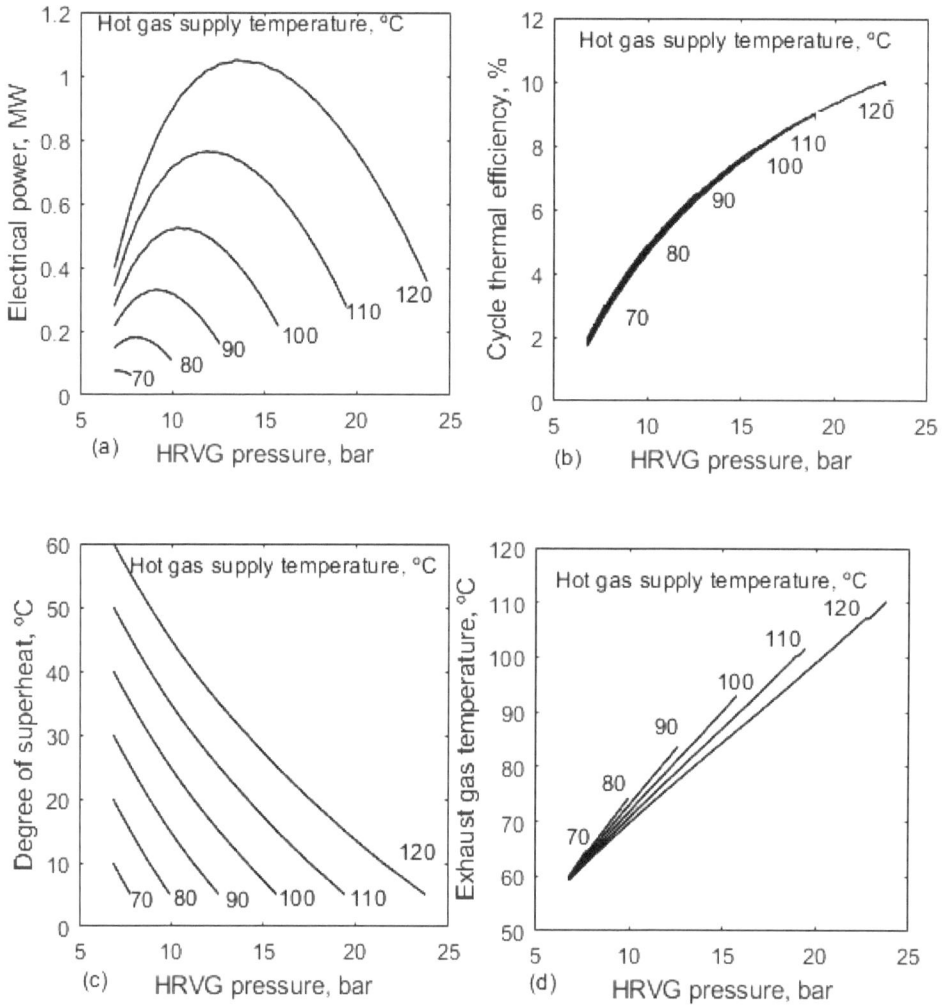

Fig. (10). Influence of source temperature with HRVG pressure on **(a)** net power generation, **(b)** cycle energy conversion efficiency **(c)** degree of superheat and **(d)** exhaust gas temperature with R124 fluid.

Fig. (**11**) shows the power, thermal efficiency, degree of superheat and exhaust gas temperature with change in HRVG pressure and turbine inlet temperature with R134a fluid. The optimum HRVG pressure is increasing with increase in source temperature.

Similarly, for R245fa, the performance and process conditions are developed in Fig. (**12**). The optimum pressures are generated as a function of source temperature.

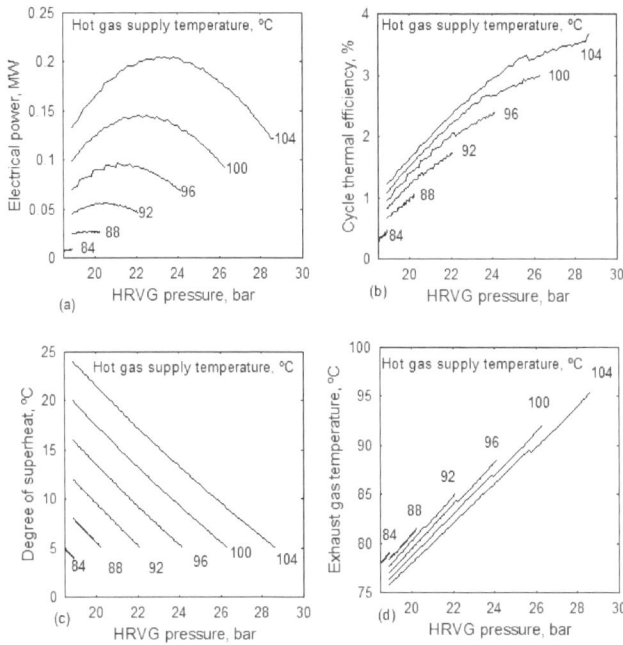

Fig. (11). Influence of source temperature with HRVG pressure on **(a)** net power, **(b)** cycle energy conversion efficiency **(c)** degree of superheat and **(d)** exhaust gas temperature with R134a fluid.

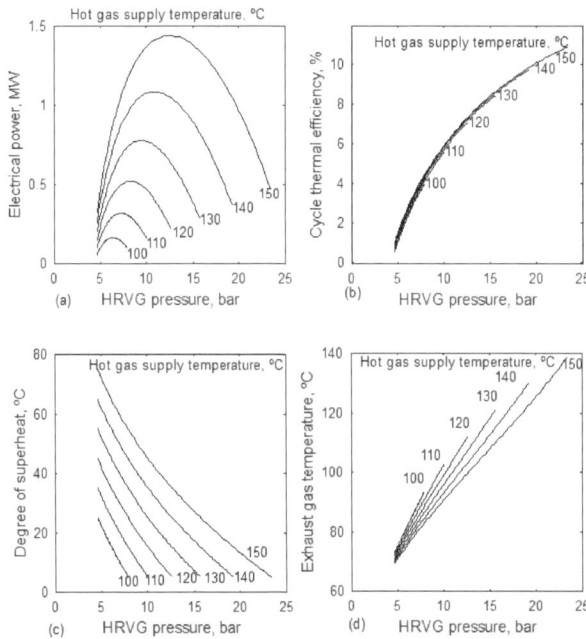

Fig. (12). Influence of source temperature with HRVG pressure on **(a)** net power generation, **(b)** cycle energy conversion efficiency **(c)** degree of superheat and **(d)** exhaust gas temperature with R245fa fluid.

Fig. (**13**) shows the power, thermal efficiency, DSH and exhaust gas temperature with R717 fluid.

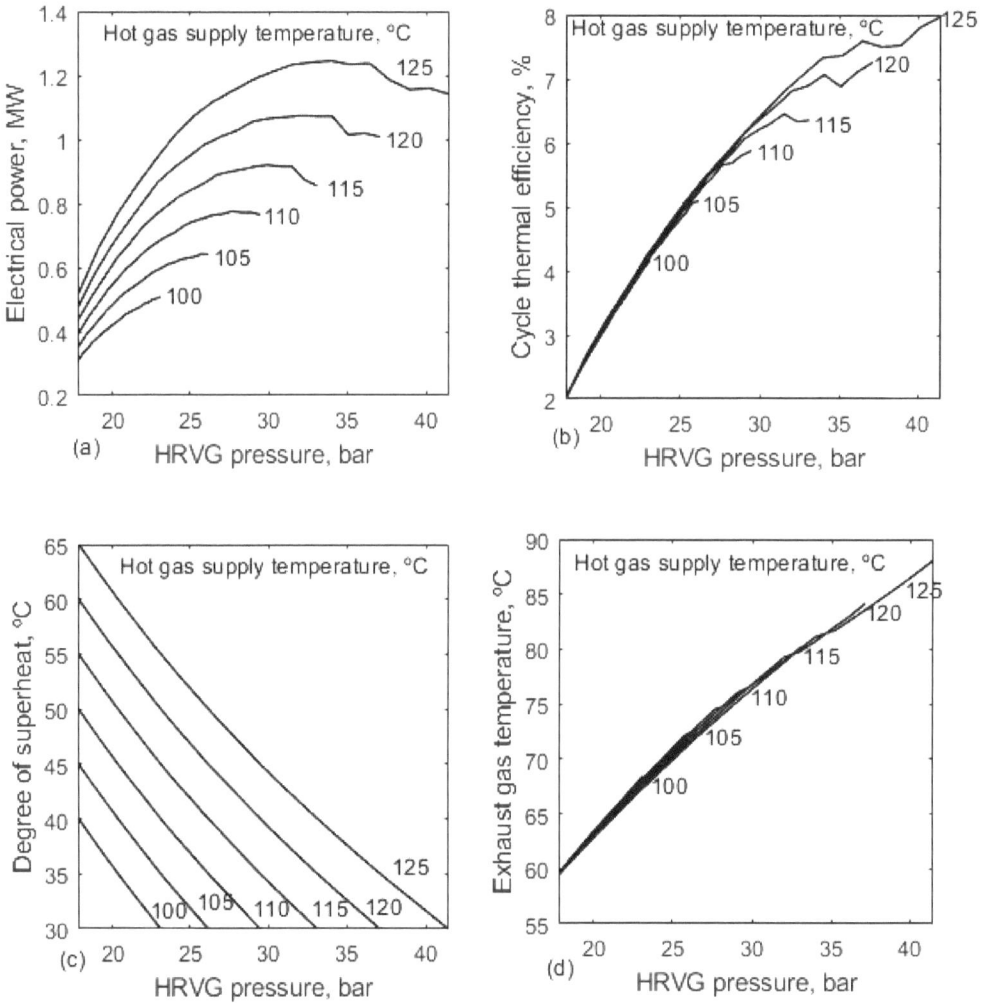

Fig. (13). Influence of source temperature with HRVG pressure on (**a**) net power generation, (**b**) cycle energy conversion efficiency (**c**) degree of superheat and (**d**) exhaust gas temperature with R717 fluid.

Fig. (**14**) analyzes the power, thermal efficiency, DSH, and exhaust gas temperature with R407C fluid. Since the fluid is zeotropic mixture, the results are differing comparing with the other single fluid OFC plants. The optimum HRVG pressure is not resulted as the power is raising continuously. Therefore, the pressure can be selected at a higher temperature.

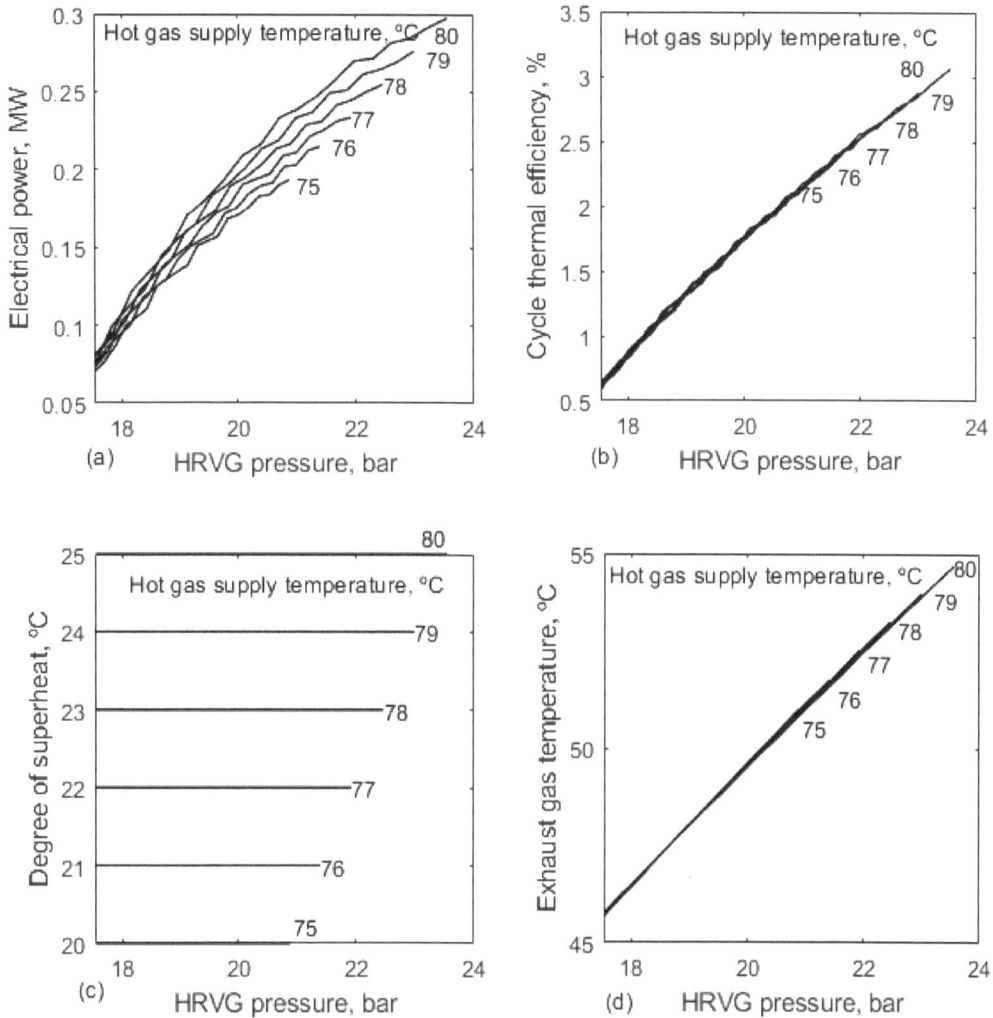

Fig. (14). Influence of source temperature with HRVG pressure on **(a)** net power generation, **(b)** cycle energy conversion efficiency **(c)** degree of superheat and **(d)** exhaust gas temperature with R407C fluid.

It is clear that the optimum HRVG pressure depends on source temperature and working fluid at a fixed condenser temperature. The data of optimum pressure at the chosen source temperature and fluid has a significant importance to design a power plant. For every source temperature and working fluid, there is an optimum HRVG pressure that can be determined for maximum power. To formulate the HRVG pressure as a function of source temperature with working fluid, nonlinear correlations are developed from the best fit method and presented in Table **1**. A non-linear quadratic fit has been developed to find the optimum pressure.

Therefore, the optimum pressure can be directly determined from these correlations without repeating the analysis and optimization.

Table 1. Coefficients to find the optimum HRVG pressure at source temperature below critical temperature and 30 °C of circulating water temperature (condensate temperature is 40 °C)..

Working fluid	Coefficients		
	a_1	a_2	a_3
R123	3.2225	-0.0491	0.0005
R124	6.5299	-0.0636	0.001
R134a	15.325	-0.0825	0.0015
R245fa	6.027	-0.0749	0.0008
R717	-43.1223	0.8483	-0.0019
R407C	9.5082	-0.2082	0.0048

The nonlinear equation to find the optimum HRVG pressure from source temperature,

$$P_{HRVG\ opt} = a_1 + a_2 T + a_3 T^2 \tag{12}$$

where T is the hot gas (source) temperature in °C.

Optimum State of LPF and HPF

Figs. (**15 - 20**) analyzes the location of LP flasher and HP flasher on power with fluids. The vapour expands in three stages in OFC turbine. In stage 1, vapour expands from HRVG pressure to HP flasher pressure. In first stage, there is no flash vapour and the superheated vapour expands in the turbine. At the exit of 1st stage, HP flash vapour is mixed with the turbine vapour. The combined vapour expands from HP flash pressure to LP flash pressure in stage 2. Now the LP flash vapour joins with the turbine fluid. The total working fluid expands in the last stage, *i.e.* 3rd stage, from LP flash pressure to condenser pressure. To study the location of LP flasher and HP flasher, the state has been expressed as a temperature ratio. The temperature of flash is expressed in a ratio of temperature difference with reference to condenser temperature and HRVG pressure. Therefore, the flash temperature can be varied between lower pressure and upper pressure. For example, the LP flash state is changed from condenser state to HP flash state. The HP state is changed between LP flash state and HRVG state. The saturation state in condenser, LP flasher, HP flasher and HRVG can be found either from saturation temperature or saturation pressure. If it is expressed in saturation temperature, the pressure can be determined from this temperature. The capacity of pump is independent of these flash operational conditions as the fluid flow in pump is constant. The condenser pressure and HRVG pressures are constant with changes in LP flasher and HP flasher conditions. With increase in LP flash temperature ratio (proportionately flash temperature), the change in stage 3 and the total change in stage 1 and stage 2 are observed. The temperature of LP flasher is between stage 3 and other two stages. The increase in LP flash temperature increases the expansion in stage 3 with subsequent drop in other stages, *i.e.* stage 1 and stage 2. Similarly, the increase in HP flash temperature, increases the stage 3 output and drops the stage 1 and stage 2 combined output. These two contrast results of stage 3 and combined stage 1 and stage 2 maximizes the power. The location of LP flasher and HP flasher are optimized for maximum power. The results are suggesting a mid-state to LP flasher and HP flasher with reference to other neighbor temperatures. For example, if the LP flasher's temperature is the mid temperature of condenser and HP flasher, a maximum power can be generated. Similarly, if the HP flasher's temperature is the mid temperature of LP flasher and HRSG, power can be maximized. The pump supply in OFC is a considerable amount with some working fluids such as R134a and R245fa. Therefore, the power augmentation with these fluids is difficult compared to the other fluids.

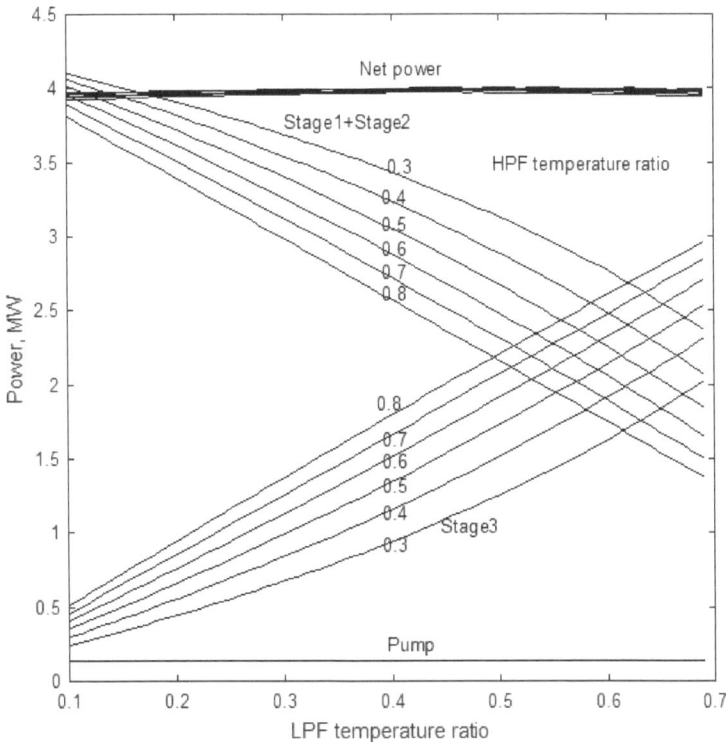

Fig. (15). Influence of LPF and HPF temperature ratios on power generation of OFC with R123 fluid.

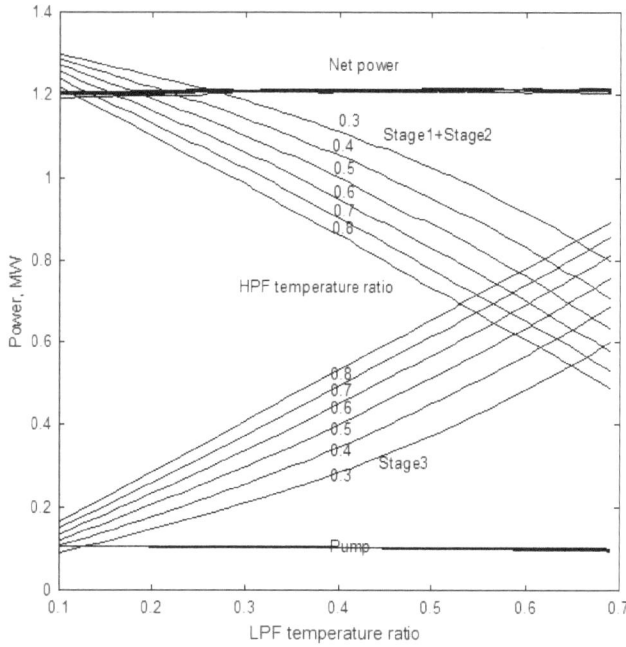

Fig. (16). Influence of LPF and HPF temperature ratios on power generation of OFC with R124 fluid.

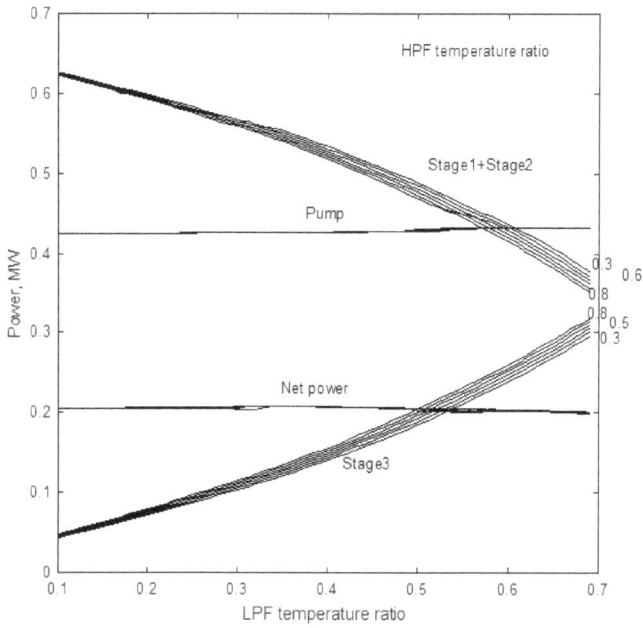

Fig. (17). Influence of LPF and HPF temperature ratios on power generation of OFC with R134a fluid.

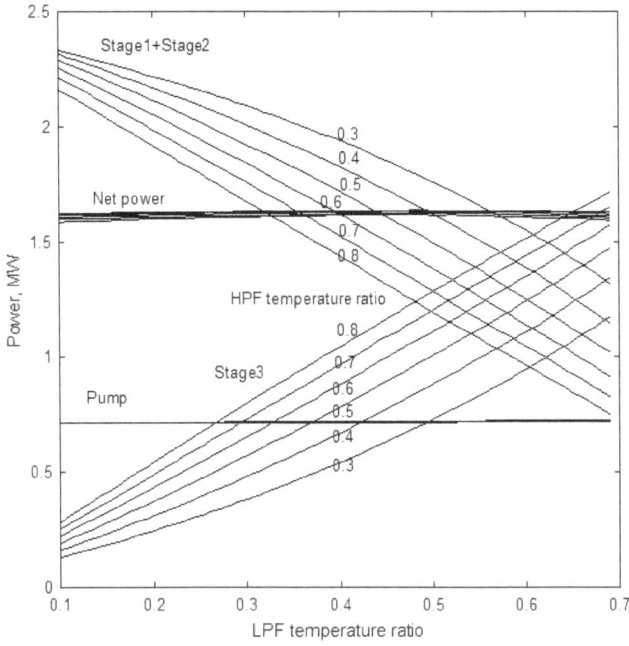

Fig. (18). Influence of LPF and HPF temperature ratios on power generation of OFC with R245fa fluid.

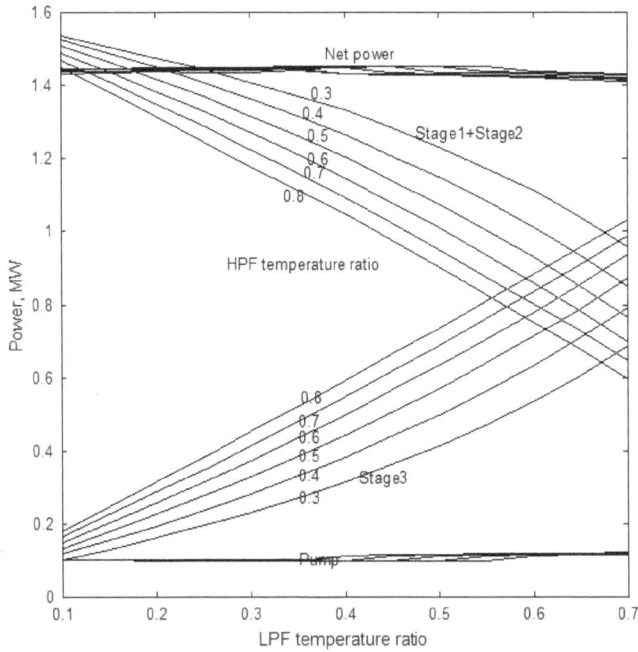

Fig. (19). Influence of LPF and HPF temperature ratios on power generation of OFC with R717 fluid.

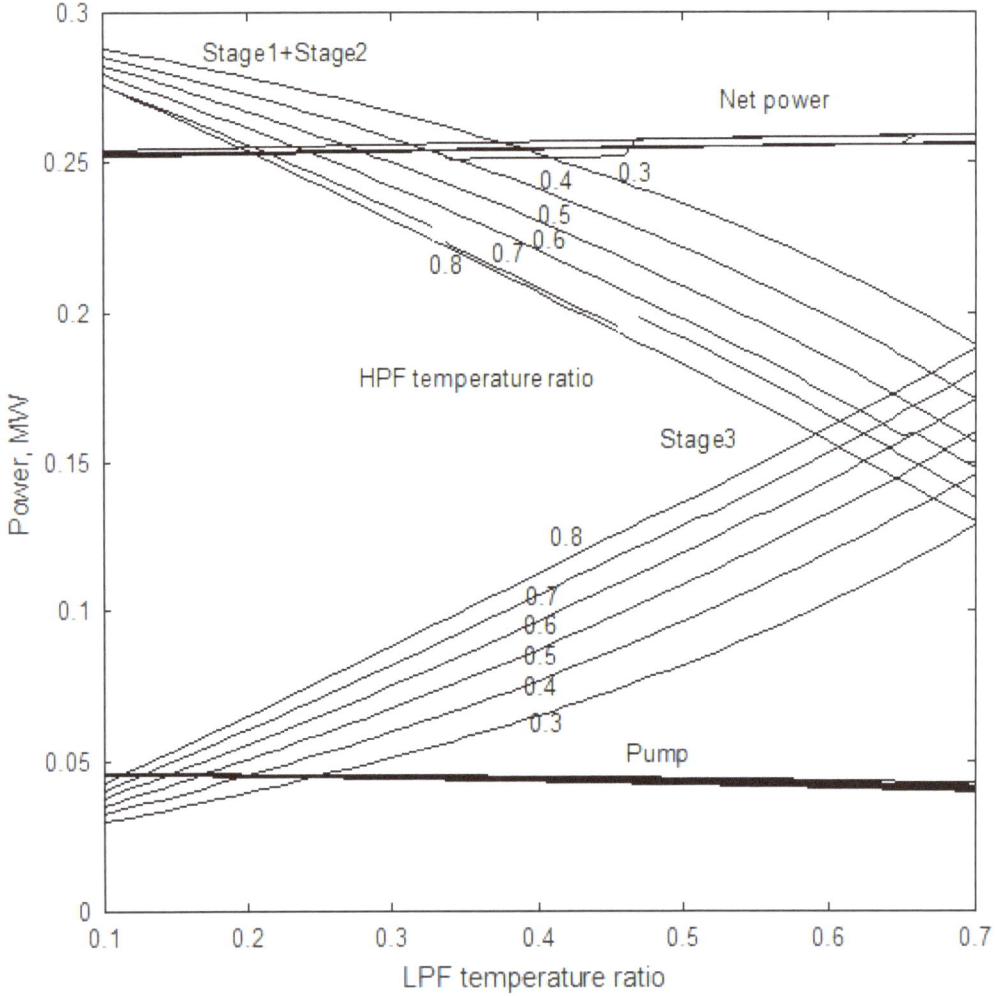

Fig. (20). Influence of LPF and HPF temperature ratios on power generation of OFC with R407C fluid.

Effect of Sink

Fig. (**21**) reveals the influence of sink temperature and HRVG pressure on power output and thermal efficiency with R123 fluid. HRVG pressure can be picked between condenser pressure (related to sink temperature) and pressure related to the source temperature. The HRVG pressure is plotted against the sink temperature. The sink temperature is changed from 20 °C to 35 °C. Now the lower limit of HRVG pressure is changed with change in sink temperature. The upper limit is same with fixed source temperature. The power characteristics Fig. (**21a**), shows that the optimum HRVG pressure is increasing with increase in sink temperature but with a fall in power production. As per the Carnot's theorem, the efficiency of the cycle decreases with increase in the sink temperature (Fig. **21b**). For maximum thermal efficiency, optimum pressure is not resulted due to lack of peak. Finally, the optimum pressure can be chosen for the maximum power production from the source and sink temperatures. Similar to R123, the other plots *i.e.* Figs. (**22 - 26**) develops the influence of sink temperature on optimum HRVG pressure with the fluids of R124, R134a, R245fa, R717 and R407C. In all the plants, the source temperature is fixed. For example, for the R123 sink characteristics, the source temperature is fixed at 180 °C. The hot gas supply temperature is 120 °C, 100 °C, 150 °C, 125 °C and 60 °C respectively maintained with R124, R134a, R245fa, R717 and R407C fluids to study the optimum HRVG pressure with a change in sink temperature.

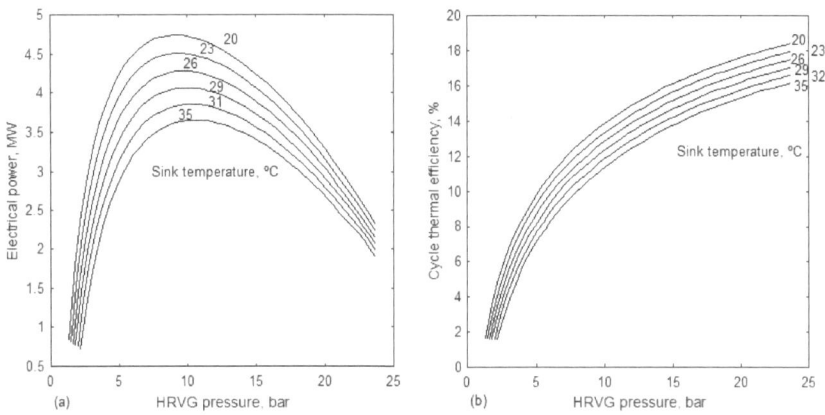

Fig. (21). Influence of sink temperature with HRVG pressure on (**a**) net power generation and (**b**) cycle energy conversion efficiency with R123.

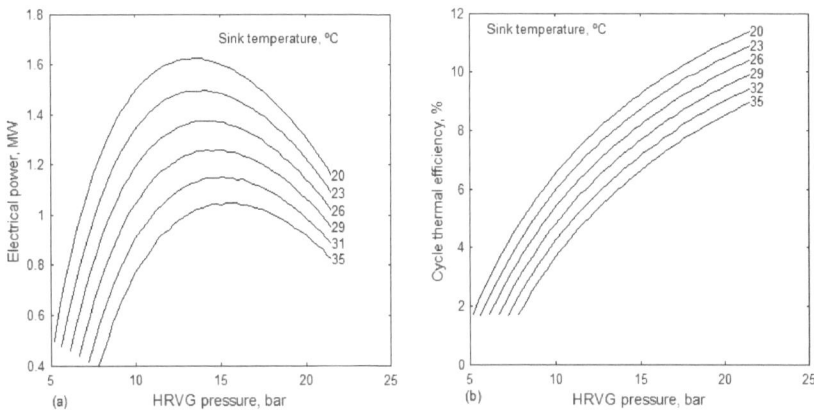

Fig. (22). Influence of sink temperature with HRVG pressure on (**a**) net power generation and (**b**) cycle energy conversion efficiency with R124.

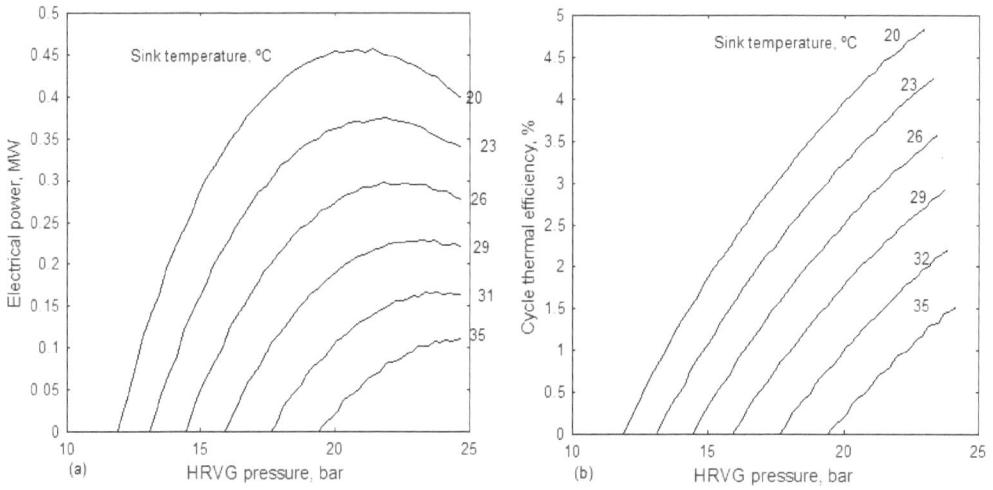

Fig. (23). Influence of sink temperature with HRVG pressure on **(a)** net power generation and **(b)** cycle energy conversion efficiency with R134a.

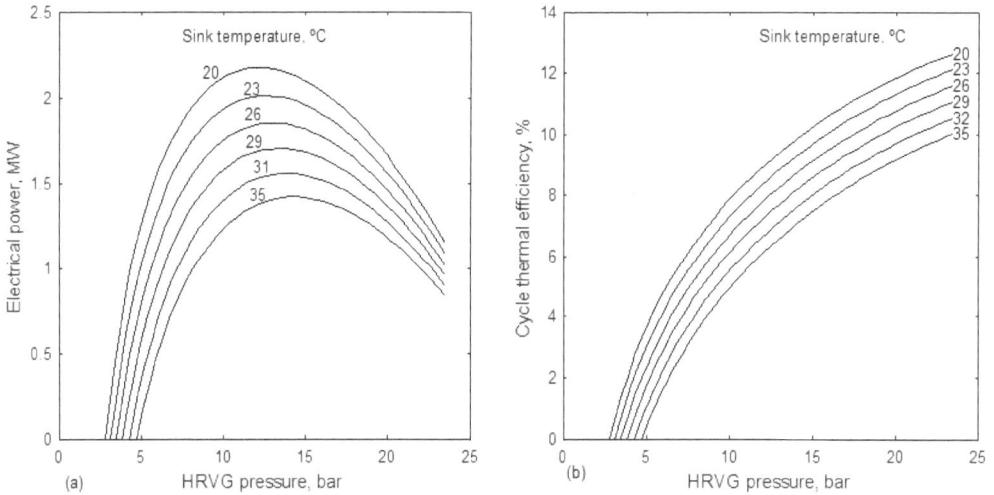

Fig. (24). Influence of sink temperature with HRVG pressure on **(a)** net power generation and **(b)** cycle energy conversion efficiency with R245fa.

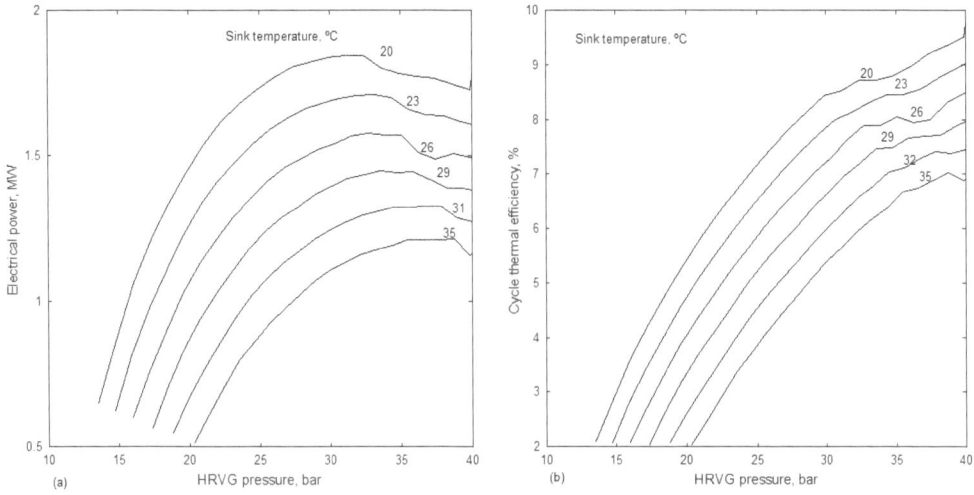

Fig. (25). Influence of sink temperature with HRVG pressure on **(a)** net power generation and **(b)** cycle energy conversion efficiency with R717.

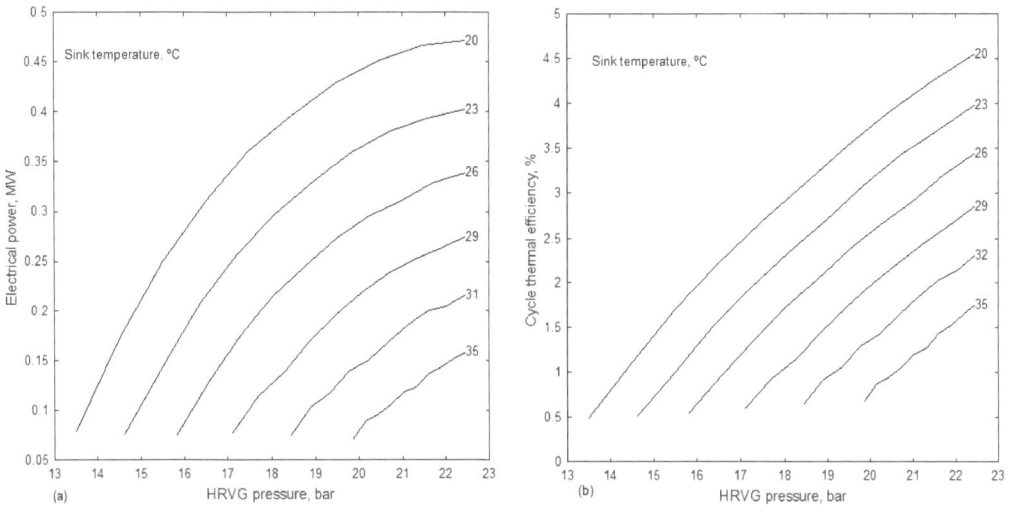

Fig. (26). Influence of sink temperature with HRVG pressure on **(a)** net power generation and **(b)** cycle energy conversion efficiency with R407C.

Property Charts and Tables at the Optimized Conditions

Figs. (**27** - **32**) shows the (a) pressure-specific enthalpy and (b) temperature-entropy plots for the regenerative ORC with the working fluids of R123, R124, R134a, R245fa, R717 and R407C. The earlier sections detailed about the developing the optimum HRVG pressure as a function of source temperature and fluid with the equation fit. The property charts and tables are prepared at the optimum pressure and temperature close to critical temperature. The optimum conditions are prepared by changing the temperature and pressure below the critical temperature and critical pressure respectively. The slope lines of phase change for R407C can be observed in Fig. (**32b**) for condensation and evaporation.

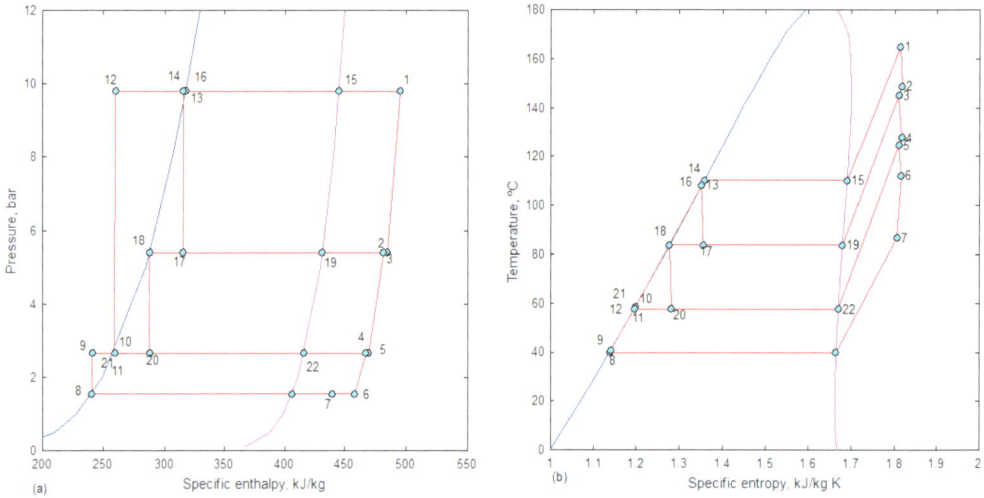

Fig. (27). (a) Pressure-specific volume and **(b)** temperature-specific entropy diagram of OFC with two flashers and R123 fluid.

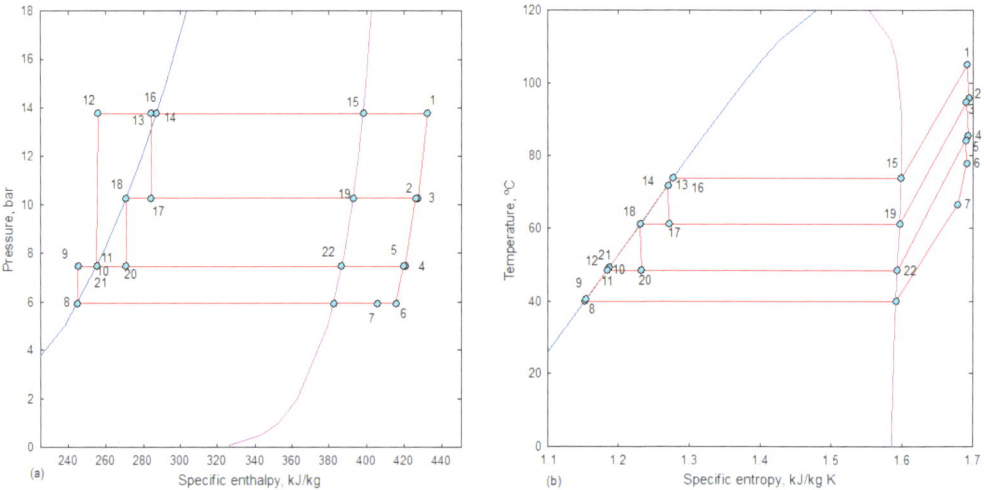

Fig. (28). (a) Pressure-specific volume and **(b)** temperature-specific entropy diagram of OFC with two flashers and R124 fluid.

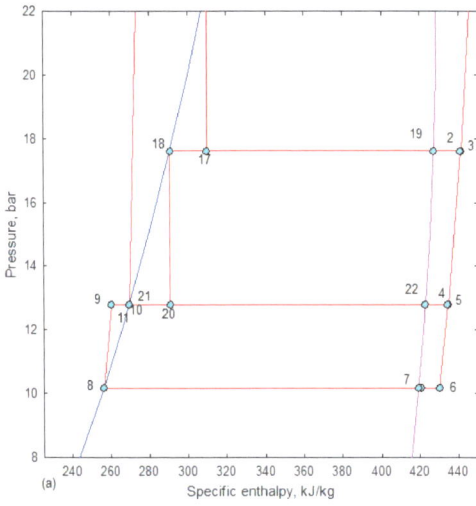

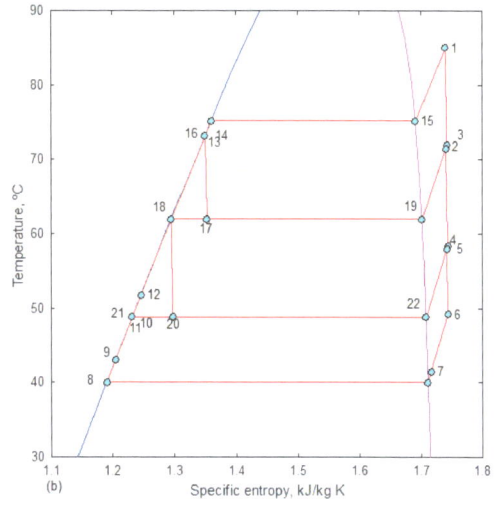

Fig. (29). (a) Pressure-specific volume and **(b)** temperature-specific entropy diagram of OFC with two flashers and R134a fluid.

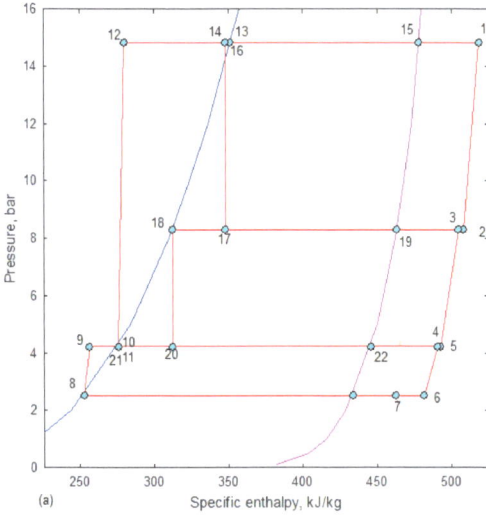

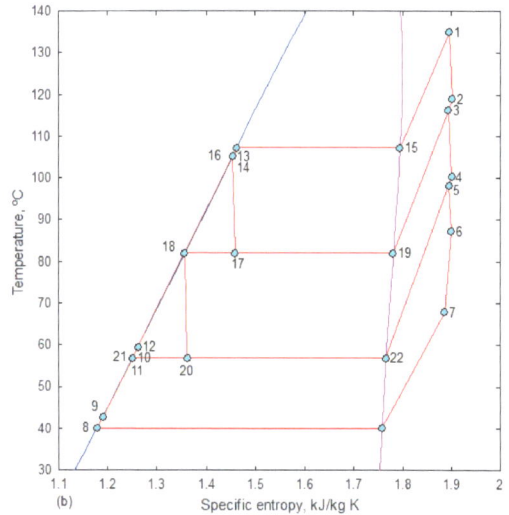

Fig. (30). (a) Pressure-specific volume and **(b)** temperature-specific entropy diagram of OFC with two flashers and R245fa fluid.

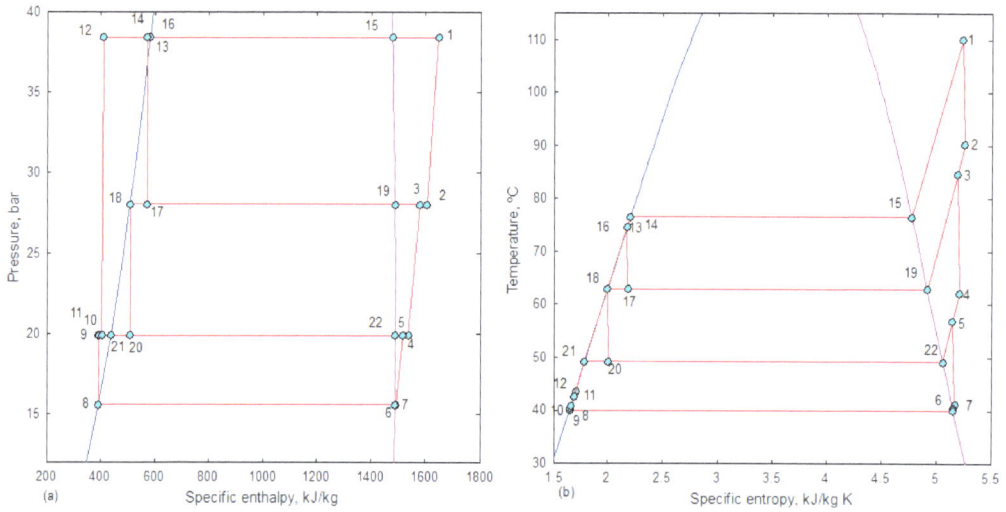

Fig. (31). (a) Pressure-specific volume and **(b)** temperature-specific entropy diagram of OFC with two flashers and R717 fluid.

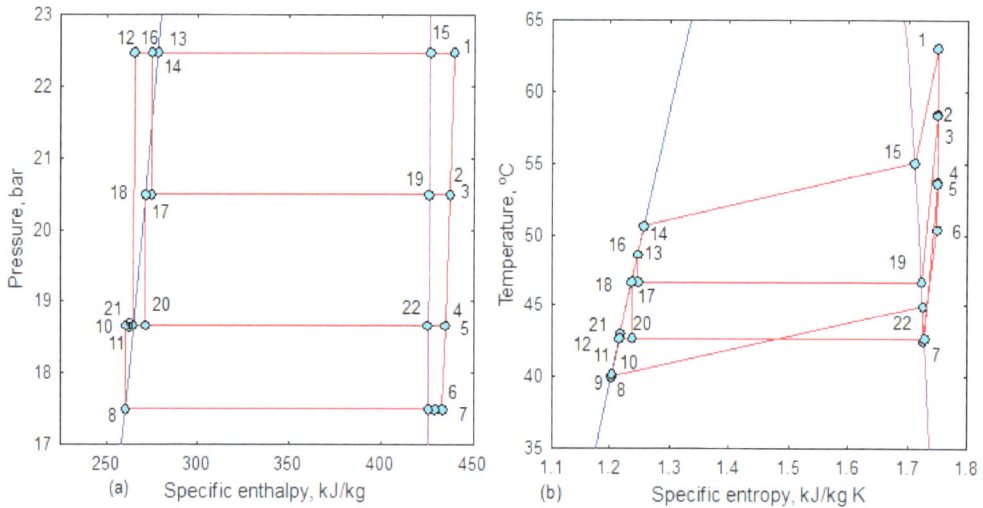

Fig. (32). (a) Pressure-specific volume and **(b)** temperature-specific entropy diagram of OFC with two flashers and R407C fluid.

The values of pressure, temperature, flow rate, specific enthalpy and specific entropy are presented from Tables **2** - **7** respectively for the working fluids of R123, R124, R134a, R245fa, R717 and R407C.

Table 2. OFC material balance results with R123 fluid at hot gas supply of 180 °C.

State	P, bar	t, °C	m, kg/s	h, kJ/kg	s, kJ/kg K
1.	10.58	165.00	114.80	494.59	1.81
2.	5.68	147.93	114.80	483.81	1.81
3.	5.68	143.96	122.68	480.50	1.81
4.	2.73	125.93	122.68	467.99	1.81
5.	2.73	122.86	128.52	465.65	1.81
6.	1.54	109.71	128.52	456.01	1.81
7.	1.54	83.08	128.52	436.73	1.76
8.	1.54	40.00	128.52	240.59	1.14
9.	2.73	40.58	128.52	240.70	1.14
10.	2.73	58.42	128.52	259.98	1.20
11.	2.73	58.42	153.07	259.98	1.20
12.	10.58	59.42	153.07	260.75	1.20
13.	10.58	111.70	153.07	319.33	1.36
14.	10.58	111.70	114.80	319.33	1.36
15.	10.58	113.70	114.80	446.65	1.69
16.	10.58	111.70	38.27	319.33	1.36
17.	5.68	86.06	38.27	319.33	1.37
18.	5.68	86.06	30.39	290.06	1.28
19.	5.68	86.06	7.88	432.26	1.68
20.	2.73	58.42	30.39	290.06	1.29
21.	2.73	58.42	24.55	259.98	1.20
22.	2.73	58.42	5.84	416.48	1.67
23.	1.01	25.00	29.72	0.00	0.00
24.	1.01	35.00	339.21	4.09	0.01
25.	1.01	900.00	368.90	1006.40	1.54
26.	1.01	180.00	368.90	162.56	0.44
27.	1.01	166.06	368.90	147.64	0.41
28.	1.01	128.70	368.90	108.02	0.31
29.	1.01	105.64	368.90	83.71	0.25
30	1.01	25.00	753.82	0.00	0.00
31.	1.01	33.00	753.82	33.44	0.11

Table 3 . OFC with regenerator material balance results with R124 fluid at hot gas supply of 120 °C.

State	P, bar	t, °C	m, kg/s	h, kJ/kg	s, kJ/kg K
1.	13.30	105.00	84.00	432.92	1.70
2.	10.02	96.37	84.00	428.07	1.70
3.	10.02	95.16	86.91	426.87	1.69
4.	7.38	86.37	86.91	421.65	1.70
5.	7.38	85.17	89.72	420.55	1.69
6.	5.93	79.09	89.72	416.73	1.69
7.	5.93	68.35	89.72	407.30	1.67
8.	5.93	40.00	89.72	245.07	1.15
9.	7.38	40.54	89.72	245.22	1.15
10.	7.38	48.07	89.72	254.65	1.18
11.	7.38	48.07	111.99	254.65	1.18
12.	13.30	49.00	111.99	255.27	1.19
13.	13.30	70.30	111.99	282.22	1.26
14.	13.30	70.30	84.00	282.22	1.26
15.	13.30	72.30	84.00	397.64	1.60
16.	13.30	70.30	28.00	282.22	1.26
17.	10.02	60.19	28.00	282.22	1.27
18.	10.02	60.19	25.09	269.43	1.23
19.	10.02	60.19	2.91	392.42	1.60
20.	7.38	48.07	25.09	269.43	1.23
21.	7.38	48.07	22.28	254.65	1.18
22.	7.38	48.07	2.81	386.56	1.59
23.	1.01	25.00	29.72	0.00	0.00
24.	1.01	35.00	339.21	4.09	0.01
25.	1.01	900.00	368.90	1006.40	1.54
26.	1.01	120.00	368.90	98.84	0.29
27.	1.01	112.39	368.90	90.81	0.27
28.	1.01	87.30	368.90	64.53	0.20
29.	1.01	79.48	368.90	56.35	0.17
30	1.01	25.00	435.25	0.00	0.00
31.	1.01	33.00	435.25	33.44	0.11

Table 4. OFC with regenerator material balance results with R134a fluid at hot gas supply of 100 °C.

State	P, bar	t, °C	m, kg/s	h, kJ/kg	s, kJ/kg K
1.	22.07	85.00	34.55	450.64	1.76
2.	16.79	73.62	34.55	445.59	1.76
3.	16.79	73.10	35.93	444.86	1.76
4.	12.52	61.56	35.93	439.30	1.76
5.	12.52	61.11	37.18	438.74	1.76
6.	10.17	53.28	37.18	434.63	1.76
7.	10.17	46.62	37.18	426.56	1.74
8.	10.17	40.00	37.18	256.41	1.19
9.	12.52	42.98	37.18	260.41	1.20
10.	12.52	47.97	37.18	268.48	1.23
11.	12.52	47.97	46.07	268.48	1.23
12.	22.07	50.94	46.07	272.60	1.24
13.	22.07	69.88	46.07	304.08	1.33
14.	22.07	69.88	34.55	304.08	1.33
15.	22.07	71.88	34.55	428.85	1.69
16.	22.07	69.88	11.52	304.08	1.33
17.	16.79	59.93	11.52	304.08	1.33
18.	16.79	59.93	10.14	287.38	1.28
19.	16.79	59.93	1.38	426.61	1.70
20.	12.52	47.97	10.14	287.38	1.29
21.	12.52	47.97	8.89	268.48	1.23
22.	12.52	47.97	1.24	422.68	1.71
23.	1.01	25.00	29.72	0.00	0.00
24.	1.01	35.00	339.21	4.09	0.01
25.	1.01	900.00	368.90	1006.40	1.54
26.	1.01	100.00	368.90	77.82	0.23
27.	1.01	98.07	368.90	75.78	0.23
28.	1.01	86.88	368.90	64.09	0.20
29.	1.01	83.14	368.90	60.16	0.18
30	1.01	25.00	189.16	0.00	0.00
31.	1.01	33.00	189.16	33.44	0.11

Table 5. OFC with regenerator material balance results with R245fa fluid at hot gas supply of 150 °C.

State	P, bar	t, °C	m, kg/s	h, kJ/kg	s, kJ/kg K
1.	12.79	135.00	73.10	521.50	1.91
2.	7.47	121.13	73.10	511.88	1.92
3.	7.47	118.38	78.06	508.60	1.91
4.	4.02	104.34	78.06	497.49	1.92
5.	4.02	102.12	81.76	495.10	1.91
6.	2.50	92.38	81.76	486.67	1.91
7.	2.50	75.34	81.76	469.44	1.87
8.	2.50	40.00	81.76	252.57	1.18
9.	4.02	42.73	81.76	256.20	1.19
10.	4.02	55.12	81.76	273.42	1.24
11.	4.02	55.12	97.47	273.42	1.24
12.	12.79	57.86	97.47	277.17	1.26
13.	12.79	98.50	97.47	337.42	1.42
14.	12.79	98.50	73.10	337.42	1.42
15.	12.79	100.50	73.10	474.54	1.79
16.	12.79	98.50	24.37	337.42	1.42
17.	7.47	77.81	24.37	337.42	1.43
18.	7.47	77.81	19.41	306.00	1.34
19.	7.47	77.81	4.96	460.29	1.78
20.	4.02	55.12	19.41	306.00	1.34
21.	4.02	55.12	15.71	273.42	1.24
22.	4.02	55.12	3.69	444.58	1.77
23.	1.01	25.00	29.72	0.00	0.00
24.	1.01	35.00	339.21	4.09	0.01
25.	1.01	900.00	368.90	1006.40	1.54
26.	1.01	150.00	368.90	130.58	0.37
27.	1.01	141.25	368.90	121.27	0.34
28.	1.01	115.50	368.90	94.10	0.28
29.	1.01	100.37	368.90	78.18	0.23
30	1.01	25.00	530.22	0.00	0.00
31.	1.01	33.00	530.22	33.44	0.11

Table 6. OFC with regenerator material balance results with R717 fluid at hot gas supply of 125 °C.

State	P, bar	t, °C	m, kg/s	h, kJ/kg	s, kJ/kg K
1.	33.23	110.00	13.59	1671.53	5.36
2.	25.44	93.48	13.59	1633.77	5.37
3.	25.44	87.93	13.81	1610.65	5.31
4.	19.07	69.42	13.81	1572.93	5.32
5.	19.07	64.06	14.04	1552.87	5.26
6.	15.55	50.53	14.04	1526.87	5.27
7.	15.55	40.07	14.04	1490.16	5.16
8.	15.55	40.00	14.04	390.64	1.64
9.	19.07	40.26	14.04	391.45	1.65
10.	19.07	47.53	14.04	428.16	1.76
11.	19.07	47.53	18.12	428.16	1.76
12.	33.23	48.28	18.12	431.49	1.77
13.	33.23	68.12	18.12	534.95	2.08
14.	33.23	68.12	13.59	534.95	2.08
15.	33.23	70.12	13.59	1483.84	4.84
16.	33.23	68.12	4.53	534.95	2.08
17.	25.44	58.82	4.53	534.95	2.08
18.	25.44	58.82	4.31	485.88	1.93
19.	25.44	58.82	0.22	1489.58	4.96
20.	19.07	47.53	4.31	485.88	1.94
21.	19.07	47.53	4.07	428.16	1.76
22.	19.07	47.53	0.23	1491.03	5.08
23.	1.01	25.00	29.72	0.00	0.00
24.	1.01	35.00	339.21	4.09	0.01
25.	1.01	900.00	368.90	1006.40	1.54
26.	1.01	125.00	368.90	104.11	0.30
27.	1.01	118.46	368.90	97.20	0.28
28.	1.01	85.12	368.90	62.25	0.19
29.	1.01	80.27	368.90	57.17	0.18
30	1.01	25.00	461.76	0.00	0.00
31.	1.01	33.00	461.76	33.44	0.11

Table 7. OFC with regenerator material balance results with R407C zeotrope at hot gas supply of 78 °C.

State	P, bar	t, °C	m, kg/s	h, kJ/kg	s, kJ/kg K
1.	22.47	63.00	53.57	439.06	1.75
2.	20.50	58.41	53.57	436.96	1.75
3.	20.50	58.36	53.96	436.88	1.75
4.	18.65	53.67	53.96	434.69	1.75
5.	18.65	53.58	54.69	434.56	1.75
6.	17.50	50.36	54.69	432.97	1.75
7.	17.50	42.45	54.69	428.82	1.73
8.	17.50	40.00	54.69	260.34	1.20
9.	18.65	40.15	54.69	260.48	1.20
10.	18.65	42.65	54.69	264.64	1.22
11.	18.65	42.65	71.43	264.64	1.22
12.	22.47	43.01	71.43	265.12	1.22
13.	22.47	48.60	71.43	274.69	1.25
14.	22.47	48.60	53.57	274.69	1.25
15.	22.47	55.02	53.57	426.29	1.71
16.	22.47	48.60	17.86	274.69	1.25
17.	20.50	46.63	17.86	274.69	1.25
18.	20.50	46.63	17.47	271.33	1.24
19.	20.50	46.63	0.39	425.69	1.72
20.	18.65	42.65	17.47	271.33	1.24
21.	18.65	42.65	16.74	264.64	1.22
22.	18.65	42.65	0.73	425.03	1.73
23.	1.01	25.00	29.72	0.00	0.00
24.	1.01	35.00	339.21	4.09	0.01
25.	1.01	900.00	368.90	1006.40	1.54
26.	1.01	78.00	368.90	54.82	0.17
27.	1.01	76.24	368.90	52.97	0.16
28.	1.01	55.02	368.90	30.95	0.10
29.	1.01	53.25	368.90	29.10	0.09
30	1.01	25.00	275.53	0.00	0.00
31.	1.01	33.00	275.53	33.44	0.11

Sankey and Grossman Diagrams

The energy analysis and exergy analysis are presented from Figs. (**33** - **38**) with the fluids. The Sankey diagram and Grassmann diagram are generated from the first law analysis and second law analysis respectively. On an average, the energy of hot gas supply is lost 60% with exhaust and 35% at condenser. Nearly 2-5% is the output from the energy supply. The balanced energy is resulted as power generation. The depicted exergy losses are HRVG, turbine, regenerator, condenser, pump, exhaust and hot water. The exergy analysis is conducted with the exergy balance of the components in the plant. Similar to first law results, the second law losses are also more in exhaust from 25% to 40%. After the exhaust exergetic decay, HRVG shares the majority of losses. The balance components offer nominal exergy losses.

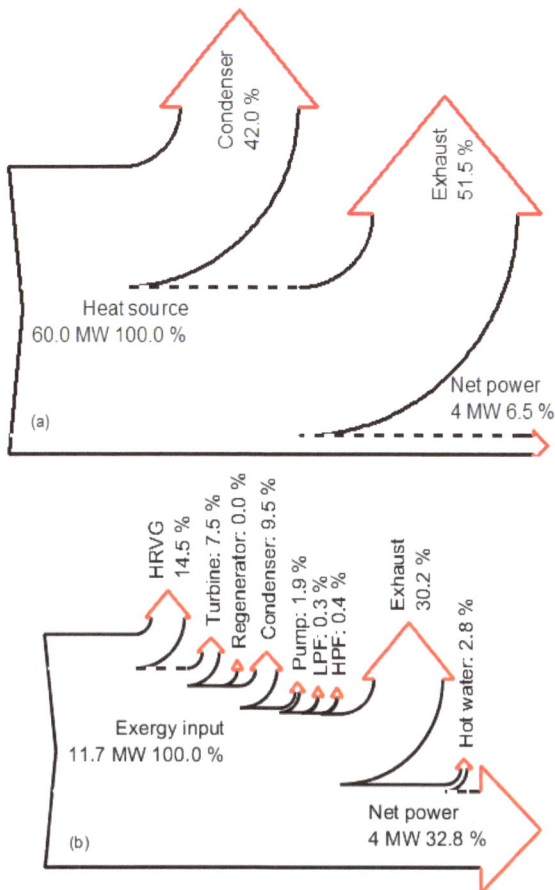

Fig. (33). (a) Sankey diagram and **(b)** Grassmann diagram of OFC with R123.

Condenser
39.9 %

Exhaust
57.0 %

Heat source
36.5 MW 100.0 %

Net power
1 MW 3.1 %

(a)

HRVG
18.2 %

Turbine: 4.4 %

Regenerator: 0.0 %
Condenser
12.4 %

Pump: 3.5 %

LPF: 0.2 %
HPF: 0.2 %

Exhaust
35.0 %

Hot water: 4.0 %

Exergy input
4.8 MW 100.0 %

Net power
1 MW 22.0 %

(b)

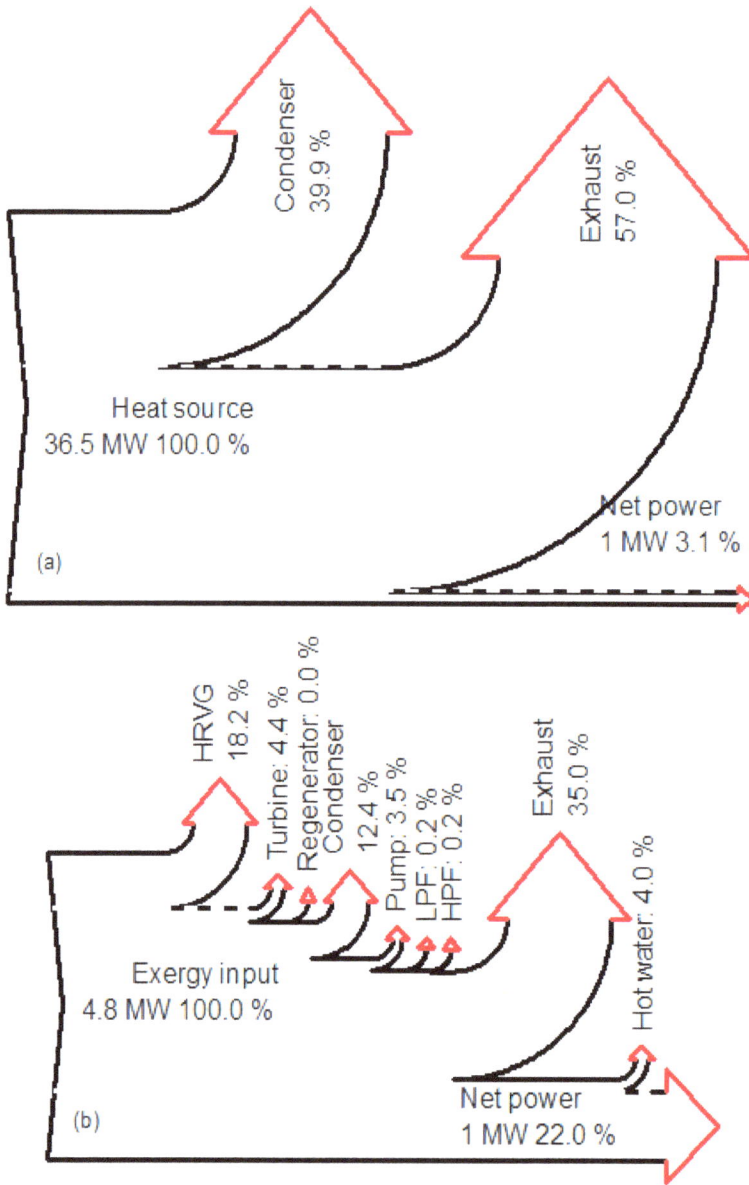

Fig. (34). (a) Sankey diagram and (b) Grassmann diagram of OFC with R124.

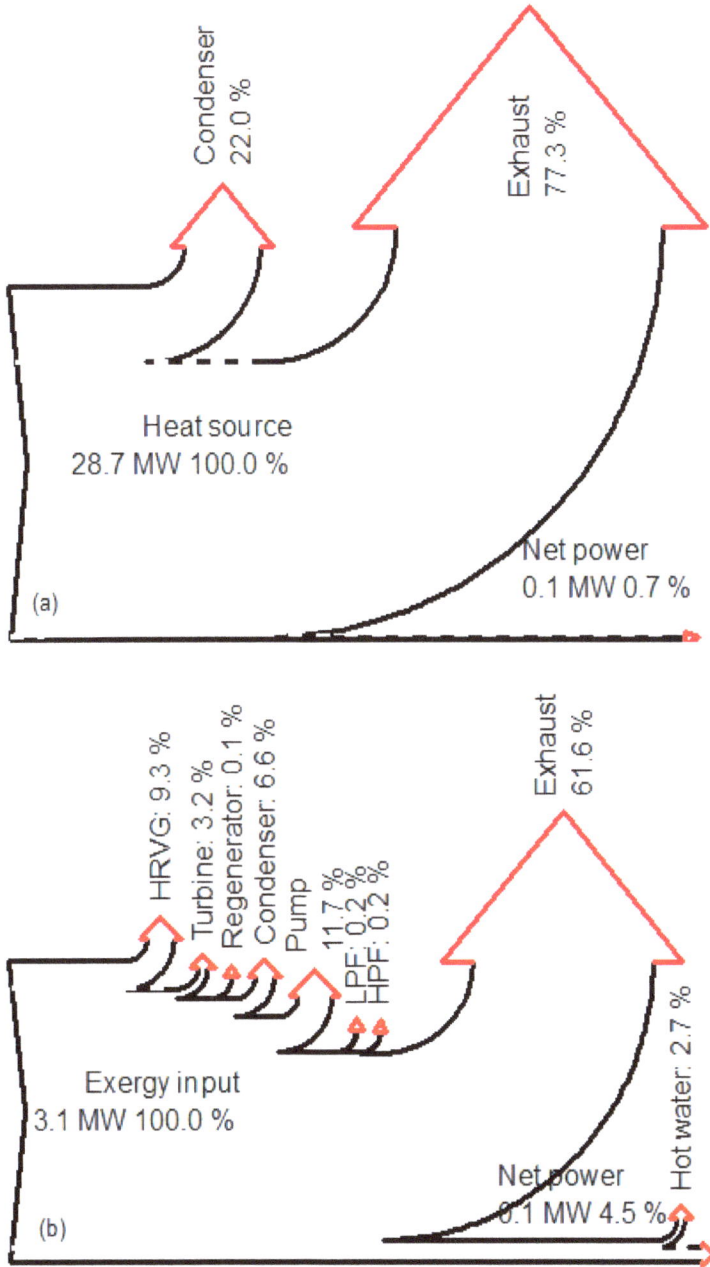

Fig. (35). (a) Sankey diagram and **(b)** Grassmann diagram of OFC with R134a.

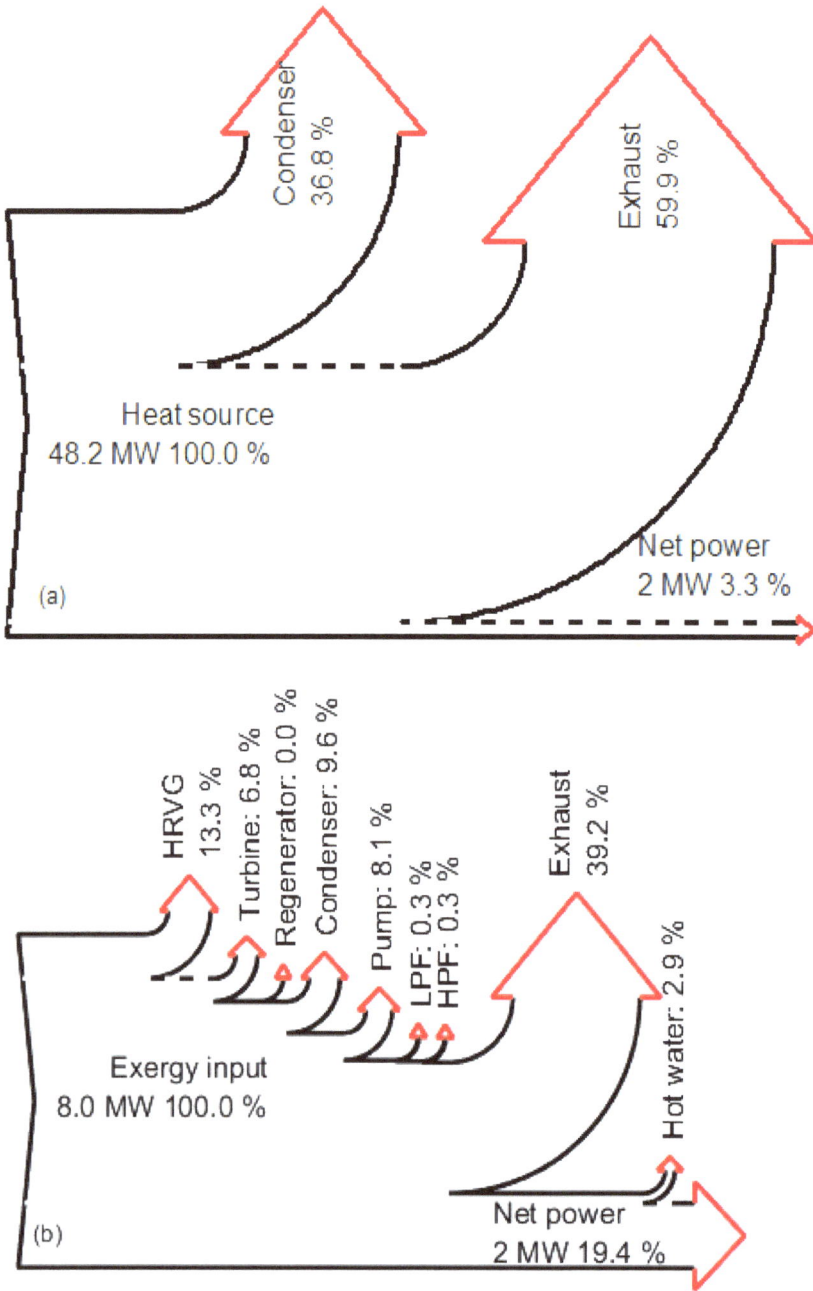

Fig. (36). (a) Sankey diagram and **(b)** Grassmann diagram of OFC with R245fa.

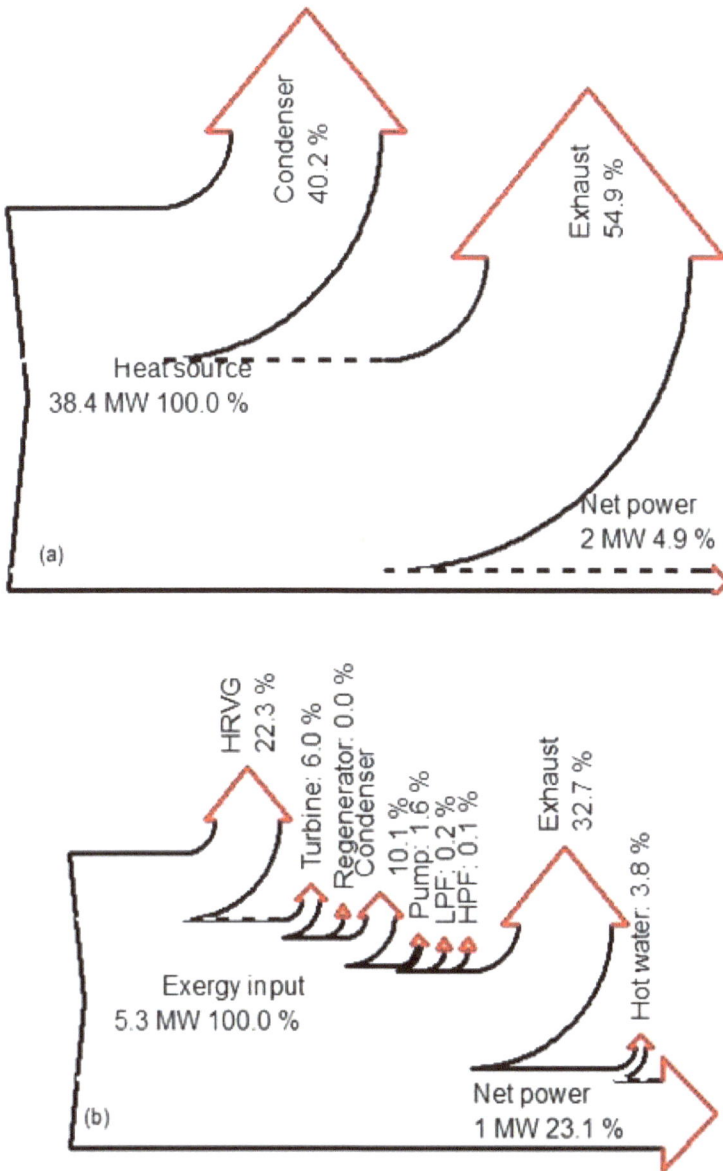

Fig. (37). (a) Sankey diagram and **(b)** Grassmann diagram of OFC with R717.

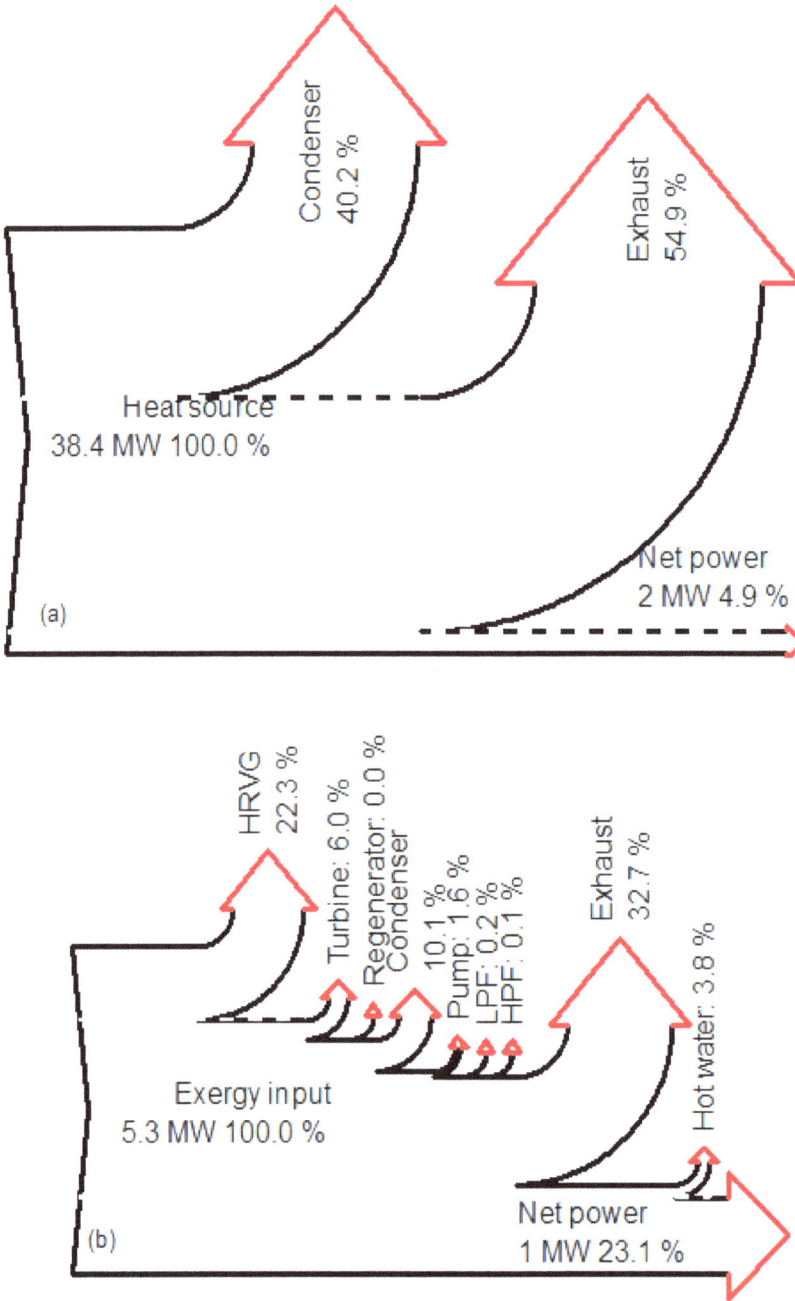

Fig. (38). (a) Sankey diagram and (b) Grassmann diagram of OFC with R407C.

Performance Comparison of OFC with Working Fluids

Fig. (**39**) compares the performance (power and efficiency) changes with change in working fluid and temperature. For each fluid, temperature range is selected from the properties of the fluids. R134a and R407C are suitable to low temperature heat. R123 has good compatibility with the high temperature. The power and efficiency increases with increase in temperature. The temperature range for R123 is more compared to the other fluids.

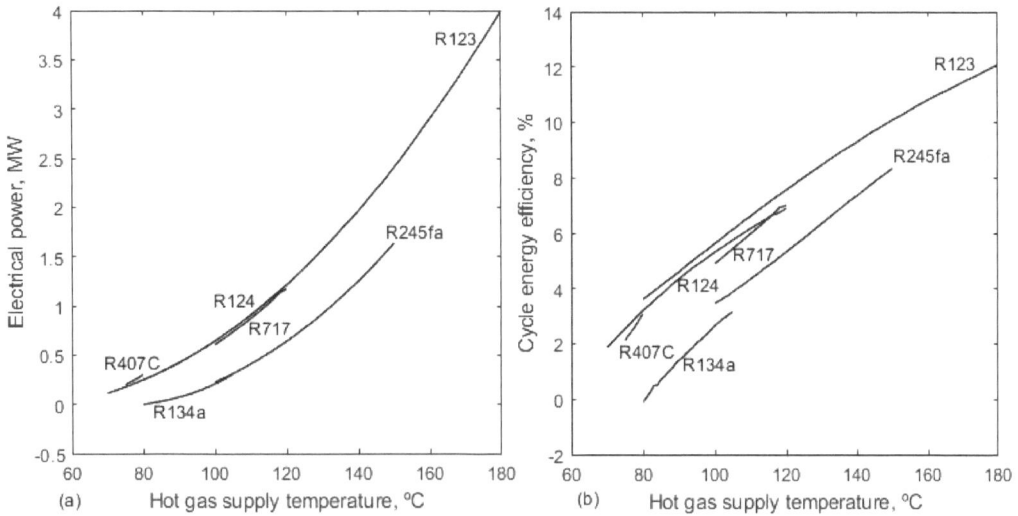

Fig. (39). Performance comparison of OFC with working fluids.

The power curves are matched with the fluids of R123, R124 and R407. Similarly, performance curves related to power in the same curve for R134a, R717 and R245fa with different temperature source (Fig. **39a**). The slope of the power curve is increasing with increase in temperature. From Fig. (**39b**), it can be observed that the slop of efficiency curve is decreasing with increase in the temperature.

Table **8** compares the specifications developed with the thermodynamic study for the OFC with regenerator. At the beginning, the critical temperature and critical pressure of the selected fluids are presented to know the deviation of the turbine inlet state with reference to critical state. The turbine inlet temperature is just below the critical temperature of the fluid. The critical temperature of R123 is high and the critical pressure of R717 is high compared to other fluids. The critical pressure of R123, R124 and R245fa is close with each other. As per the order, the lowest turbine inlet temperature is maintained with R407C. The maximum and minimum HRVG pressure are resulted with R717 and R123 respectively. The maximum and minimum condenser pressure are with R407C and R123 respectively. In addition to the HRVG pressure and condenser pressure, the flash pressure in HP flasher and LP flasher are also reported. The HP flash pressure is between LP flash pressure and HRVG pressure. Similarly, the LP flash pressure lies between condenser pressure and HP flash pressure. The higher and lower HP flash pressure are with R717 and R407C respectively. The same fluids resulted the maximum and minimum LP flash pressure in the same order. As per the earlier discussion, the high specific power is observed with R717. R134a is producing low specific power.

Table 8. Comparison of OFC specifications with working fluids at optimized HRVG pressure and source temperature below the critical temperature..

Working Fluid	R123	R124	R134a	R245fa	R717	R407C
Critical temperature, °C	183.68	122.47	101.08	154.10	132.40	86.03
Critical pressure, bar	36.68	36.34	40.60	36.40	113.40	46.30
Hot fluid inlet temperature, °C	180.00	120.00	100.00	150.00	125.00	78.00
Turbine inlet temperature, °C	165.00	105.00	85.00	135.00	110.00	63.00
Optimized HRVG pressure, bar	10.58	13.30	22.07	12.79	33.23	22.47
HP flash pressure, bar	5.68	10.02	16.79	7.47	25.44	20.49
LP flash pressure, bar	2.73	7.37	12.52	4.02	19.07	18.65
Condenser pressure, bar	1.54	5.93	10.17	2.50	15.55	17.49
Specific power, kW/kg	31.68	12.46	4.15	19.67	91.38	4.75
Plant power, MW	3.63	1.05	0.14	1.43	1.24	0.25
Cycle thermal efficiency, %	12.50	6.67	2.20	7.44	7.17	2.68
Relative cycle efficiency, %	43.83	38.83	17.53	31.96	39.24	39.23
Plant exergy efficiency, %	30.96	21.68	4.62	18.01	23.39	15.79

ORGANIC FLASH CYCLE WITH SOURCE TEMPERATURE ABOVE THE CRITICAL TEMPERATURE

In the earlier sections, the source temperature is maintained below the critical temperature of the fluid. This section is presented on the development of performance characteristics with the source temperature above the critical temperature of the working fluid. A separate correlation is developed for below and above the critical temperature of the fluid. In both the cases, the HRVG pressure is selected below the critical pressure. Therefore, the studied power plant cycles are sub-critical power plants.

Optimum Boiler Pressure with Source Temperature

Figs. (**40** - **45**) shows the optimum boiler pressure of OFC with source temperature above the critical temperature respectively with R123, R124, R134a, R245fa, R717 and R407C. Fig. (**40**) depicts the change in hot gas supply temperature and HRVG pressure on power, thermal efficiency, degree of superheat and exhaust gas temperature. Suitable temperature range (200 °C to 250 °) and pressure range (2 bar to 30 bar) are selected to operate the plant with source temperature above the critical temperature of R123. The HRVG pressure is increased from condenser to close to critical pressure. The condenser pressure is fixed with the sink temperature. Maximum power is resulting at the optimum HRVG pressure. The optimum HRVG or boiler pressure differs with source temperature. It increases with increase in source temperature or turbine inlet temperature. The rise in turbine inlet pressure increases the vapor expansion in turbine which causes power boost. On other side, the pump supply also increases with rise in boiler pressure. The trade-off results the optimum HRVG pressure with the objective of maximum net power. The efficiency is the function of power and heat supply. The rise in pressure, increases the power and drops the heat supply with pinch point. These two factors, favors to thermal efficiency with diminishing rate as shown in Fig. (**40b**). The degree of superheat decreases with increase in pressure and also with source temperature (Fig. **40c**). The exhaust gas temperature increases with increase in pressure and decreases with increase in source temperature (Fig. **40d**).

Fig. (**41**) is developed to study the performance and process conditions of OFC with R124 fluid. Fig. (**41a**) results can be used to select the best HRVG pressure at a suitable/available source temperature. Similar to earlier plant, the optimum HRVG pressure is increasing with increase in source temperature at maximum power condition. Fig. (**41c**) is prepared to study the variations in DSH resulted with changes in temperature and pressure. The DSH is increasing with increase in source temperature and decreasing with rise in HRVG pressure.

Similarly, for R245fa, the performance and process conditions are developed in Fig. (**43**).

Fig. (**42**) shows the power, thermal efficiency, degree of superheat and exhaust gas temperature with changes in HRVG pressure and turbine inlet temperature with the fluid R134a. The optimum HRVG pressure is increasing with increase in source temperature.

It is clear that the optimum HRVG pressure depends on source temperature and working fluid at a fixed condenser temperature. The data of best optimum pressure at the chosen source temperature and fluid has a significant role on design of a power plant. For every source temperature and working fluid, there will be a one optimum HRVG pressure to maximize the power. To formulate the HRVG pressure as a function of source temperature with working fluid, correlations are developed from the best fit method and presented in Table **9**.

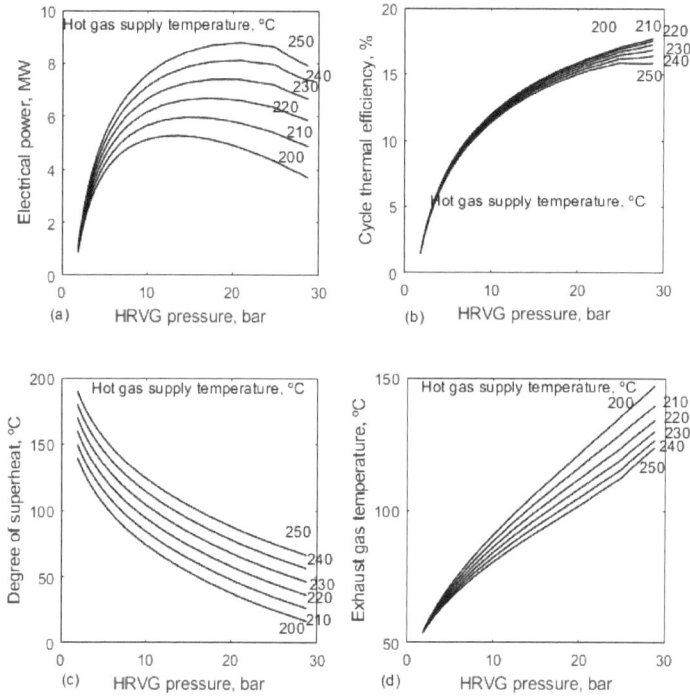

Fig. (40). Influence of source temperature with HRVG pressure on **(a)** net power, **(b)** cycle energy conversion efficiency **(c)** degree of superheat and **(d)** exhaust gas temperature with R123 fluid.

Table 9. Developed coefficients to find the optimum HRVG pressure at source temperature above the critical temperature and 30 °C of circulating water temperature (condensate temperature is 40 °C)..

Working Fluid	Coefficients		
	A_1	A_2	A_3
R123	-109.579	0.9781	-0.0018
R124	-76.0972	1.1357	-0.0032
R134a	-3.8816	0.5372	-0.002
R245fa	-135.029	1.5352	-0.0037
R717	-412.2922	5.1655	-0.0140
R407C	-2.302	0.6942	-0.0031

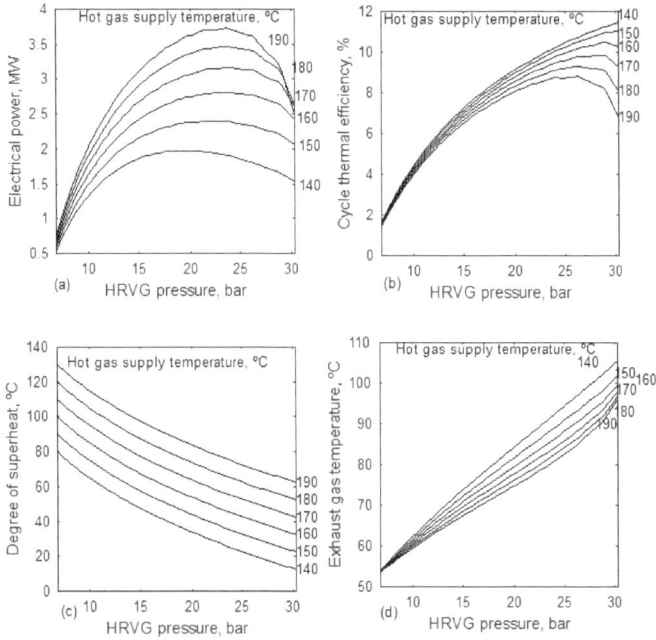

Fig. (41). Influence of source temperature with HRVG pressure on **(a)** net power generation, **(b)** cycle energy conversion efficiency **(c)** degree of superheat and **(d)** exhaust gas temperature with R124 fluid.

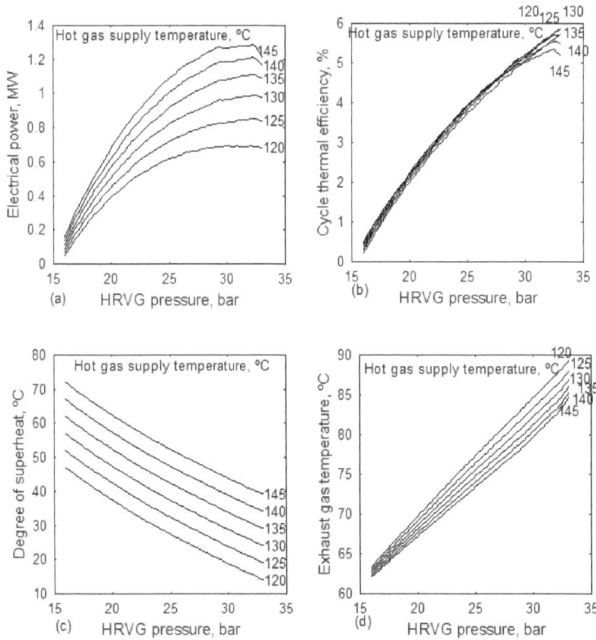

Fig. (42). Influence of source temperature with HRVG pressure on **(a)** net power generation, **(b)** cycle energy conversion efficiency **(c)** degree of superheat and **(d)** exhaust gas temperature with R134a fluid.

Fig. (**44**) shows the power, thermal efficiency, DSH and exhaust gas temperature with R717 fluid.

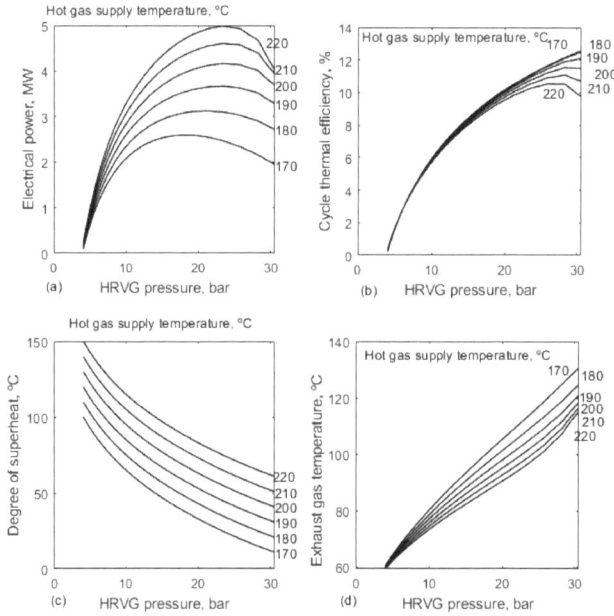

Fig. (43). Influence of source temperature with HRVG pressure on (**a**) net power generation, (**b**) cycle energy conversion efficiency (**c**) degree of superheat and (**d**) exhaust gas temperature with R245fa fluid.

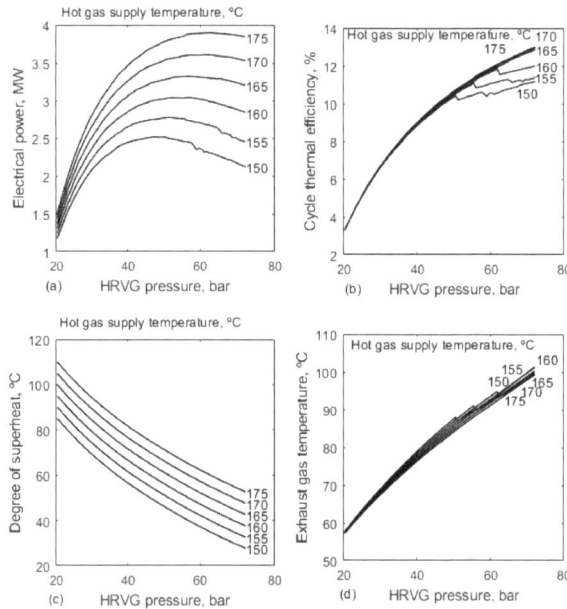

Fig. (44). Influence of source temperature with HRVG pressure on (**a**) net power generation, (**b**) cycle energy conversion efficiency (**c**) degree of superheat and (**d**) exhaust gas temperature with R717 fluid.

Fig. (**45**) analyzes the power, thermal efficiency, DSH, and exhaust gas temperature with R407C fluid.

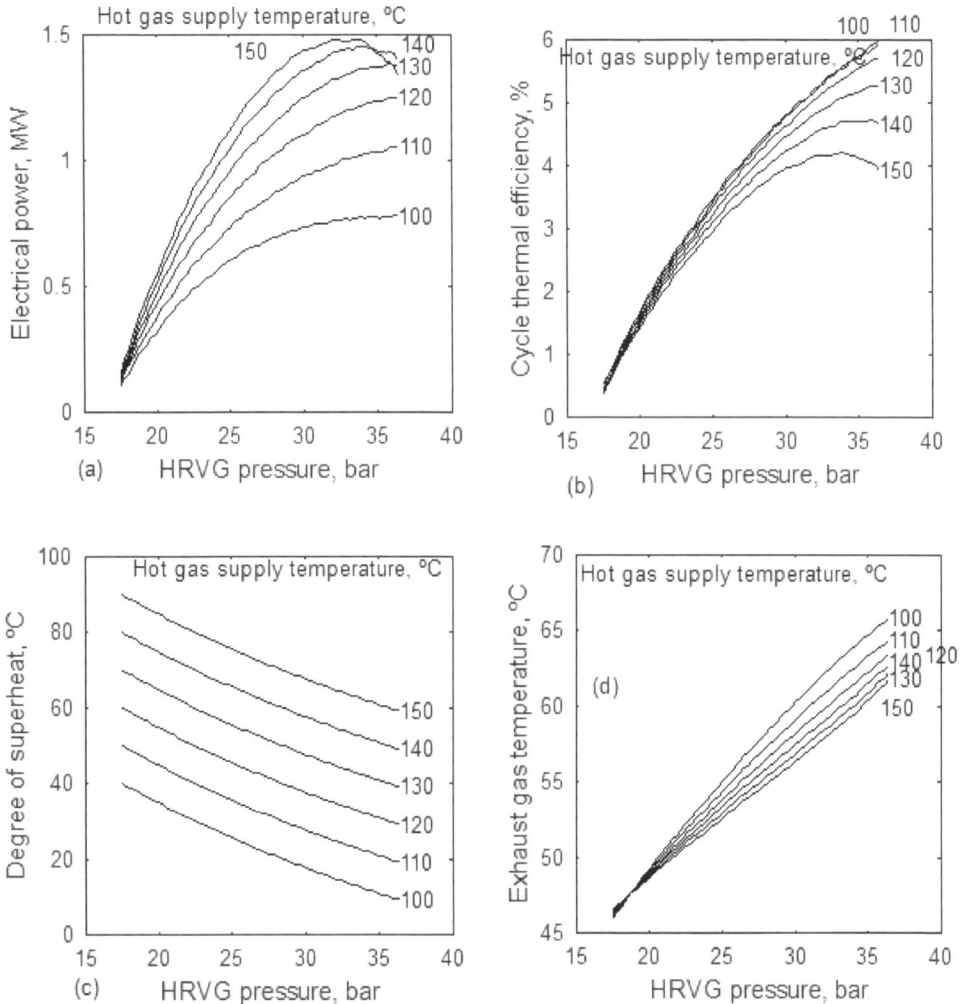

Fig. (45). Influence of source temperature with HRVG pressure on (**a**) net power generation, (**b**) cycle energy conversion efficiency (**c**) degree of superheat and (**d**) exhaust gas temperature with R407C fluid.

Optimum State of LPF and HPF

Figs. (**45 - 51**) shows changes in the OFC plant's power with HPF temperature ratio and LPF temperature ratio on power with the six working fluids. At the beginning of the LPF temperature ratio and HPF temperature ratio, an improvement in net power has been found and latter it is dropping. The vapor supplied to turbine from flasher is decreasing with increase in LPF and HPF temperature ratios. Approximately at the mid of HPF and LPF, power is maximum. With increase in LPF and HPF temperature ratio the exhaust gas temperature is increasing because of low heat recovery.

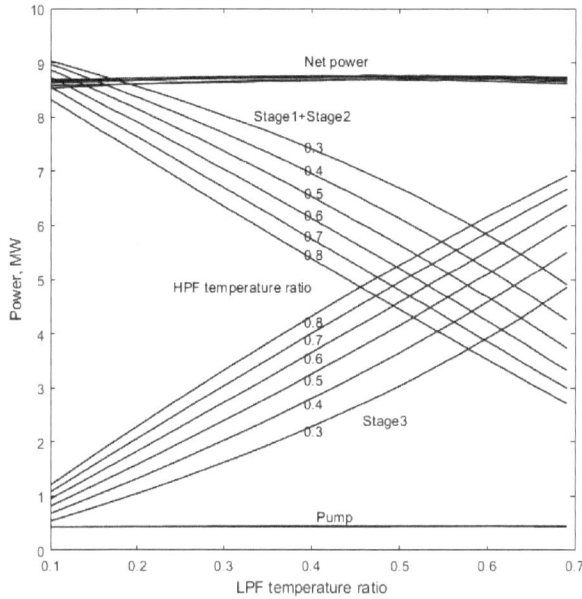

Fig. (46). Influence of LPF and HPF temperature ratios on power generation of OFC with R123 fluid.

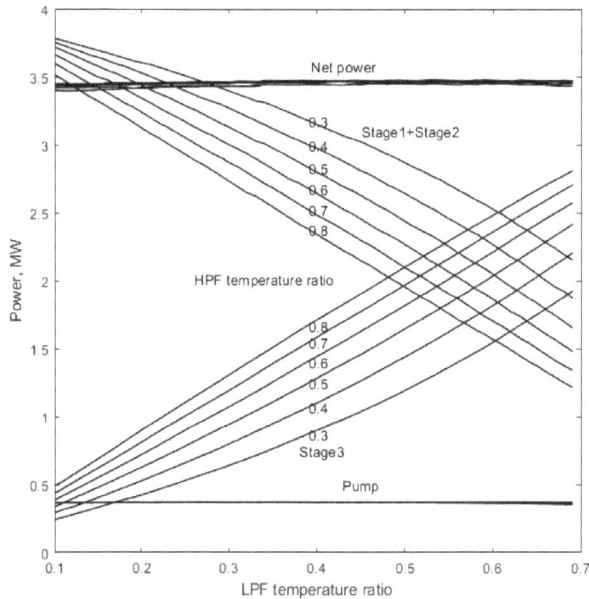

Fig. (47). Influence of LPF and HPF temperature ratios on power generation of OFC with R124 fluid.

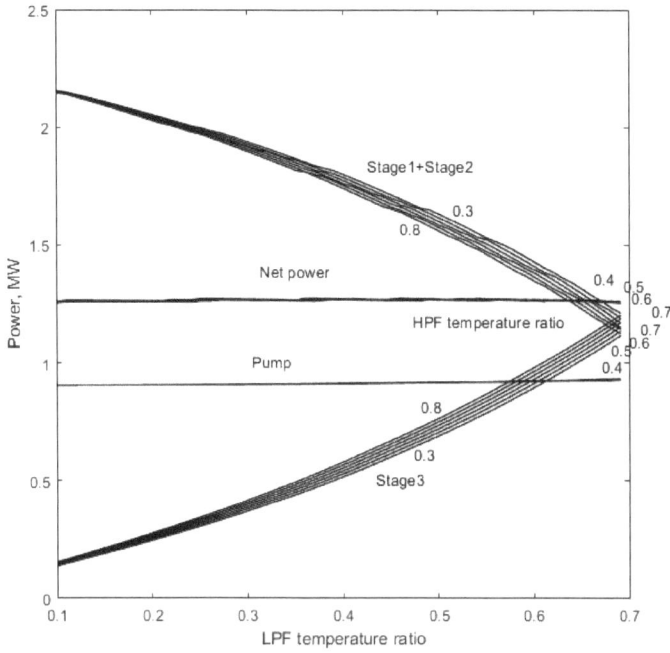

Fig. (48). Influence of LPF and HPF temperature ratios on power generation of OFC with R134a fluid.

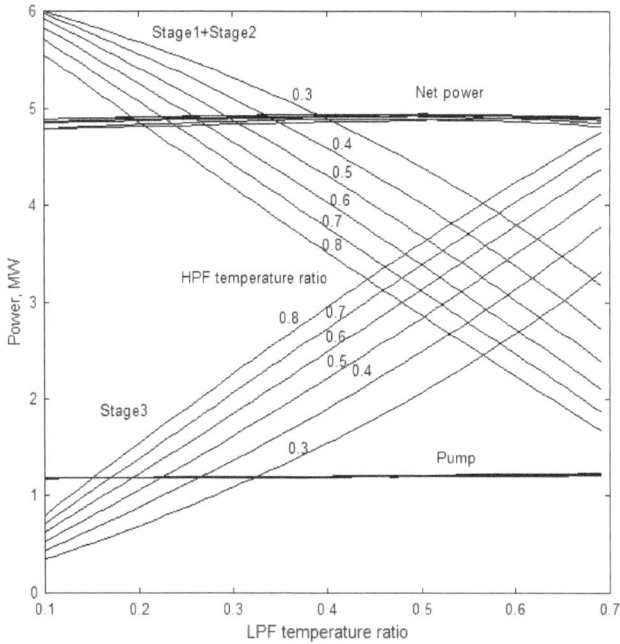

Fig. (49). Influence of LPF and HPF temperature ratios on power generation of OFC with R245fa fluid.

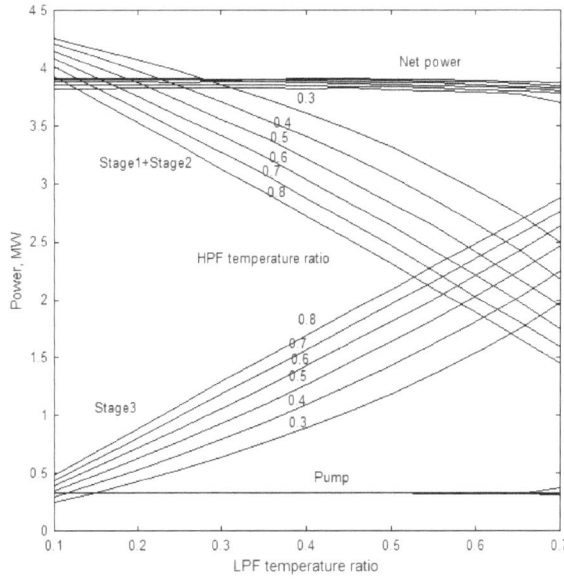

Fig. (50). Influence of LPF and HPF temperature ratios on power generation of OFC with R717 fluid.

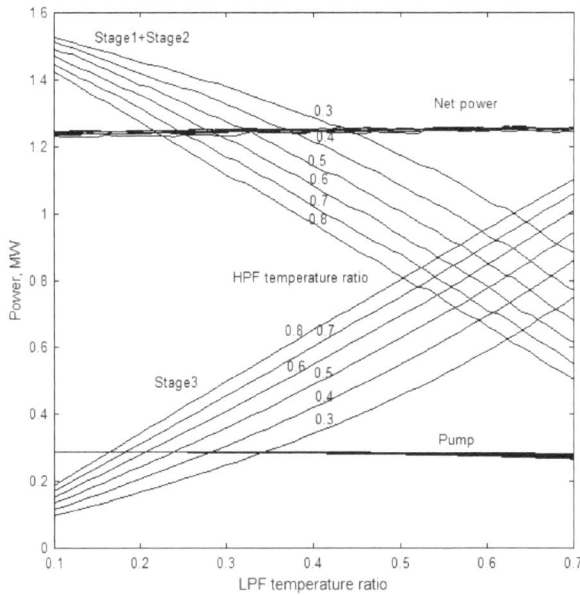

Fig. (51). Influence of LPF and HPF temperature ratios on power generation of OFC with R407C fluid.

Effect of Sink

Figs. (**52 - 57**) shows the effect of sink temperature on plant performance with the above stated working fluids. Fig. (**52**) shows the effect of sink temperature and heat recovery pressure on power and thermal efficiency of OFC-R123. HRVG pressure lies between condenser pressure and critical pressure. The sink temperature is changed from 20 °C to 35 °C. The results show that optimum pressure increases with increase in sink temperature (Fig. **52a**). Similar to R123, the other plots *i.e.* Figs. (**53 - 57**) develops the influence of sink temperature with R124, R134a, R245fa, R717 and R407C. The source temperature is fixed in this analysis. For example, for the R123 sink characteristics, the source temperature is fixed at 250 °C. The hot gas supply temperature is 180 °C, 130 °C, 200 °C, 175 °C and 120 °C respectively maintained with R124, R134a, R245fa, R717 and R407C fluids to study the optimum HRVG pressure with a change in sink temperature.

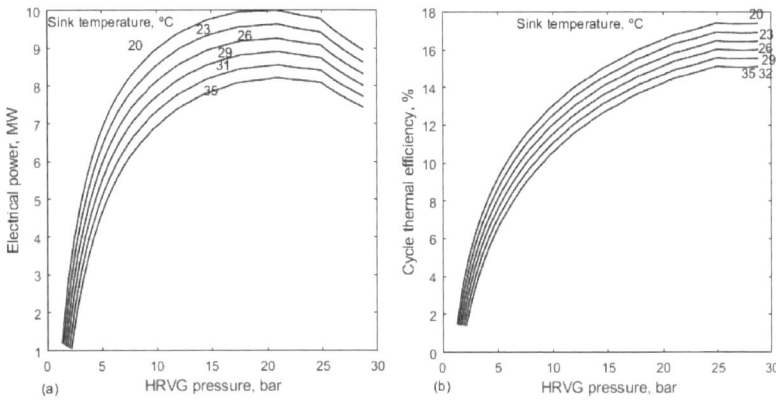

Fig. (52). Influence of sink temperature with HRVG pressure on (**a**) net power generation and (**b**) cycle energy conversion efficiency with R123.

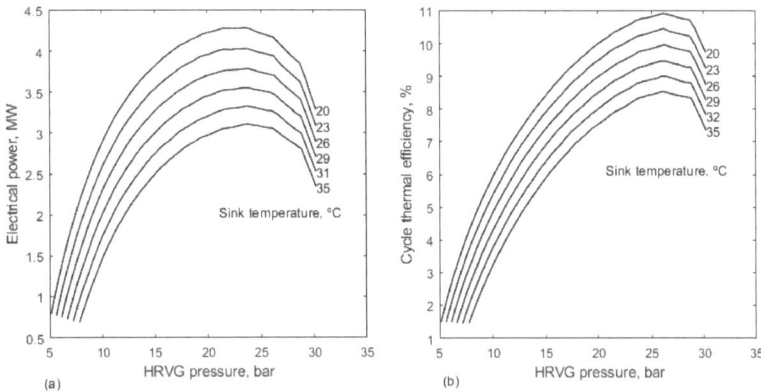

Fig. (53). Influence of sink temperature with HRVG pressure on (**a**) net power generation and (**b**) cycle energy conversion efficiency with R124.

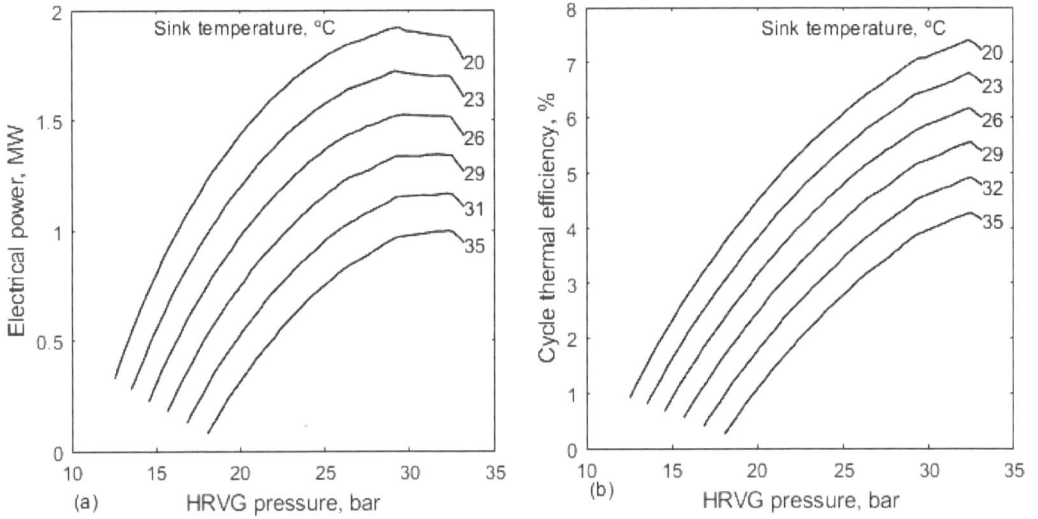

Fig. (54). Influence of sink temperature with HRVG pressure on **(a)** net power generation and **(b)** cycle energy conversion efficiency with R134a.

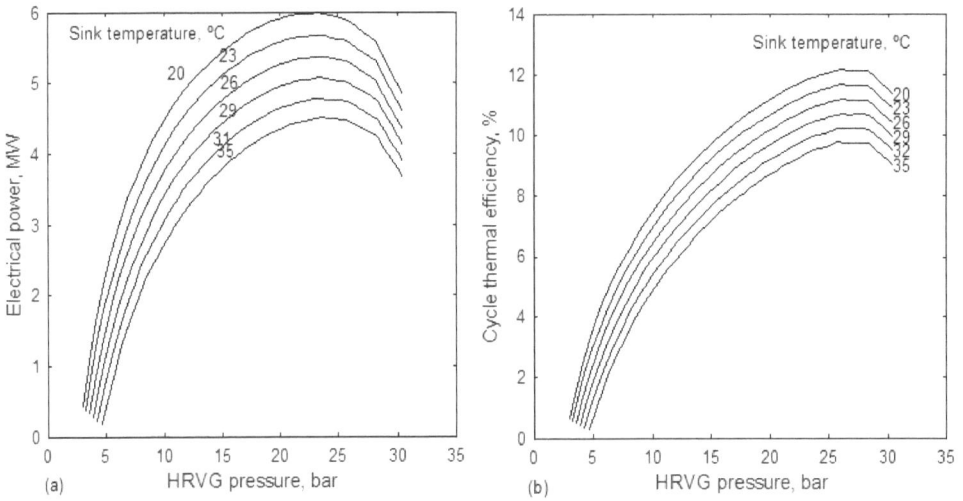

Fig. (55). Influence of sink temperature with HRVG pressure on **(a)** net power generation and **(b)** cycle energy conversion efficiency with R245fa.

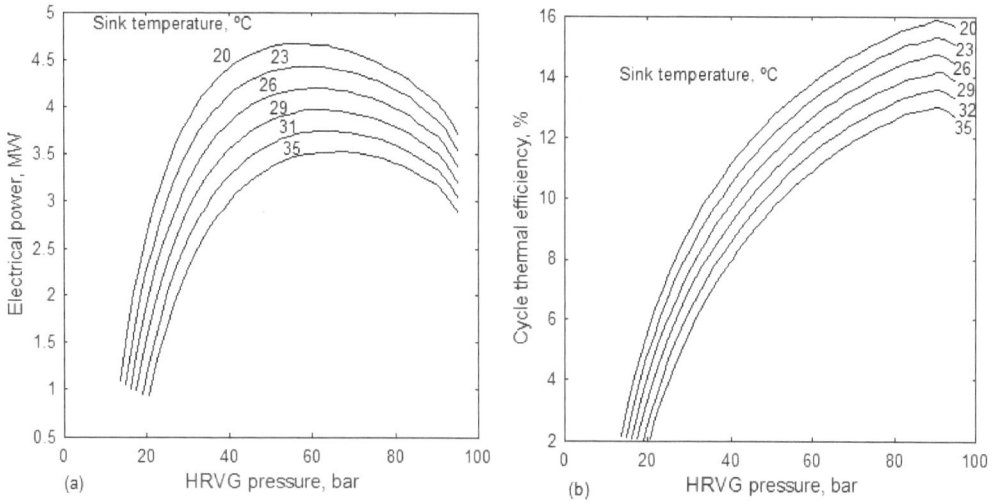

Fig. (56). Influence of sink temperature with HRVG pressure on **(a)** net power generation and **(b)** cycle energy conversion efficiency with R717.

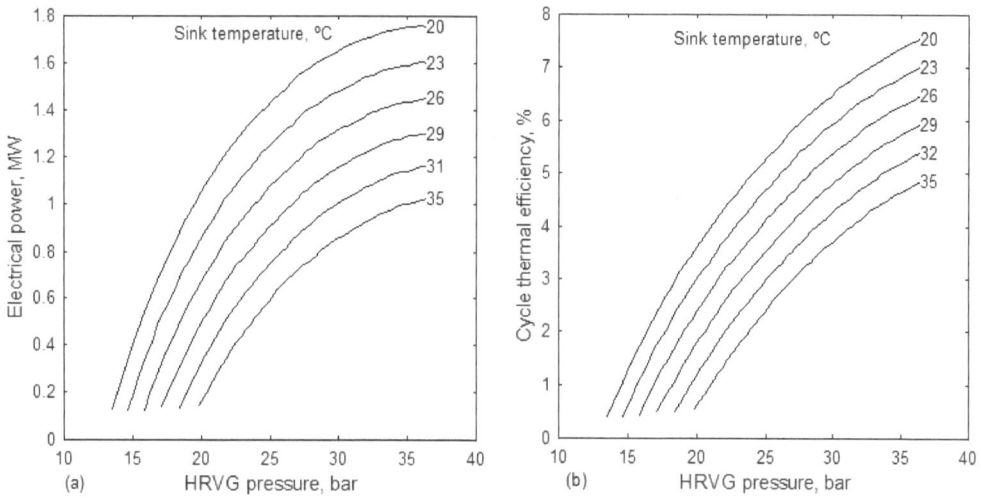

Fig. (57). Influence of sink temperature with HRVG pressure on **(a)** net power generation and **(b)** cycle energy conversion efficiency with R407C.

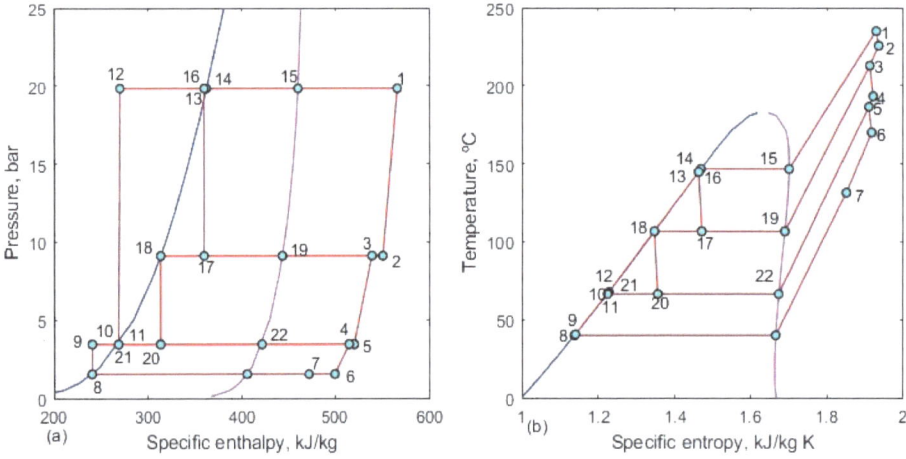

Fig. (58). (a) Pressure-specific volume and **(b)** temperature-specific entropy diagram of OFC with two flashers and R123 fluid.

Property Charts and Tables at the Optimized Conditions

Figs. (**58 - 63**) shows the (a) pressure-specific enthalpy and (b) temperature-entropy plots for the regenerative OFC with the working fluids of R123, R124, R134a, R245fa, R717 and R407C. These are developed at the optimum pressure and the temperature stated in the earlier section.

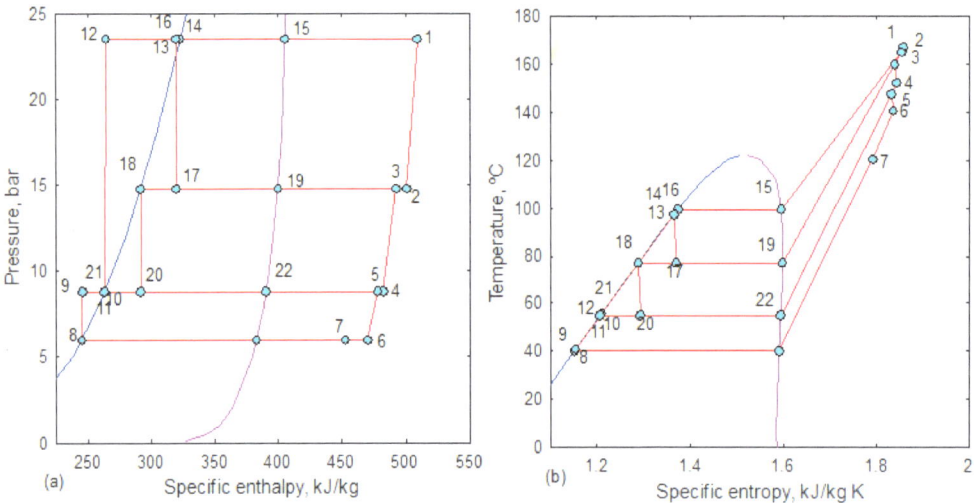

Fig. (59). (a) Pressure-specific volume and **(b)** temperature-specific entropy diagram of OFC with two flashers and R124 fluid.

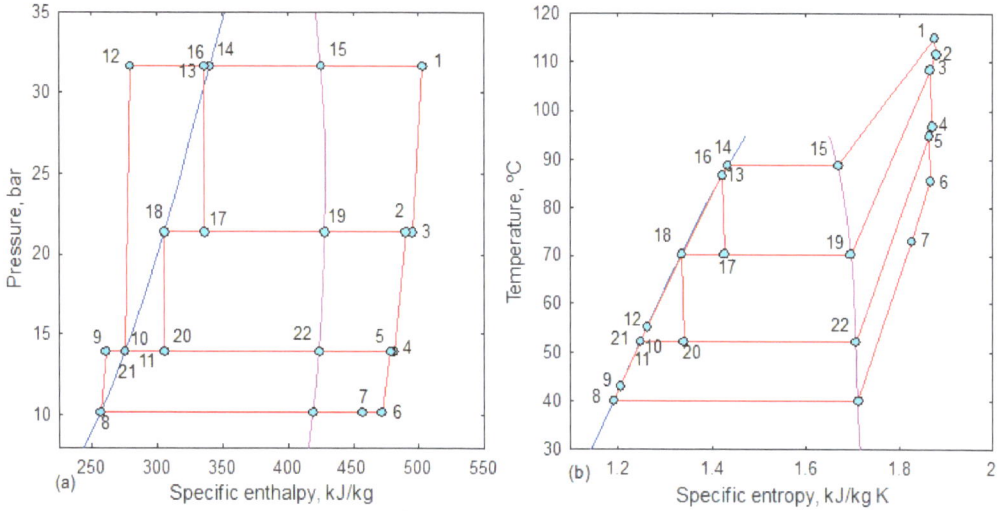

Fig. (60). (a) Pressure-specific volume and **(b)** temperature-specific entropy diagram of OFC with two flashers and R134a fluid.

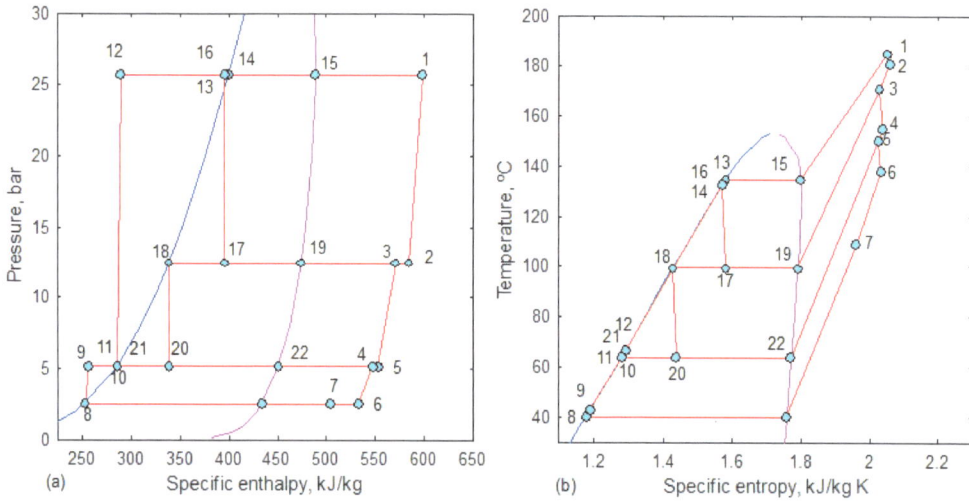

Fig. (61). (a) Pressure-specific volume and **(b)** temperature-specific entropy diagram of OFC with two flashers and R245fa fluid.

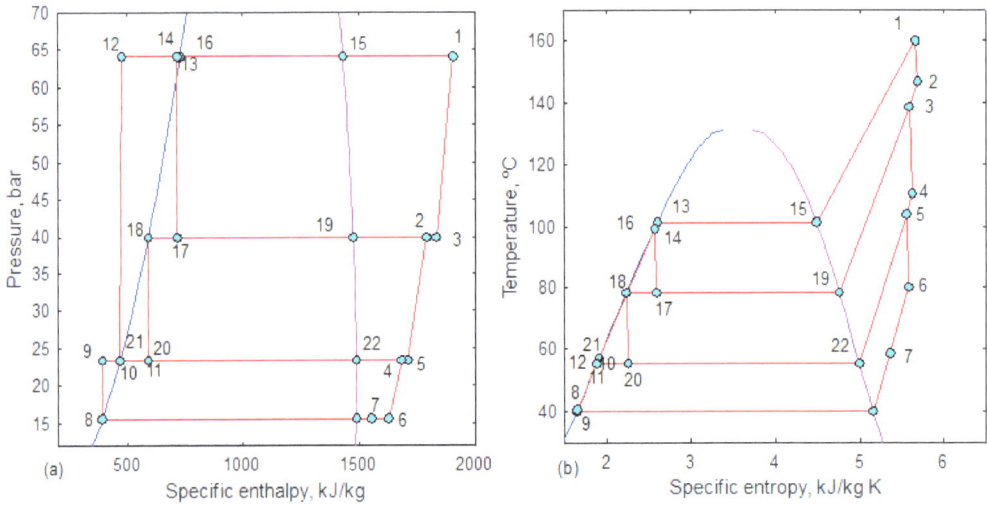

Fig. (62). (a) Pressure-specific volume and **(b)** temperature-specific entropy diagram of OFC with two flashers and R717 fluid.

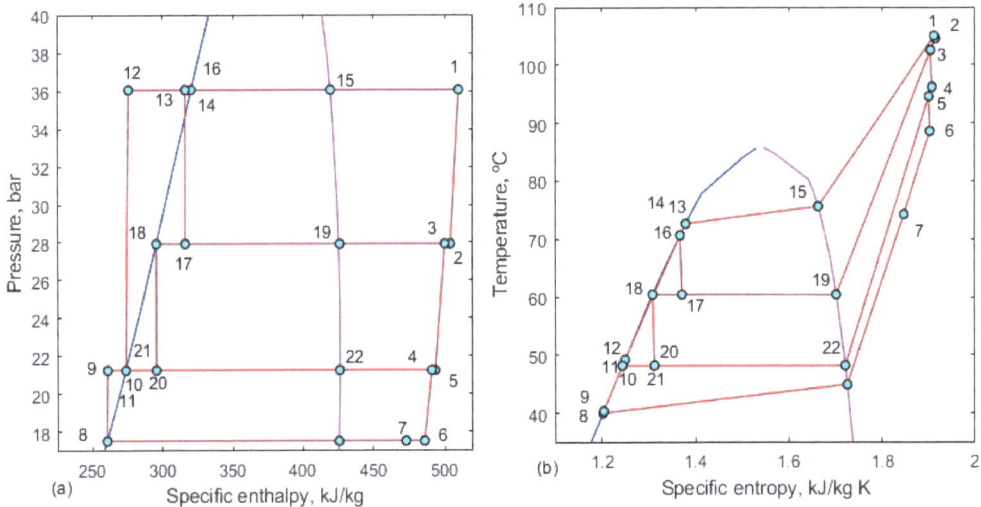

Fig. (63). (a) Pressure-specific volume and **(b)** temperature-specific entropy diagram of OFC with two flashers and R407C fluid.

The pressure, temperature, flow rate, specific enthalpy and specific entropy are given from Tables **10 - 15** respectively with R123, R124, R134a, R245fa, R717 and R407C.

Table 10. OFC material balance results with R123 fluid at hot gas supply of 250 °C.

State	P, bar	t, °C	m, kg/s	h, kJ/kg	s, kJ/kg K
1.	22.45	235.00	171.35	569.62	1.93
2.	10.00	229.15	171.35	554.27	1.94
3.	10.00	215.21	194.28	541.42	1.91
4.	3.62	194.82	194.28	521.65	1.92
5.	3.62	188.17	205.08	516.42	1.91
6.	1.54	170.78	205.08	500.23	1.92
7.	1.54	129.43	205.08	470.28	1.85
8.	1.54	40.00	205.08	240.59	1.14
9.	3.62	40.66	205.08	240.78	1.14
10.	3.62	68.43	205.08	270.73	1.23
11.	3.62	68.43	228.46	270.73	1.23
12.	22.45	69.70	228.46	272.59	1.23
13.	22.45	151.73	228.46	369.52	1.48
14.	22.45	151.73	171.35	369.52	1.48
15.	22.45	153.73	171.35	461.78	1.70
16.	22.45	151.73	57.12	369.52	1.48
17.	10.00	111.08	57.12	369.52	1.49
18.	10.00	111.08	34.18	318.60	1.36
19.	10.00	111.08	22.94	445.40	1.69
20.	3.62	68.43	34.18	318.60	1.37
21.	3.62	68.43	23.38	270.73	1.23
22.	3.62	68.43	10.79	422.30	1.67
23.	1.01	25.00	29.72	0.00	0.00
24.	1.01	35.00	339.21	4.09	0.01
25.	1.01	900.00	368.90	1006.40	1.54
26.	1.01	250.00	368.90	238.12	0.59
27.	1.01	203.76	368.90	188.03	0.49
28.	1.01	163.73	368.90	145.18	0.40
29.	1.01	107.01	368.90	85.15	0.25
30	1.01	25.00	1408.66	0.00	0.00
31.	1.01	33.00	1408.66	33.44	0.11

Table 11. OFC with regenerator material balance results with R124 fluid at hot gas supply of 180 °C.

State	P, bar	t, °C	m, kg/s	h, kJ/kg	s, kJ/kg K
1.	24.65	165.00	140.79	512.80	1.86
2.	15.30	169.93	140.79	504.50	1.86
3.	15.30	162.10	153.99	495.54	1.84
4.	8.91	154.77	153.99	485.13	1.85
5.	8.91	149.94	161.85	480.52	1.84
6.	5.93	142.61	161.85	472.50	1.84
7.	5.93	121.87	161.85	454.29	1.80
8.	5.93	40.00	161.85	245.07	1.15
9.	8.91	40.67	161.85	245.38	1.16
10.	8.91	55.44	161.85	263.58	1.21
11.	8.91	55.44	187.72	263.58	1.21
12.	24.65	56.45	187.72	265.27	1.21
13.	24.65	99.78	187.72	323.16	1.38
14.	24.65	99.78	140.79	323.16	1.38
15.	24.65	101.78	140.79	405.22	1.59
16.	24.65	99.78	46.93	323.16	1.38
17.	15.30	78.61	46.93	323.16	1.38
18.	15.30	78.61	33.73	293.10	1.29
19.	15.30	78.61	13.20	400.03	1.60
20.	8.91	55.44	33.73	293.10	1.30
21.	8.91	55.44	25.87	263.58	1.21
22.	8.91	55.44	7.86	390.18	1.59
23.	1.01	25.00	29.72	0.00	0.00
24.	1.01	35.00	339.21	4.09	0.01
25.	1.01	900.00	368.90	1006.40	1.54
26.	1.01	180.00	368.90	162.56	0.44
27.	1.01	141.47	368.90	121.50	0.34
28.	1.01	111.78	368.90	90.18	0.27
29.	1.01	83.68	368.90	60.73	0.19
30	1.01	25.00	1012.64	0.00	0.00
31.	1.01	33.00	1012.64	33.44	0.11

Table 12. OFC with regenerator material balance results with R134a fluid at hot gas supply of 130 °C.

State	P, bar	t, °C	m, kg/s	h, kJ/kg	s, kJ/kg K
1.	32.15	115.00	71.34	503.40	1.87
2.	21.63	111.87	71.34	495.58	1.88
3.	21.63	108.63	77.48	490.29	1.86
4.	13.99	96.83	77.48	480.95	1.87
5.	13.99	94.84	81.10	478.42	1.86
6.	10.17	85.38	81.10	471.39	1.87
7.	10.17	72.70	81.10	456.45	1.82
8.	10.17	40.00	81.10	256.41	1.19
9.	13.99	42.98	81.10	260.41	1.20
10.	13.99	52.39	81.10	275.35	1.25
11.	13.99	52.39	95.12	275.35	1.25
12.	32.15	55.36	95.12	279.55	1.26
13.	32.15	87.55	95.12	337.68	1.42
14.	32.15	87.55	71.34	337.68	1.42
15.	32.15	89.55	71.34	425.63	1.67
16.	32.15	87.55	23.78	337.68	1.42
17.	21.63	70.97	23.78	337.68	1.43
18.	21.63	70.97	17.64	305.97	1.34
19.	21.63	70.97	6.14	428.75	1.69
20.	13.99	52.39	17.64	305.97	1.34
21.	13.99	52.39	14.01	275.35	1.25
22.	13.99	52.39	3.63	424.28	1.71
23.	1.01	25.00	29.72	0.00	0.00
24.	1.01	35.00	339.21	4.09	0.01
25.	1.01	900.00	368.90	1006.40	1.54
26.	1.01	130.00	368.90	109.39	0.31
27.	1.01	115.76	368.90	94.35	0.28
28.	1.01	99.55	368.90	77.35	0.23
29.	1.01	85.25	368.90	62.36	0.19
30	1.01	25.00	485.17	0.00	0.00
31.	1.01	33.00	485.17	33.44	0.11

Table 13. OFC with regenerator material balance results with R245fa fluid at hot gas supply of 220 °C.

State	P, bar	t, °C	m, kg/s	h, kJ/kg	s, kJ/kg K
1.	23.64	205.00	127.75	633.45	2.13
2.	11.69	207.69	127.75	619.43	2.14
3.	11.69	195.02	144.18	602.65	2.10
4.	4.98	183.83	144.18	584.42	2.11
5.	4.98	177.50	152.11	577.40	2.10
6.	2.50	167.81	152.11	562.93	2.10
7.	2.50	140.33	152.11	535.15	2.04
8.	2.50	40.00	152.11	252.57	1.18
9.	4.98	42.73	152.11	256.20	1.19
10.	4.98	62.60	152.11	283.98	1.28
11.	4.98	62.60	170.33	283.98	1.28
12.	23.64	65.34	170.33	287.78	1.29
13.	23.64	128.41	170.33	387.52	1.55
14.	23.64	128.41	127.75	387.52	1.55
15.	23.64	130.41	127.75	487.77	1.80
16.	23.64	128.41	42.58	387.52	1.55
17.	11.69	96.51	42.58	387.52	1.56
18.	11.69	96.51	26.15	334.31	1.42
19.	11.69	96.51	16.43	472.20	1.79
20.	4.98	62.60	26.15	334.31	1.43
21.	4.98	62.60	18.22	283.98	1.28
22.	4.98	62.60	7.93	449.86	1.77
23.	1.01	25.00	29.72	0.00	0.00
24.	1.01	35.00	339.21	4.09	0.01
25.	1.01	900.00	368.90	1006.40	1.54
26.	1.01	220.00	368.90	205.57	0.53
27.	1.01	173.07	368.90	155.13	0.42
28.	1.01	140.41	368.90	120.41	0.34
29.	1.01	96.72	368.90	74.36	0.22
30	1.01	25.00	1285.40	0.00	0.00
31.	1.01	33.00	1285.40	33.44	0.11

Table 14. OFC with regenerator material balance results with R717 fluid at hot gas supply of 175 °C.

State	P, bar	t, °C	m, kg/s	h, kJ/kg	s, kJ/kg K
1.	62.92	160.00	21.21	1903.08	5.65
2.	39.42	146.18	21.21	1832.39	5.69
3.	39.42	138.17	22.19	1790.80	5.59
4.	23.15	110.61	22.19	1713.14	5.62
5.	23.15	103.87	22.91	1686.11	5.55
6.	15.55	80.20	22.91	1631.02	5.58
7.	15.55	59.10	22.91	1556.94	5.36
8.	15.55	40.00	22.91	390.64	1.64
9.	23.15	40.45	22.91	392.39	1.65
10.	23.15	55.07	22.91	466.47	1.88
11.	23.15	55.07	28.28	466.47	1.88
12.	62.92	56.83	28.28	476.04	1.90
13.	62.92	98.29	28.28	710.11	2.55
14.	62.92	98.29	21.21	710.11	2.55
15.	62.92	100.29	21.21	1435.76	4.49
16.	62.92	98.29	7.07	710.11	2.55
17.	39.42	77.68	7.07	710.11	2.57
18.	39.42	77.68	6.09	587.36	2.22
19.	39.42	77.68	0.98	1476.82	4.76
20.	23.15	55.07	6.09	587.36	2.25
21.	23.15	55.07	5.38	466.47	1.88
22.	23.15	55.07	0.72	1490.55	5.00
23.	1.01	25.00	29.72	0.00	0.00
24.	1.01	35.00	339.21	4.09	0.01
25.	1.01	900.00	368.90	1006.40	1.54
26.	1.01	175.00	368.90	157.21	0.43
27.	1.01	149.80	368.90	130.34	0.36
28.	1.01	110.29	368.90	88.62	0.26
29.	1.01	93.20	368.90	70.68	0.21
30	1.01	25.00	798.89	0.00	0.00
31.	1.01	33.00	798.89	33.44	0.11

Table 15. OFC with regenerator material balance results with R407C zeotrope at hot gas supply of 120 °C.

State	P, bar	t, °C	m, kg/s	h, kJ/kg	s, kJ/kg K
1.	36.36	105.00	88.07	510.59	1.91
2.	28.08	105.03	88.07	504.82	1.92
3.	28.08	102.98	92.82	500.78	1.91
4.	21.29	96.68	92.82	493.96	1.91
5.	21.29	95.04	96.41	491.43	1.90
6.	17.50	89.03	96.41	486.34	1.90
7.	17.50	74.49	96.41	473.04	1.85
8.	17.50	40.00	96.41	260.34	1.20
9.	21.29	40.36	96.41	260.81	1.20
10.	21.29	48.25	96.41	274.11	1.24
11.	21.29	48.25	117.43	274.11	1.24
12.	36.36	49.37	117.43	276.07	1.25
13.	36.36	71.02	117.43	317.22	1.37
14.	36.36	71.02	88.07	317.22	1.37
15.	36.36	75.94	88.07	418.99	1.66
16.	36.36	71.02	29.36	317.22	1.37
17.	28.08	60.64	29.36	317.22	1.37
18.	28.08	60.64	24.61	296.23	1.31
19.	28.08	60.64	4.75	425.92	1.70
20.	21.29	48.25	24.61	296.23	1.31
21.	21.29	48.25	21.02	274.11	1.24
22.	21.29	48.25	3.59	425.90	1.72
23.	1.01	25.00	29.72	0.00	0.00
24.	1.01	35.00	339.21	4.09	0.01
25.	1.01	900.00	368.90	1006.40	1.54
26.	1.01	120.00	368.90	98.84	0.29
27.	1.01	99.21	368.90	76.97	0.23
28.	1.01	75.94	368.90	52.68	0.16
29.	1.01	63.36	368.90	39.58	0.12
30	1.01	25.00	613.25	0.00	0.00
31.	1.01	33.00	613.25	33.44	0.11

Sankey and Grossman Diagrams

Figs. (**64 - 69**) analyzes the energy and exergy results with the fluids. The Sankey charts and Grassmann diagrams are showing the first law output and second law output respectively.

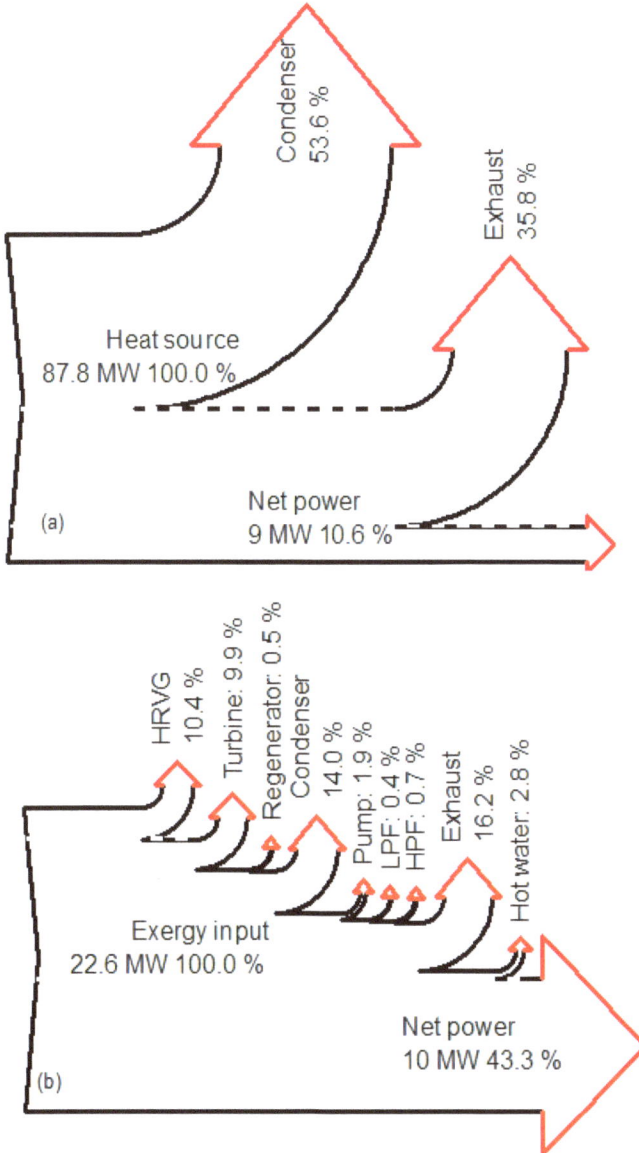

Fig. (**64**). (**a**) Sankey diagram and (**b**) Grassmann diagram of OFC with R123.

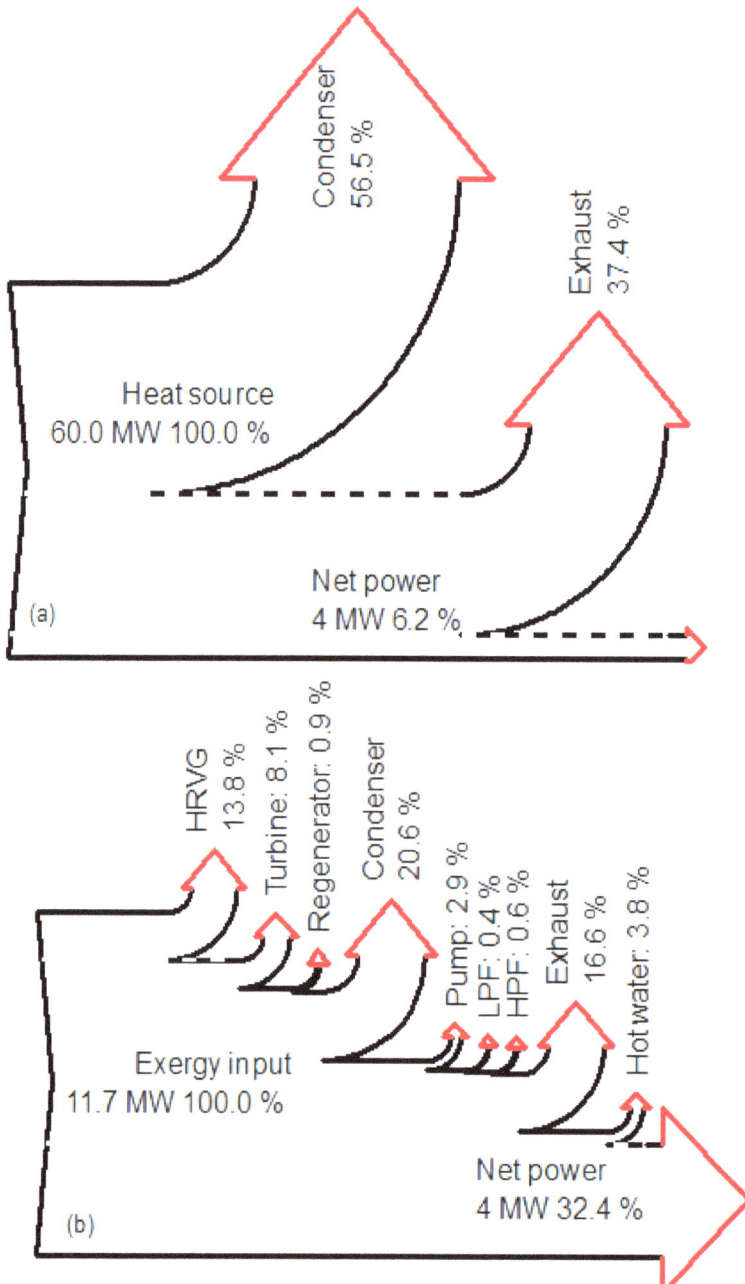

Fig. (65). (a) Sankey diagram and **(b)** Grassmann diagram of OFC with R124.

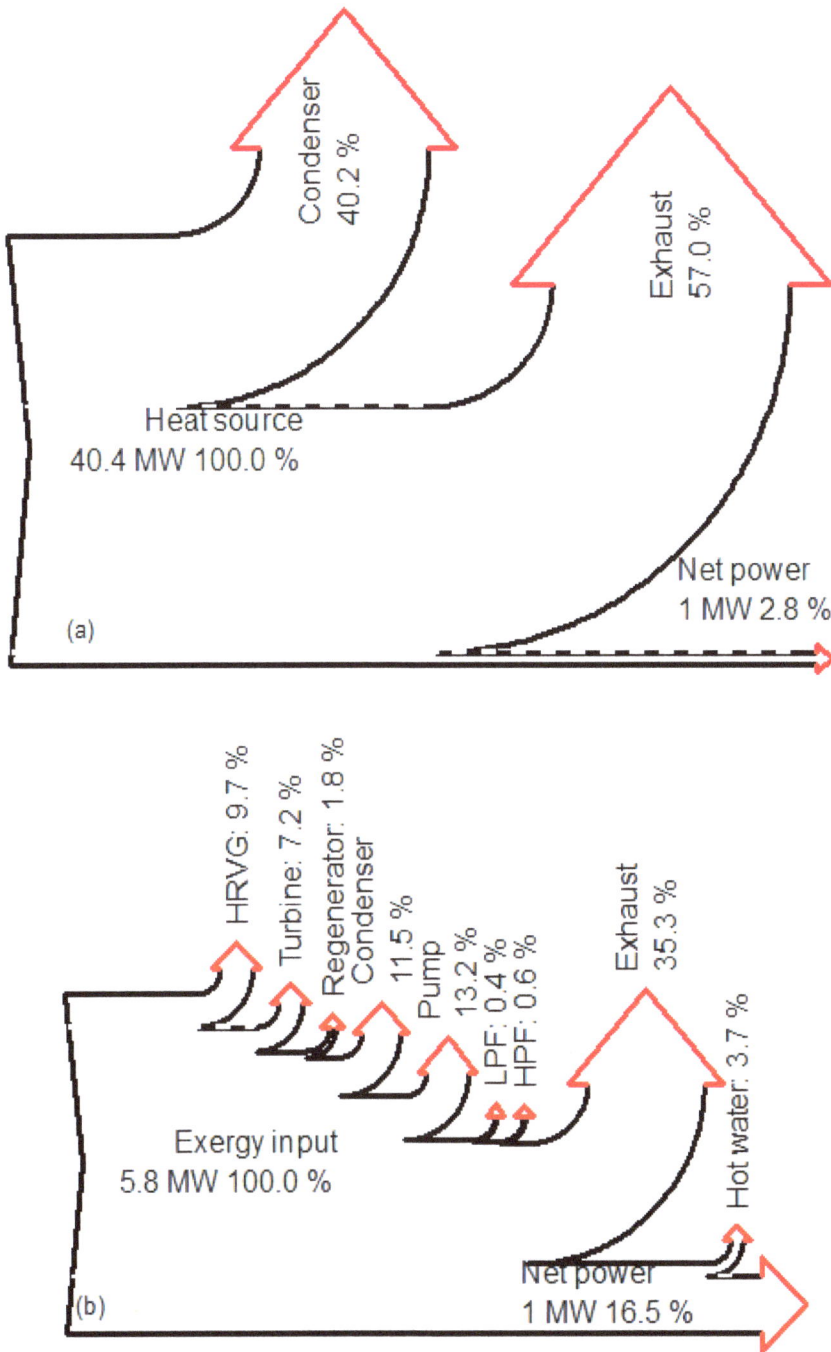

Fig. (66). (a) Sankey diagram and **(b)** Grassmann diagram of OFC with R134a.

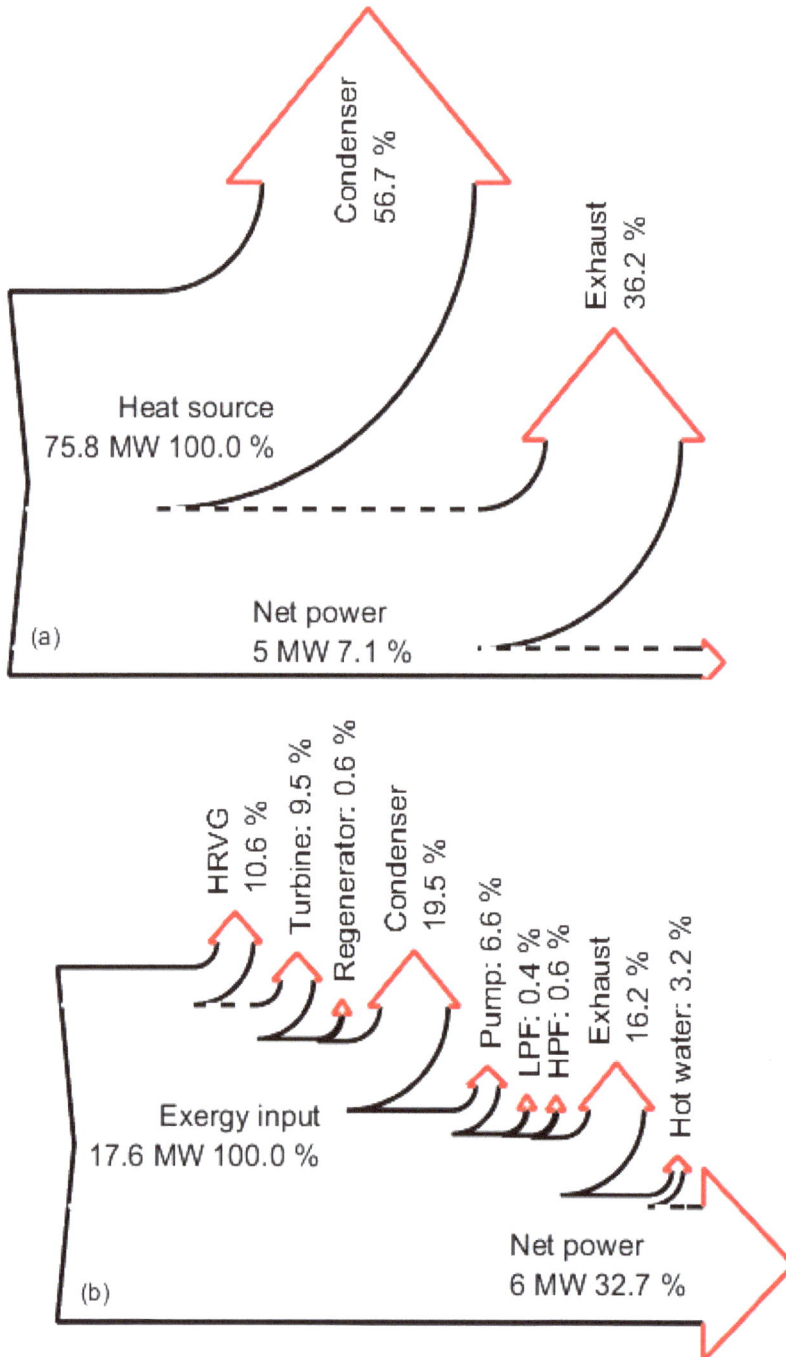

Fig. (67). (a) Sankey diagram and **(b)** Grassmann diagram of OFC with R245fa.

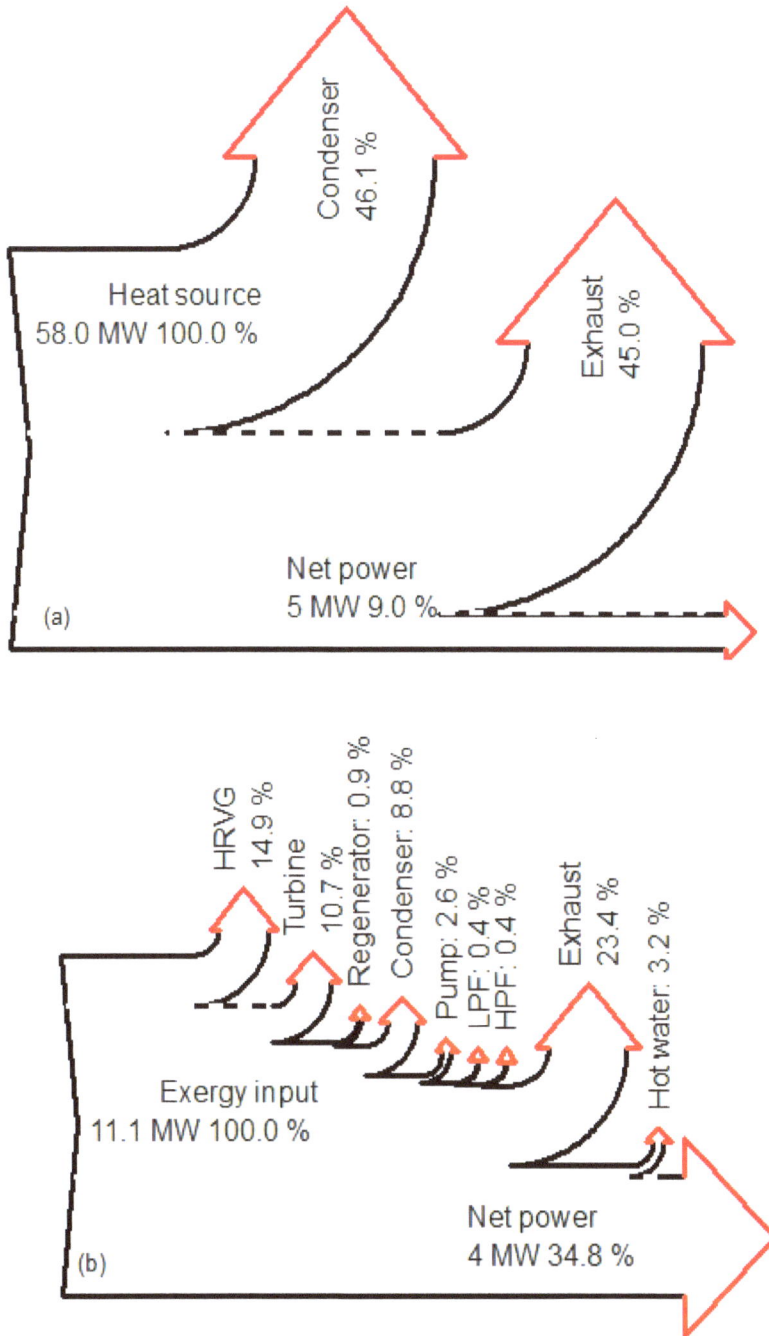

Fig. (68). (a) Sankey diagram and **(b)** Grassmann diagram of OFC with R717.

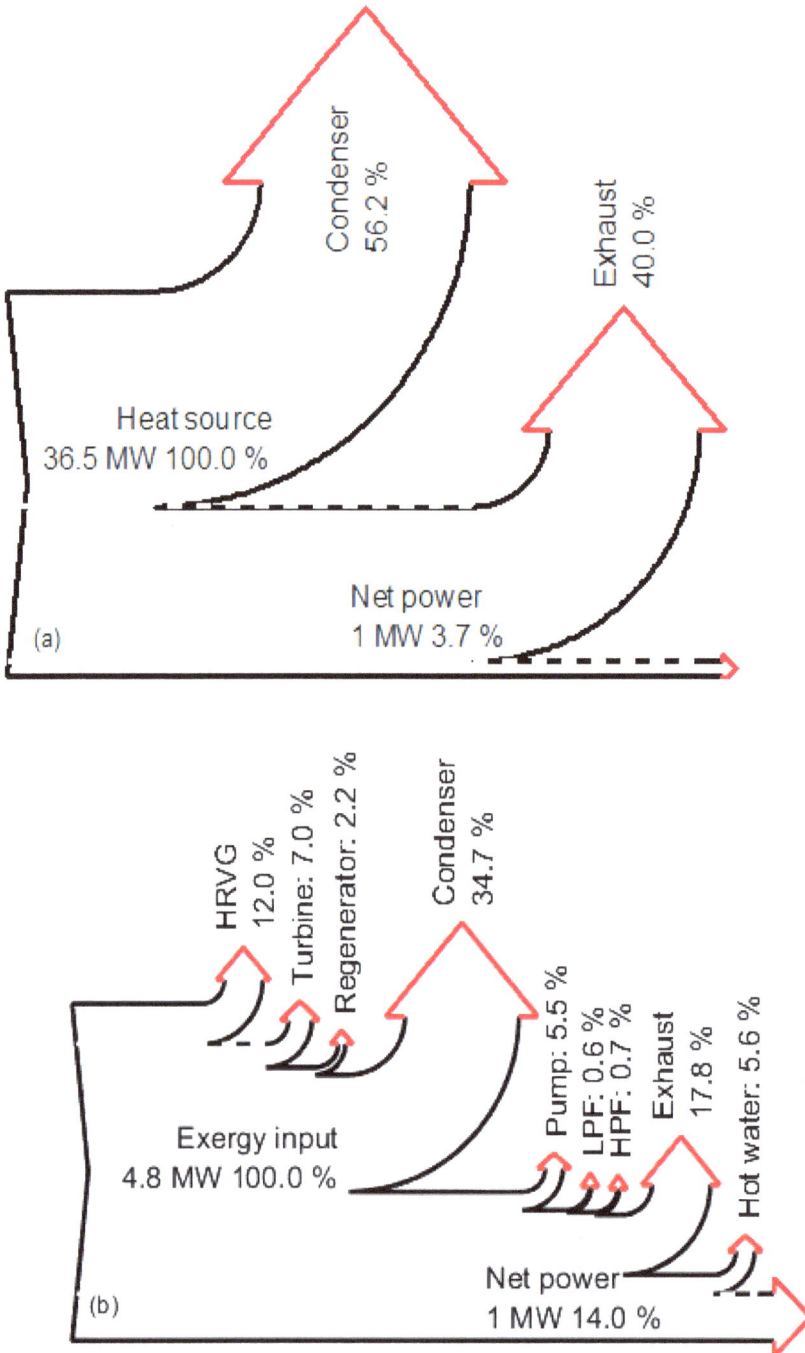

Fig. (69). (a) Sankey diagram and **(b)** Grassmann diagram of OFC with R407C.

Performance Comparison of OFC with Working Fluids

Fig. (**70**) compares the changes in (a) power and (b) thermal efficiency of thermal power cycle with the fluid and temperature. The results are plotted at the optimum HRVG pressure determined from the source temperature. The optimum pressure differs with the fluid and temperature. The power curves are changes linearly with temperature. An optimum source temperature has been found to result maximum thermal efficiency for the fluids. This optimum temperature is resulted only above the critical temperature but not below the critical temperature of the fluid.

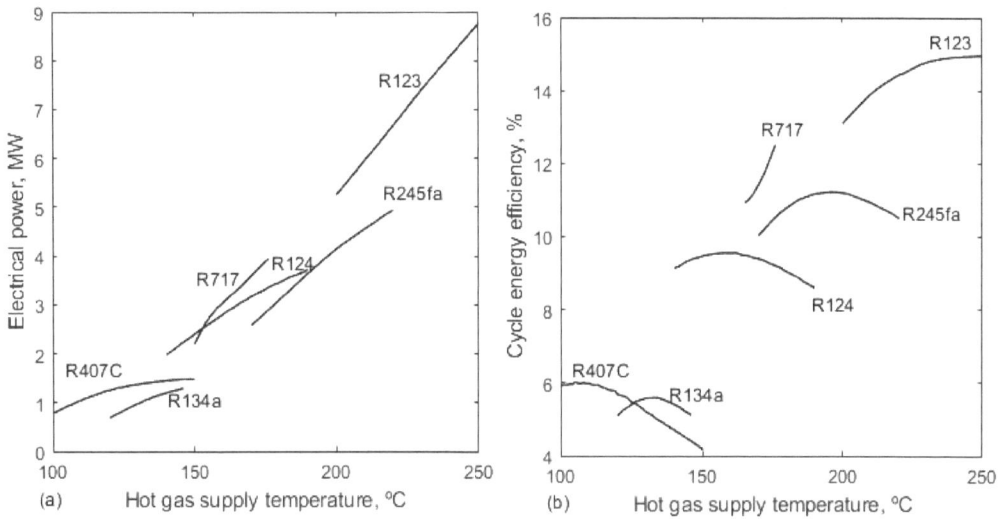

Fig. (70). Performance comparison of OFC with working fluids.

Table **16** is listed with the cycle specification developed at the optimum process conditions. The higher and lower turbine inlet temperatures are respectively with the fluids of R123 and R407C. The higher specific power is with R717. The lowt specific power is with R134a. The maximum and minimum energy efficiency and exergy efficiency are R123 and R134a respectively. But the maximum relative efficiency is with R717.

Table 16. Comparison of OFC specifications with working fluids at optimized HRVG pressure and source temperature above the critical temperature..

Working Fluid	R123	R124	R134a	R245fa	R717	R407C
Hot fluid inlet temperature, °C	250.00	180.00	130.00	220.00	175.00	120.00
Turbine inlet temperature, °C	235	165.00	115.00	205.00	160.00	105.00
Optimized HRVG pressure, bar	22.45	24.64	32.15	23.63	62.92	36.36

(Table 16) cont....

HP flash pressure, bar	10.00	15.30	21.63	11.69	39.42	28.08
LP flash pressure, bar	3.61	8.91	13.99	4.98	23.15	21.29
Condenser pressure, bar	1.54	5.93	10.17	2.50	15.55	17.49
Specific power, kW/kg	50.92	24.48	13.83	38.97	183.63	14.16
Plant power, MW	8.73	3.45	0.99	4.98	3.89	1.25
Cycle thermal efficiency, %	15.46	9.17	5.68	10.29	12.20	5.70
Relative cycle efficiency, %	40.30	32.16	29.43	29.81	44.04	33.19
Plant exergy efficiency, %	38.66	29.34	16.99	28.27	35.15	25.83

SUMMARY

ORC has been improved by adding the fluid separators between condenser pressure and boiler pressure after the flasher. The performance characteristics of OFC are generated at the source temperature below and above the critical temperature of fluid with single flasher and two flashers. The optimum boiler pressure is developed at the supply temperature with the fluids under the consideration. The optimum boiler pressure is increasing with increase in source temperature and sink temperature. The property charts, tables, Sankey diagrams and Grossmann diagrams are generated with the working fluids, R123, R124, R134a, R245fa, R717 and R407C. The plant specifications are compared at the optimized working parameters. For higher performance, R123 has been recommended due to its higher critical temperature. At the lower source temperature, R407C and R134a are recommended. OFC resulting a maximum of 3.63 MW and 12.50% thermal efficiency at the source temperature of 180 °C (below critical temperature). The same plant is generating 8.73 MW and 15.46% efficiency at the source temperature of 250 °C which is above the critical temperature of R123. Since the critical pressure of R717 is high, the highest specific power is observed with this fluid at the source temperature below and above the critical temperature. If the source temperature is below the critical temperature, R123 gave highest energy and exergy efficiencies. At the source temperature above the critical temperature, R717 resulting higher relative efficiency.

CHAPTER 6

Kalina Cycle

Abstract: The plant layout and working of Kalina cycle (KC) differs with the organic Rankine Cycle (ORC), steam Rankine cycle (SRC) and flash cycle. Separation and absorption of binary working fluid makes this cycle a different heat engine cycle. This chapter is aimed at the description, formulation and performance characteristics of KC suitable to a heat recovery. The strong solution concentration and vapour concentration are studied to maximize the power. A suitable KC configuration for the low temperature heat recovery has been selected and studied.

Keywords: Exergy analysis, Heat recovery, Kalina cycle, Modified rankine cycle, Power generation, Thermal efficiency, Thermal power cycles.

INTRODUCTION

Kalina cycle (KC) is a binary vapour absorption power cycle. The heat addition and rejection occurs with the variable temperature. Ammonia-water mixture is used as a working fluid in KC. The addition of ammonia to water permits the operation at low temperature (100 °C – 150 °C). The power plant layout is configured to suit the source temperatures. Water is an ideal fluid in power plant. However, this fluid is not suitable for low temperature heat source. Ammonia cycle is suitable for low temperature heat source. Condensing ammonia in condenser (absorber) is difficult (except at high pressures) as it needs very low temperature (below 0 °C). A feasible condenser can be designed by diluting the ammonia with water. For this process, the plant needs some modifications to the existing Rankine cycle. The process consists of absorption of vapour into the liquid. Therefore, the cycle is called as vapour absorption cycle. The changes in layout need additional components such as separator and mixture. The turbine, condenser (absorber in KC), pump and boiler are the common components in Rankine cycle and KC.

KALINA CYCLE

Fig. (**1**) shows the KC plant layout with industrial waste heat recovery. The cycle consists of external heat recovery, internal heat recovery, machines such as turbine, pump and other heat exchangers *i.e.* absorber, dephlegmator *etc.* The external heat recovery in a heat exchanger is called as heat recovery vapour generator (HRVG). The internal heat recovery consists of low temperature regenerator (LTR) and high temperature regenerator (HTR).

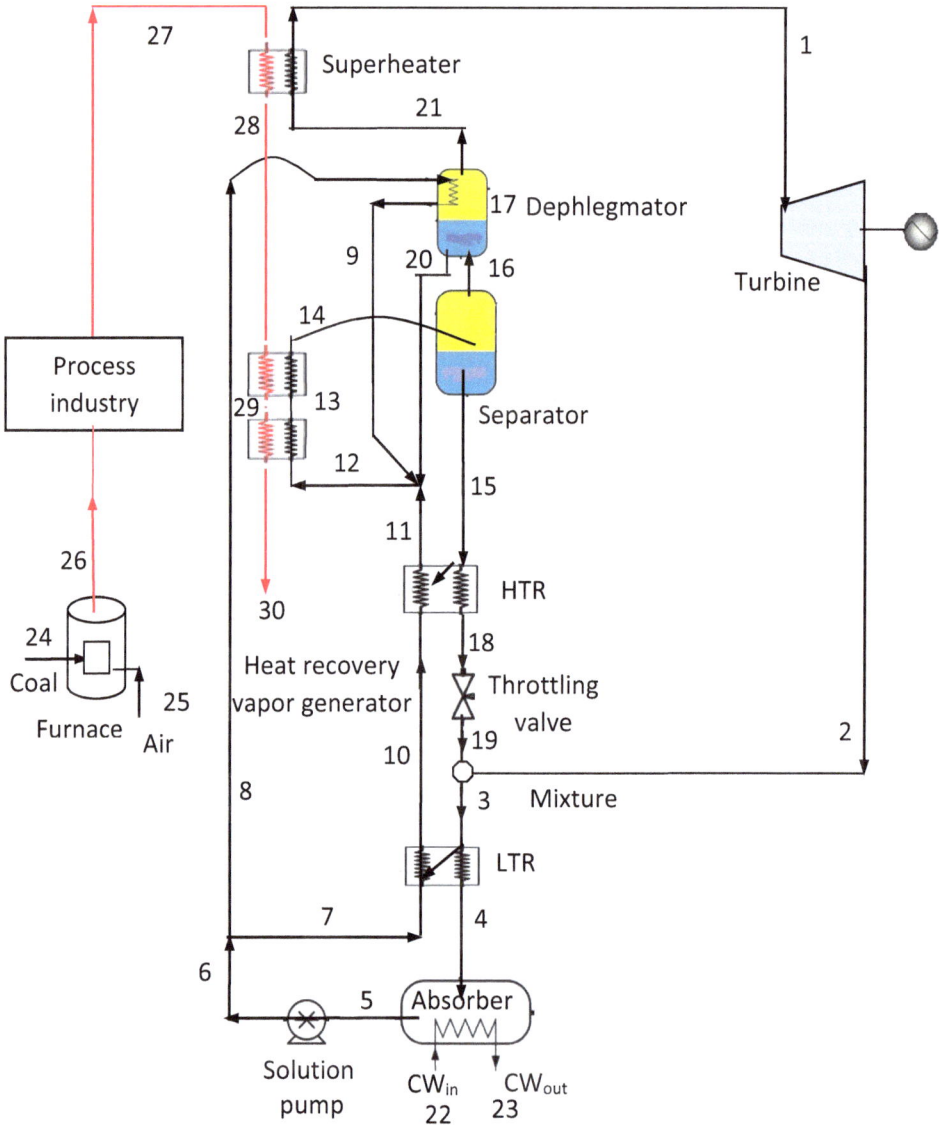

Fig. (1). Power plant layout of Kalina cycle with heat recovery arrangement.

Since the internal heat recovery shares the economizer's load, HRVG has partial section of economizer. The saturated liquid at the inlet of the evaporator is converted into liquid-vapour mixture. The vapour from the separator is superheated and supplied to turbine. The superheated vapour expands in turbine to generate electricity in generator. The condensation of vapour at the exit of turbine is not possible as it demands low temperature sink. Therefore, to condense the vapour at the available sink temperature, the vapour should be diluted by mixing with the weak solution. The weak solution is a separated liquid from the separator. The weak solution after mixing with vapour is capable of internal heat recovery. Therefore, LTR is located between mixture and absorber. The absorber is an air cooled or water cooled condenser to condense the mixture into saturated liquid. The liquid is pumped to HRVG to generate the vapour for the cycle. HRVG converts the preheated liquid from LTR and HTR into a liquid-vapour mixture. The turbine demands vapour without any liquid traces. Therefore, the vapour and liquid are separated after HRVG. The vapour is supplied to dephlegmator and weak solution is passed through internal heat recovery and throttling. Dephlegmator is a heat and mass transfer unit located between separator and superheater. The dephlagmator increases the concentration of vapour by condensing the water from the vapour. The incorporation of dephlegmator permits the engineer to design the plant at the low pressure.

(Fig. **2**) shows pressure-enthalpy diagram of KC with all the processes. The concentration curves are shown for liquid side (left side) and vapour side (right side). The high pressure and low pressure are the two horizontal lines. The expansion of vapour has been depicted with a drop in pressure and temperature. The mixing before absorber is a horizontal line. Similarly, the mixing of fluids at the inlet of HRVG is a horizontal line. The throttling from high pressure to low pressure and pumping from low pressure to high pressure are shown as vertical lines.

Table **1** lists the thermal properties of KC shown in (Fig. **1**). The two pressures in the cycle are high pressure (13.78 bar) and low pressure (2.58 bar). The combustion and process states are shown from state 25 to state 27. The hot gas temperature is dropped from state 27 to state 30. The strong solution concentration is 0.4. The weak solution concentration is 0.26. The vapour concentration at the exit of separator is 0.85. The concentration is increased from 0.85 to 0.87 in the dephlegmator. As per the assumptions, at the hot gas supply temperature of 150 °C, the resulted turbine inlet temperature is 135 °C.

Table 1. KC material balance results.

State	P, bar	t, °C	x	m, kg/s	h, kJ/kg	s, kJ/kg K
1	13.78	135.00	0.87	7.62	1689.12	5.29
2	2.58	65.60	0.87	7.62	1449.59	5.40
3	2.58	66.28	0.40	32.96	484.22	2.11

(Table 1) cont.....

State	P, bar	t, °C	x	m, kg/s	h, kJ/kg	s, kJ/kg K
4	2.58	62.56	0.40	32.96	420.74	1.93
5	2.58	35.00	0.40	32.96	-72.57	0.39
6	13.78	36.78	0.40	32.96	-63.75	0.42
7	13.78	36.78	0.40	32.87	-63.75	0.42
8	13.78	36.78	0.40	0.09	-63.75	0.42
9	13.78	110.00	0.40	0.09	476.99	1.93
10	13.78	51.08	0.40	32.87	-0.69	0.62
11	13.78	83.75	0.40	32.87	146.16	1.05
12	13.78	84.34	0.40	33.26	149.07	1.06
13	13.78	93.96	0.40	33.26	193.54	1.18
14	13.78	125.00	0.40	33.26	733.75	2.58
15	13.78	125.00	0.26	25.34	384.39	1.61
16	13.78	125.00	0.85	7.92	1686.15	5.27
17	13.78	120.57	0.85	7.92	1680.15	5.25
18	13.78	83.92	0.26	25.34	193.37	1.10
19	2.58	67.77	0.26	25.34	195.80	1.12
20	13.78	120.57	0.28	0.30	356.36	1.55
21	13.78	120.57	0.87	7.62	1651.48	5.20
22	1.01	25.00	- -	486.20	0.00	0.00
23	1.01	33.00	- -	486.20	33.44	0.11
24	1.01	25.00	- -	38.62	0.00	0.00
25	1.01	35.00	- -	333.63	4.16	0.01
26	1.01	900.00	- -	372.24	1021.19	1.56
27	1.01	150.00	- -	372.24	131.55	0.37
28	1.01	149.31	- -	372.24	130.78	0.37
29	1.01	103.96	- -	372.24	82.51	0.25
30	1.01	100.24	- -	372.24	78.54	0.23

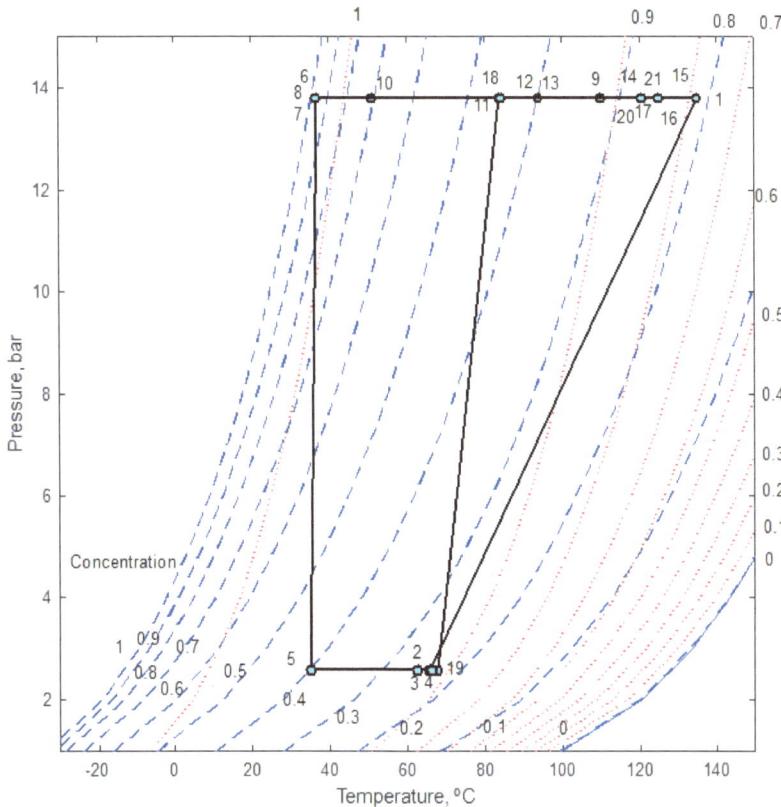

Fig. (2). Pressure-temperature-concentration diagram of Kalina cycle.

THERMODYNAMIC EVALUATION OF KALINA CYCLE

Thermal evaluation of KC has been conducted with the thermodynamic properties of ammonia-water mixture. The development of properties of working fluid is beyond the scope of this book. Compared to single fluid system such as organic Rankine cycle (ORC) or steam Rankine cycle (SRC), the solutions of KC are complex because of extra property, concentration. Concentration of fluid is expressed in the mass ratio of ammonia to the mixture. The assumptions involved in this thermodynamic work are as follows:

The terminal temperature difference (TTD) in superheater is 15 °C. TTD is the temperature difference between hot gas supply temperature and turbine inlet temperature. The isentropic efficiency of pump is 75%. Similarly, the isentropic efficiency of turbine is 85%. The hot gas flow rate is 1000000 m³/h taken from a typical cement factory. The hot gas temperature supplied to process industry is 900 °C. Coal is used as a fuel in furnace. The hot gas supply temperature is 150 °C. Degree of superheat (DSH) is 10 °C. DSH is the temperature difference between superheated fluid and dew point temperature (DPT) of working fluid. The end of saturation in a binary fluid system is DPT. Similarly, the beginning of saturation is bubble point temperature (BPT). The pinch point (PP) in evaporator is 10 °C. PP is the minimum temperature difference between hot fluid and cold fluid in evaporator. The strong solution concentration is 0.4. In separator, the vapour concentration is 0.85. Circulating water supply temperature in absorber is 30 °C. The maximum concentration in dephlegmator is 100%. Therefore, with reference to this state, the effectiveness of dephlegmator is defined and 0.15 is used in this simulation. The mechanical efficiency of machines such as pump and turbine is 96%. The generator efficiency is 98%.The turbine inlet temperature is determined from hot gas supply temperature and TTD.

$$T_{t\ in} = T_{gas\ in} - TTD \tag{1}$$

Separator temperature is the HRVG exit temperature. Since the boiling is not complete, the separator temperature is below the DPT. The addition of separator temperature and DSH is the turbine inlet temperature.

$$T_{sep} = T_{t\ in} - DSH \tag{2}$$

The turbine concentration is higher than the separator vapour concentration with dephlegmator action. The turbine concentration can be determined from separator vapour concentration and dephlegmator's effectiveness as follows:

$$x_t = x_{sep\ v} + \varepsilon_{deph}\ (1\text{-}x_{sep\ v}) \tag{3}$$

High pressure can be determined at the vapour state in separator. The temperature and vapour concentration can be used to find the high pressure in cycle. At this temperature and pressure, the liquid concentration can be determined. The resulted liquid concentration is the weak solution concentration. Similarly, the low pressure is determined from the saturated liquid state at the absorber exit. The low pressure can be evaluated at the absorber exit condition from the saturated liquid temperature and strong solution concentration. The iteration of low pressure in KC results the mixture state at the inlet of LTR.

(Fig. **3**) shows the temperature-heat transferred diagram of heat source in KC. Heat has been supplied at three sections *viz.* economizer, evaporator and superheater. Since dephlegmator is placed between evaporator and superheater, the temperature of fluid drops at the entry of superheater as shown.

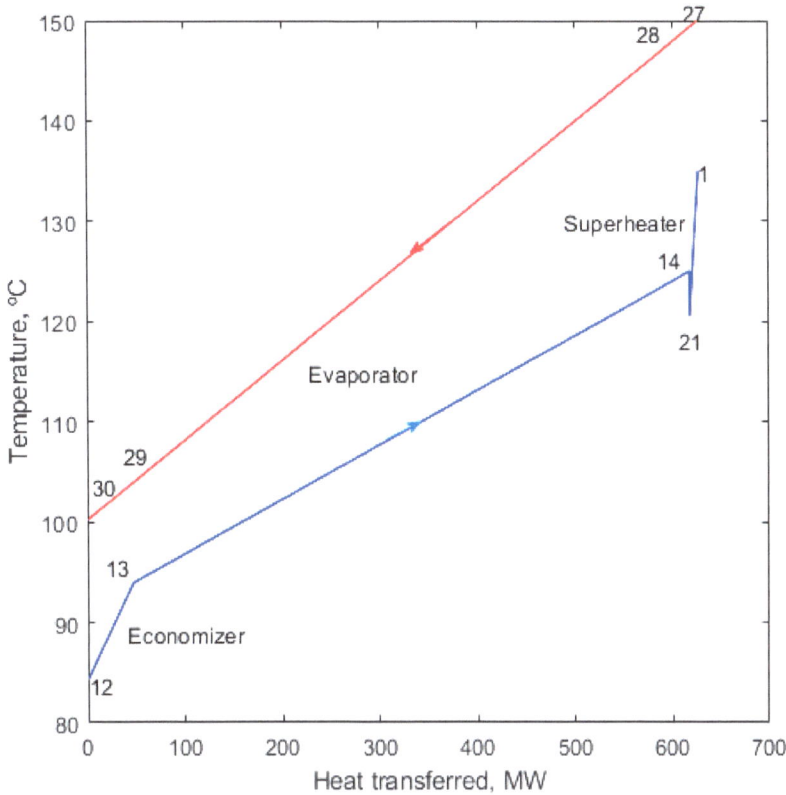

Fig. (3). Temperature-heat transferred diagram of HRVG in Kalina cycle.

With reference to the (Fig. **3**), the hot gas temperature between economizer and evaporator can be evaluated using pinch point (PP). PP is the minimum temperature difference between hot fluid (gas) and cold fluid (working fluid) to ensure the positive heat transfer from hot fluid to cold fluid. The sum of BPT and PP gives the hot gas temperature at this junction.

The hot fluid temperature at the exit of evaporator,

$$T_{hot\,gas} = T_{bp} + PP \tag{4}$$

The hot gas temperatures at other states are determined from the energy balance equations.

KC is solved with unit mass of fluid in absorber. The specific properties are converted into total values by multiplying the fluid mass obtained from energy balance equation.

The dryness fraction (d) of liquid-vapour mixture is a mass ratio of vapour and total mass. The liquid-vapour mixture state can be observed in the separator and dephlegmator.

As per the lever rule,

$$d_{sep} = \frac{x_m - x_l}{x_v - x_l} \tag{5}$$

Similarly, dryness fraction at dephlegmator is determined.

The mass flow rates in the cycle can be solved through the simplifications of mass balance equations.

The mass of fluid in turbine per unit mass of fluid in absorber,

$$m_t = \frac{d_{deph} d_{sep}}{1 + d_{sep} d_{deph} - d_{sep}} m_{abs} \tag{6}$$

The mass of fluid at the entry of HRVG,

$$m_{HRVG} = \frac{m_t}{d_{sep} d_{deph}} \tag{7}$$

The mass of weak solution,

$$m_{ws} = \frac{\left(1 - d_{sep}\right)}{d_{sep} d_{deph}} m_t \tag{8}$$

The exit condition of pump and turbine are iterated from the inlet and isentropic efficiency.

Output from the turbine,

$$W_t = m_t \left(h_{in} - h_{out}\right) \eta_{m,t} \, \eta_{ge} \tag{9}$$

Work input to pump,

$$W_p = \frac{m_p (h_{out} - h_{in})}{\eta_{m,\ p}} \tag{10}$$

Net power output,

$$W_{net} = (W_t - W_p) \tag{11}$$

For exergy analysis, the irreversibilities in all the components have been determined from the exergy balance.

Specific exergy,

$$e = h - T_0\, s \tag{12}$$

The irreversitility in HRVG,

$$
\begin{aligned}
I_{HRVG} = {} & m_{gas}\, e_{gas} + m_{ex}\, e_{ex} + m_{HRVG,\ in}\, e_{HRVG,\ in} - m_{HRVG,\ out}\, e_{HRVG,\ out} \\
& + m_{SH,\ in}\, e_{SH,\ in} - m_{SH,\ out}\, e_{SH,\ out}
\end{aligned} \tag{13}
$$

The irreversibility in separator,

$$I_{sep} = m_{in}\, e_{in,} - m_v\, e_v - m_{ws}\, e_{ws} \tag{14}$$

The irreversibility in dephlegmator,

$$
\begin{aligned}
I_{deph} = {} & m_{deph}\, (e_{deph,\ in} - e_{deph,\ out}) \\
& + m_{coolant}\, (e_{coolant,\ in} - e_{coolant,\ out})
\end{aligned} \tag{15}
$$

The irreversibility in turbine,

$$I_t = m_t\, (e_{in} - e_{out}) - W_t \tag{16}$$

The irreversibility in pump,

$$I_p = m_p (e_{in} - e_{out}) + W_p \tag{17}$$

The irreversibility in LTR,

$$I_{LTR} = m_{ss} (e_{ss,\, in} - e_{ss,\, out}) + m_{ws} (e_{ws,i\, n} - e_{ws,\, out}) \tag{18}$$

The irreversibility in HTR also can be determined in a similar manner.

The irreversibility in mixture,

$$I_m = m_v\, e_v + m_{ws}\, e_{ws} - m_{ss}\, e_{ss} \tag{19}$$

The irreversibility in absorber,

$$I_{cond} = m_{ss} (e_{ss,\, in,} - e_{ss,\, out}) + m_{ciw} (e_{cw,\, in} - e_{cw,\, out}) \tag{20}$$

The irreversibility in exhaust gas from the HRVG,

$$I_{ex} = m_{ex}\, e_{ex} \tag{21}$$

The irreversibility in throttling device,

$$I_{throttling} = m_{throttling} (e_{in} + e_{out}) \tag{22}$$

The irreversibility in hot water from absorber,

$$I_{hw} = m_{hw}\, e_{hw} \tag{23}$$

The total irreversibility of the KCS,

$$I_{total} = I_{HRVG} + I_{sep} + I_{deph} + I_t + I_p + I_{LTR} + I_{HTR} + I_m + I_{abs} + I_{throttling} + I_{ex} + I_{hw} \tag{24}$$

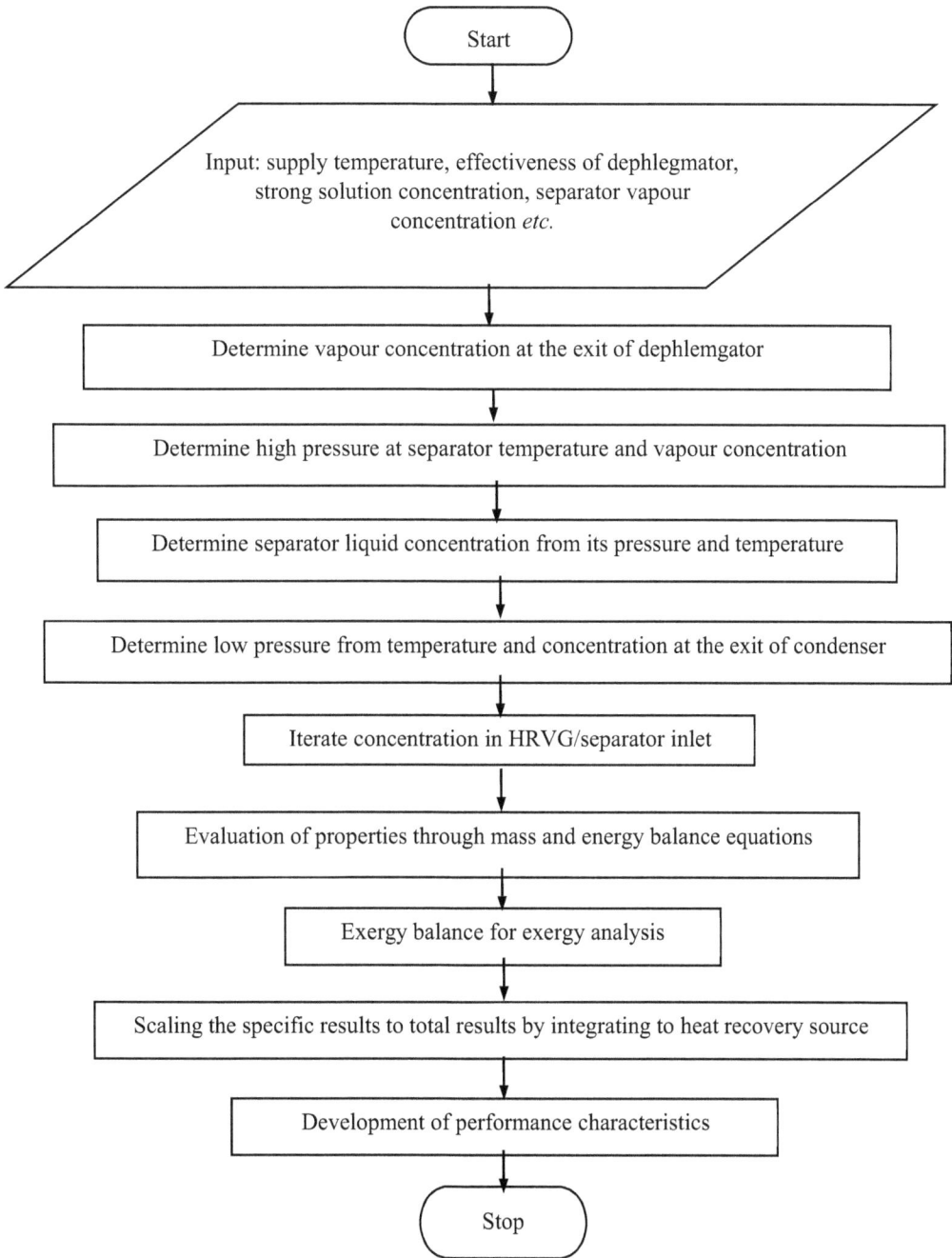

Fig. (4). Methodology developed to solve the Kalina cycle.

The energy (thermal) efficiency of cycle,

$$\eta_1 = \frac{W_{net}}{Q_{HRVG} + Q_{SH}} \times 100 \tag{25}$$

The exergy (second law) efficiency of cycle,

$$\eta_2 = \frac{E_{hot\ gas} - \sum I_{components}}{E_{hot\ gas}} \times 100 \tag{26}$$

The solutions of KC are arranged in the order and depicted in (Fig. **4**).

PERFORMANCE CHARACTERISTICS OF KALINA CYCLE

The key operational conditions in KC are strong solution concentration, separator vapour concentration and source temperature. The weak solution concentration depends on the strong solution concentration and vapour concentration. The increase in strong solution concentration indicates the movement of the cycle away from the steam Rankine cycle. In other words, the drop in strong solution concentration shows the influence of steam power plant. Therefore, the strong solution concentration indicates the balance between the ORC and SRC.

(Figs. **5** - **7**) studies the role of strong solution concentration and vapour concentration with the hot gas supply temperature. In each figure, strong solution concentration and separator vapour concentration are changed at a fixed supply temperature. But the hot gas temperature is changed from Figs. (**5-7**). The hot gas temperatures are 140 °C, 150 °C and 160 °C respectively in Figs. (**5 - 7**). The characteristics of maximum power are different compared to the maximum efficiency characteristics. For maximum power, the optimum separator's vapour concentration changes with strong solution concentration. The optimum vapour concentration increases with increase in strong solution concentration. The power increases with increase in strong solution concentration. Therefore, vapour concentration also increases with increase in strong solution concentration. A low strong solution concentration with low vapour concentration or high strong solution concentration with high vapour concentration results high thermal efficiency. But the high concentration results more power compared to the low concentration.

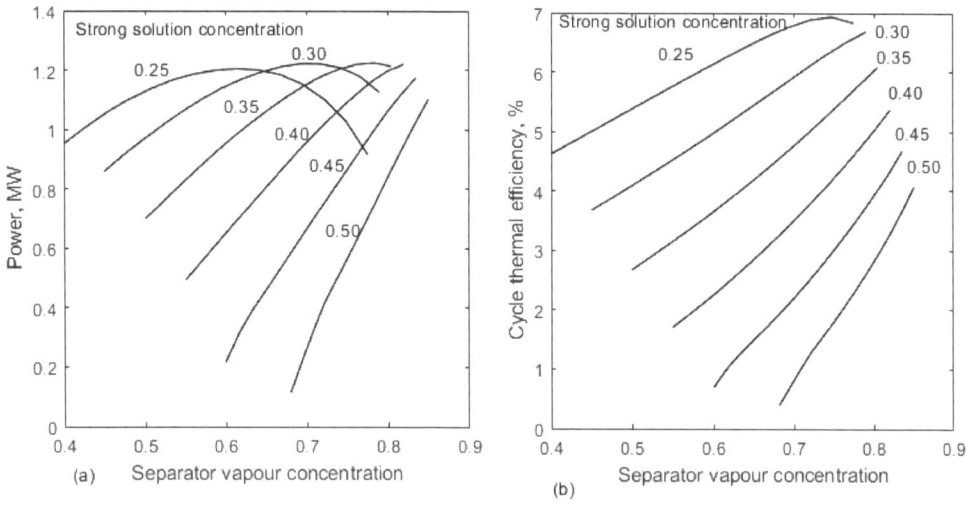

Fig. (5). Kalina cycle performance characteristics with strong solution concentration and separator vapour concentration at the hot gas supply temperature of 140 °C.

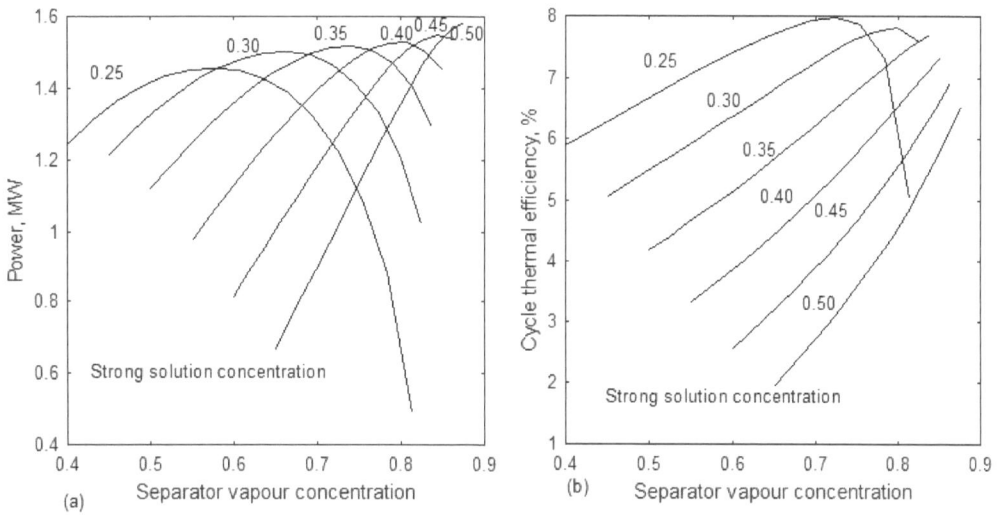

Fig. (6). Kalina cycle performance characteristics with strong solution concentration and separator vapour concentration at the hot gas supply temperature of 150 °C.

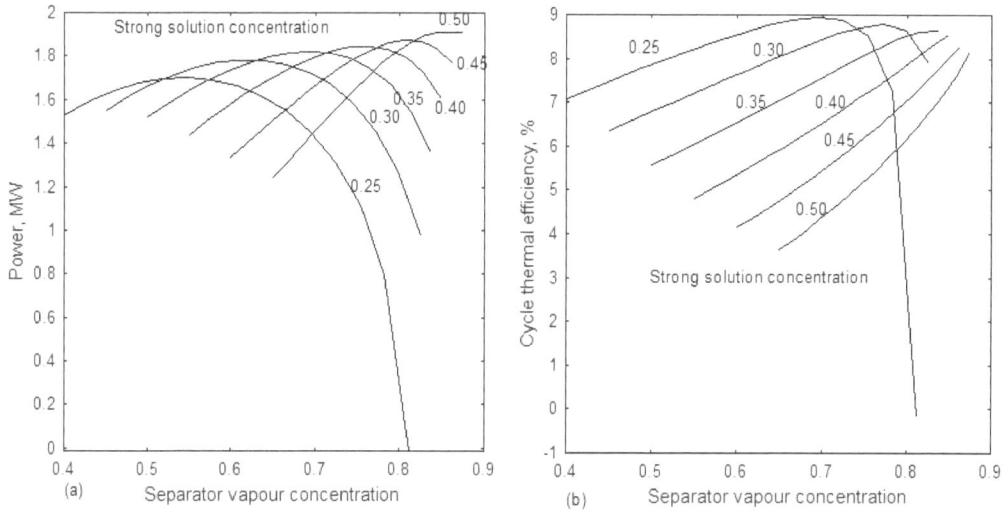

Fig. (7). Kalina cycle performance characteristics with strong solution concentration and separator vapour concentration at the hot gas supply temperature of 160 °C.

Table **2** summarizes the plant specifications at a typical strong solution concentration of 0.4, vapour concentration of 0.85 and 140 °C of hot gas supply temperature.

(Figs. **8 - 10**) also analyzes the influence of strong solution concentration, separator vapour fraction and hot gas supply temperature. In each figure, the separator vapour concentration is studied with the change in hot gas supply temperature. The strong solution concentrations are 0.3, 0.4 and 0.5 respectively in Figs. (**8 - 10**). At fixed strong solution concentration, the optimum vapour concentration decreases with increase in source temperature. The optimum vapour concentration at maximum power is less than the optimum vapour concentration at maximum thermal efficiency. These optimum concentrations are increasing with increase in strong solution concentration.

Table 2. Specifications of Kalina cycle at 0.4 strong solution concentration, 0.85 separator's vapour concentration and 150 °C hot gas supply temperature.

S. No.	Description	Result
1.	HRVG pressure, bar	13.78
2.	Heat supply in HRVG, kW	19445.00
3.	Heat supply in superheater, kW	286.92
4.	Exhaust temperature, °C	100.23
5.	Turbine output, kW	17179.00

(Table 2) cont.....

6.	Pump input, kW	268.48
7.	Net output, kW	1449.40
8.	Cycle energy efficiency, %	7.34
9.	Exergy efficiency of cycle, %	17.85
10.	Dryness fraction at exit of turbine	0.87
11.	Circulating water in condenser, kg/s	486.20
12.	Condenser heat rejection, kW	16259.00

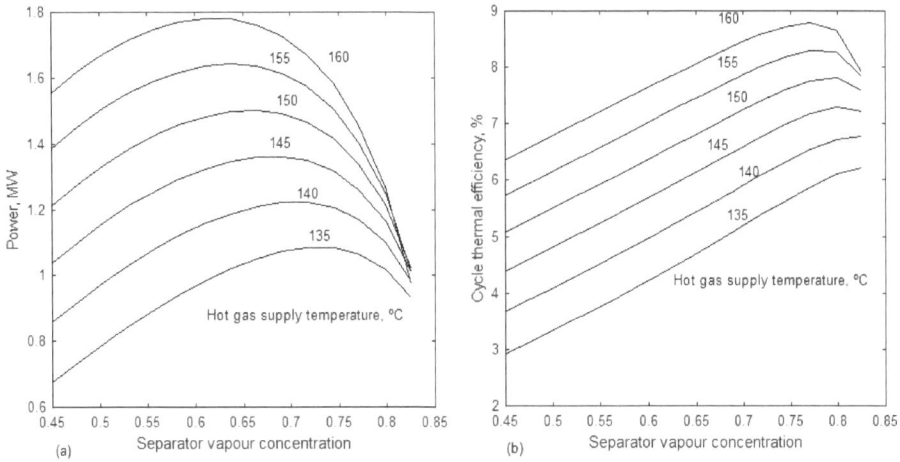

Fig. (8). Kalina cycle performance characteristics with separator vapour concentration and hot gas supply temperature at the strong solution concentration of 0.3.

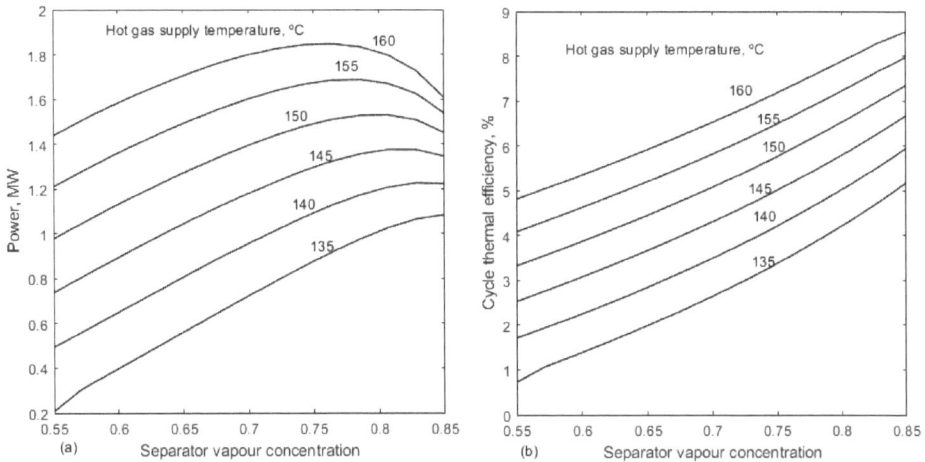

Fig. (9). Kalina cycle performance characteristics with separator vapour concentration and hot gas supply temperature at the strong solution concentration of 0.4.

(Fig. **11a**) is a Sankey diagram shown for the results of energy balance of the cycle. Similarly, (Fig. **11b**) is a Grassmann diagram for the exergy balance of the cycle. The Sankey diagram shows that the thermal efficiency of the KC is 7.1% after deducting the 33.2% of condenser loss and 59.7% of exhaust loss. The second law analysis has been shown in Grassmann diagram. After the exhaust loss, the major exergetic loss is in absorber (condenser). After, deducting the exergetic losses, the exergy efficiency is 19.4%.

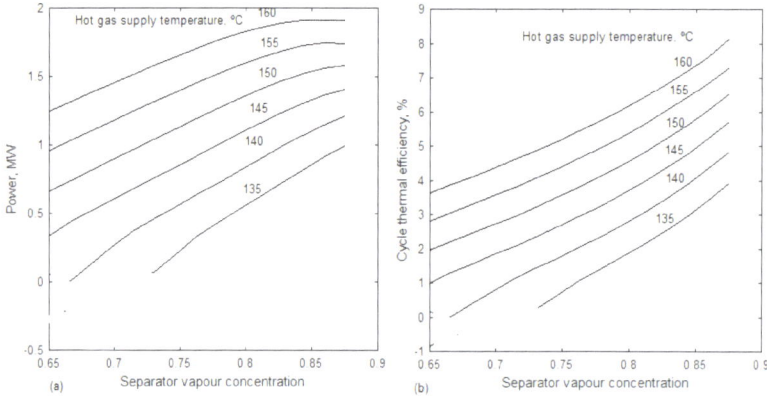

Fig. (10). Kalina cycle performance characteristics with separator vapour concentration and hot gas supply temperature at the strong solution concentration of 0.5.

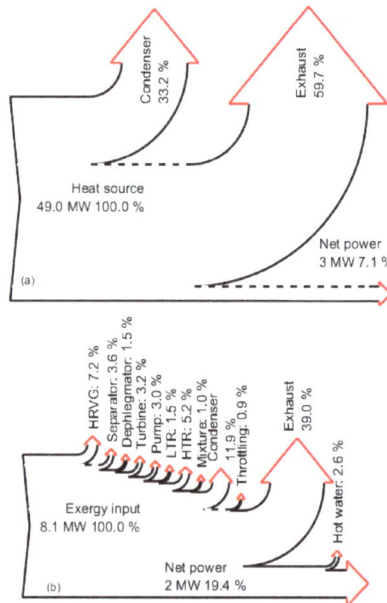

Fig. (11). (a) Sankey diagram and **(b)** Grassmann diagram of Kalina cycle plant at 0.4 strong solution concentration, 0.85 separator's vapour concentration and 150 ° hot gas supply temperature.

SUMMARY

The three main operational conditions of KC are strong solution concentration, vapour concentration and source temperature. The energy losses, exergy losses and power are compared to identify the strong and weak sections of KC. The optimum results are analyzed with combination of three variables. The optimum vapour concentration is increasing with increase in source temperature at a fixed strong solution concentration. The optimum vapour concentration is increasing with increase in strong solution concentration.

CHAPTER 7

Steam Rankine Cycle

Abstract: Steam Rankine cycle (SRC) is a widely used thermal power cycle due to its ideal properties of the fluid. Compared to organic Rankine cycle (ORC) and organic flash cycle (OFC), the working fluid and equipment are not expensive. A water treatment plant is associated with a steam thermal power plant. SRC is a power plant cycle in thermal power plant, bottoming cycle in a combined cycle power plant or power generation cycle for a waste heat recovery plant. In this chapter, the performance characteristics of SRC have been developed and analyzed. A single pressure heat recovery steam generator (HRSG) with a deaerator is considered to draw the heat from the industrial waste heat. The conditions for maximum power and maximum efficiency have been developed. Correlations are presented to find the HRSG pressure at the available hot gas supply temperature. The work is extended to a case study on SRC operating with the heat recovery of a cement factory. A power plant layout suitable to the identified waste heat sources from a cement factory has been developed. From the evaluation, it has been concluded that approximately 15 MW of electricity can be generated from its waste heat to meet the electrical load of 15 MW. It indicates that at the optimized conditions, it is possible to self-generate the electrical demand of the cement factory from its own waste heat recovery. The analysis recommended a low pressure in HRSG for maximum power generation.

Keywords: Heat recovery, Rankine cycle, Steam power cycle, Thermal efficiency, Thermal power plant.

INTRODUCTION

In majority of countries, more than 50% of electricity is from thermal power plant with steam as a working fluid. The power cycle is a steam Rankine cycle (SRC). Water is the ideal fluid for power. To operate SRC, a water treatment plant is required. The thermal properties of water/steam are superior than the others. Large size power plants can be suitably built with water as the working fluid. The equipment and systems are well developed and available around the world. The main challenges of thermal power plants are the handling of dust, ash and environmental pollution caused by the burning of coal. The ash and emission issues can be addressed by shifting from coal firing to heat recovery option or biomass firing. SRC is not suitable for a power plant with low temperature heat source. Organic Rankine cycle (ORC) is similar to SRC layout.

Tangellapalli Srinivas

ORC is suitable for low temperature heat sources. The SRC is suitable for the source above this temperature and up to 600 °C with the limitation in material properties. Above this temperature, erosion of steam turbine occurs. SRC can be operated by fuel firing, solar thermal collectors and waste heat recovery. In this chapter, waste heat has been selected. Majority of components in the basic layout of SRC and ORC are same, except for a few changes. The vapor temperature at the exit of turbine differs in SRC and ORC. In SRC, it is close to sink temperature, except back pressure turbine. In back pressure turbine steam expands to 100 °C for space heating or any process heat. The turbine exit in ORC is above the sink temperature. Therefore, at the exit of ORC turbine, regenerator is used and it is not possible in SRC. The focus of this chapter is the study and analysis of SRC.

STEAM RANKINE CYCLE

Fig. (**1**) shows the thermal power cycle of SRC with a deaerator. In a furnace, coal is fired to generate heat for process industry (here cement factory) and power plant. Fig. (**2**) shows the temperature-entropy diagram of the SRC. The hot gas is supplied to the cement processing and later to power plant. SRC shown in Fig. (**1**) consists of heat recovery steam generator (HRSG), turbine, water cooled condenser, two pumps and a deaerator. Steam is generated in the HRSG from the heat of hot gas supplied from the factory. The superheated steam is expanded in turbine and condensed to saturated water. The cycle has three pressures *viz.* HRSG pressure, deaerator pressure and condenser pressure. After the condenser, the water is pumped from the condenser pressure to deaerator pressure. In the deaerator, the steam from the turbine and pumped water mix with each other and turned into saturated liquid. Deaerator is an open feedwater heater used to remove the dissolved gases from working fluid (deaeration) and increase the feedwater temperature.

The dissolved gases in water causes corrosion of the boiler parts. They also reduce the heat transfer capacity between hot fluid and cold fluid. For example, oxygen presented in water creates pitting at the local regions of heat exchanger. Water with ammonia corrodes the copper based material in the heat recovery and bearings. The vacuum pressure in deaerator can be used in the power plant system. Vacuum ejectors are used in the vacuum deaeration. To remove the dissolved gases from the deaerator, a vacuum pump is used in vacuum deaeration. However, the deaerator pressure above the atmospheric pressure is most commonly used method in the power plant. In closed feedwater heater, heat transfers from steam to feed water without mixing. The preheating of water before HRSG in a deaerator increases the thermal efficiency with a drop in turbine output. The feedwater is pumped from the deaerator pressure to HRSG pressure.

HRSG consists of economizer, evaporator and superheater. This cycle repeats on steady state for continuous power generation.

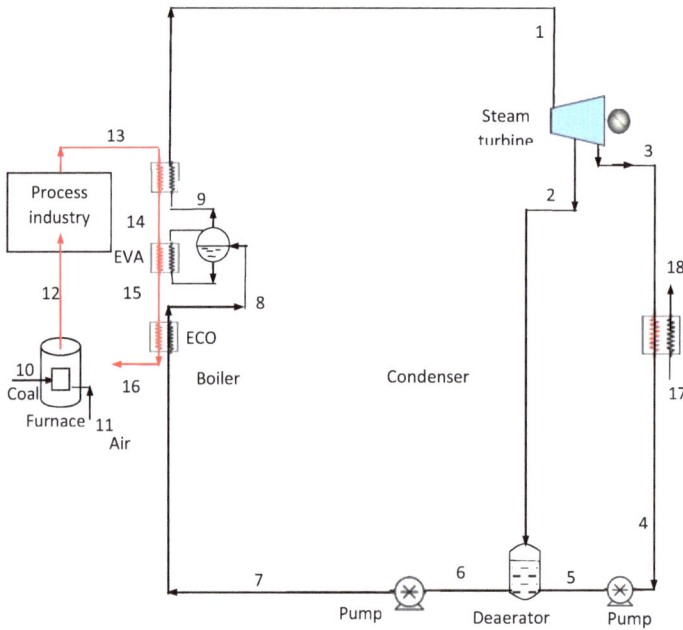

Fig. (1). Schematic plant layout of SRC with a deaerator.

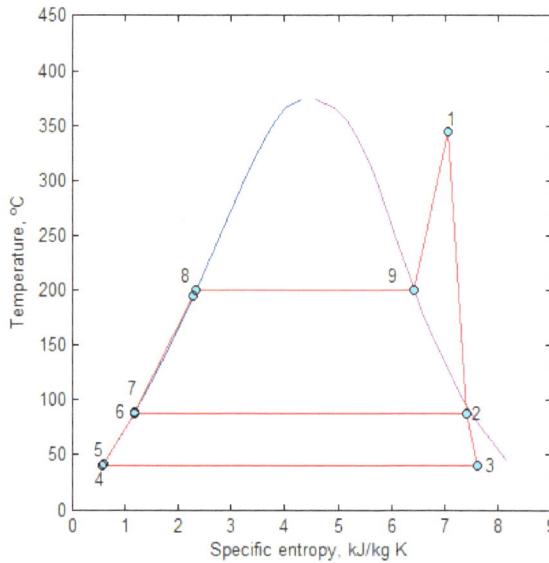

Fig. (2). Temperature-entropy diagram for SRC with a deaerator.

With reference to (Fig. **2**), the exit state of economizer is below the saturation temperature of the evaporator. It is equal to the approach point (AP) in evaporator. This design constraint avoids the sudden flashing of water into the steam between the economizer and evaporator. The temperature and enthalpies before and after pumping are nearly same.

THERMODYNAMIC EVALUATION OF STEAM RANKINE CYCLE

This section formulates the evaluation of SRC to develop and analyze the performance characteristics. The following are the assumptions used in SRC. Atmospheric condition is taken as 1.01325 bar and 25 °C. The fuel used in the furnace is coal with the combustion temperature of 900 °C. The adiabatic temperature of coal is more than this combustion temperature. Terminal temperature difference (TTD) of HRSG is 15 K. The AP in evaporator is 5 K. The Pinch point (PP) in HRSG is 10 K. The supply gas flow rate is 10^6 m³/h. The steam turbine inlet temperature is determined from the hot gas supply temperature and TTD. The HRSG optimum pressure has been selected from the result of pressure optimization. The pressure of open feed water heater (deaerator) has been taken from the deaerator's temperature ratio (0.3) which is developed with the formulation. The isentropic efficiency of steam turbine is 80%. The isentropic efficiency for the pump is 75%. The mechanical efficiency for pump and turbine is considered as 96%. Electrical generator efficiency (η_{eg}) is 98%. The pressure losses and energy losses are neglected.

Fig. (**3**) shows the heat recovery from the hot gas to the plant's working fluid to generate the superheat steam from water. The hot gas temperature in the evaporator can be determined from the PP of the HRSG. AP also maintained in the HRVG to control the sharp transition of water into steam.

The turbine inlet temperature is determined from the hot gas supply temperature and TTD,

$$T_1 = T_{13} - TTD \tag{1}$$

where TTD is the temperature difference between hot gas supply temperature and turbine inlet temperature.

The water temperature at the exit of evaporator,

$$T_8 = T_9 - AP \tag{2}$$

The temperature of hot gas in the evaporator,

$$T_{15} = T_8 + AP + PP \tag{3}$$

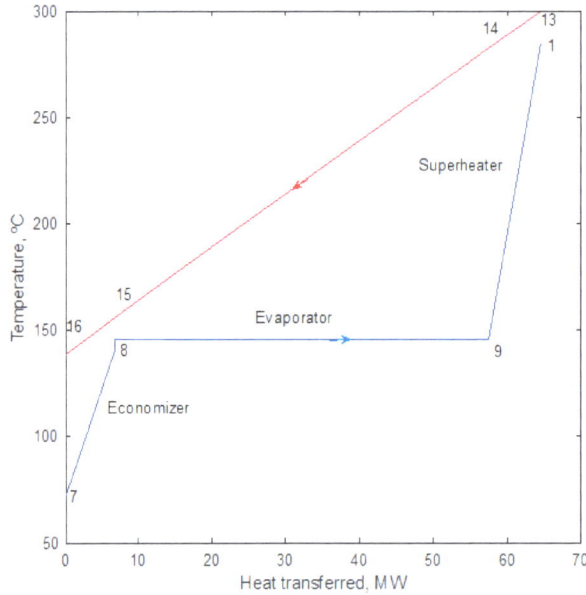

Fig. (3). Temperature-heat transferred diagram of HRSG of SRC.

The local gas temperatures are determined from the heat balance of the heat exchanger.

The deaerator location is selected from its temperature ratio. It is defined as the temperature difference between deaerator to condenser and boiler saturation to condenser.

The deaerator temperature ratio,

$$\theta_{deaerator} = \frac{T_{bled\ steam} - T_{cond}}{T_{HRSG,\ sat} - T_{cond}} \tag{4}$$

The steam capacity from the gas flow and its temperature,

$$m_1 = \frac{m_g c_{pg} \left(T_{13} - T_{15} \right)}{h_1 - h_8} \tag{5}$$

In Eq 5, the mass of hot gas is determined from the volume follow (V) and molecular weight (*M*).

$$m_g = \frac{V_g M_g}{22.4 \times 3600} \tag{6}$$

Let m_2 is the amount of steam to be removed from the turbine at the steam flow supply of m_1 for feedwater heating.

The mass balance in open feedwater heater (deaerator),

$$m_3 = m_1 - m_2 \tag{7}$$

Similarly, the heat balance in deaerator,

$$m_2\, h_2 + (m_1 - m_2)\, h_5 = m_6\, h_6 \tag{8}$$

From the simplification of these two equations,

$$m_2 = \frac{m_1\left(h_6 - h_5\right)}{h_2 - h_5} \tag{9}$$

The gross work from the turbine,

$$W_t = m_1\, (h_1 - h_2) + (m_1 - m_2)\, (h_2 - h_3) \tag{10}$$

The work supply to first pump,

$$W_{p1} = (m_1 - m_2)\, (h_5 - h_4) \tag{11}$$

where,

$$h_5 = h_4 + v\, (p_5 - p_4) \tag{12}$$

Similarly work supply to second pump *i.e.* boiler feed pump,

$$W_{p2} = m_1 (h_7 - h_6) \tag{13}$$

where,

$$h_7 = h_6 + v (p_7 - p_6) \tag{14}$$

The net output from the thermal power plant,

$$W_{net} = W_t - W_{p1} - W_{p2} \tag{15}$$

The heat supply to the power plant,

$$Q_{supply} = m_1 (h_1 - h_7) \tag{16}$$

Thermal efficiency of SRC,

$$\eta_{thermal} = \frac{W_{net}}{Q_{supply}} \times 100 \tag{17}$$

For exergy analysis, the irreversibilities are evaluated from the exergy balance of plant's components.

Irreversibility of HRSG,

$$I_{HRSG} = m_g (\varepsilon_{13} - \varepsilon_{16}) + m_1 (\varepsilon_1 - \varepsilon_7) \tag{18}$$

where ε (= h - $T_0 s$) is the specific exergy, kJ/kg.

$$I_t = m_1 (\varepsilon_1 - \varepsilon_2) + m_3 (\varepsilon_3 - \varepsilon_4) + W_t \tag{19}$$

Irreversibility in deaerator,

$$I_{deaerator} = m_2 \varepsilon_2 + m_5 \varepsilon_5 - m_6 \varepsilon_6 \qquad (20)$$

From the energy balance in steam condenser, the circulating cooling water,

$$m_w = \frac{m_3(h_3 - h_4)}{c_{pw}(T_{18} - T_{17})} \qquad (21)$$

The irreversibility of water cooled condenser,

$$I_{cond} = m_3(\varepsilon_3 - \varepsilon_4) + m_{17}(\varepsilon_{17} - \varepsilon_{18}) \qquad (22)$$

Irreversibility of two pumps,

$$I_p = m_4(\varepsilon_4 - \varepsilon_5) + m_6(\varepsilon_6 - \varepsilon_7) + W_p \qquad (23)$$

The irreversibility of exhaust gas is equal to the exergy loss by exhaust gas.

$$I_{exhaust} = m_g(\varepsilon_{12} - \varepsilon_0) \qquad (24)$$

where ε_0 is the exergy at the reference state.

Similarly, irreversibility of hot water from the condenser is equal to its exergy loss.

$$I_{hw} = m_{18}(\varepsilon_{18} - \varepsilon_0) \qquad (25)$$

Total irreversibility losses in SRC,

$$I_{total} = I_{HRSG} + I_t + I_{deaerator} + I_{cond} + I_p + I_{ex} + I_{hw} \qquad (26)$$

Exergy of hot gas source,

$$E_g = m_{38}(\varepsilon_{13} - \varepsilon_0) \qquad (27)$$

Exergy (Second law) efficiency,

$$\eta_2 = \frac{E_g - I_{total}}{E_g} \times 100 \tag{28}$$

PERFORMANCE CHARACTERISTICS

The power generation and thermal efficiency of SRC have been analyzed with hot gas supply temperature and deaerator state. HRSG pressure plays a key role in performance variations. Fig. (4) shows the variation in power, efficiency and degree of superheat (DSH) with change in hot gas supply temperature and HRSG pressure. The HRSG pressure has been changed from deaerator pressure to a pressure below the critical point. The performance characteristics are developed with the source temperature below and above the critical temperature.

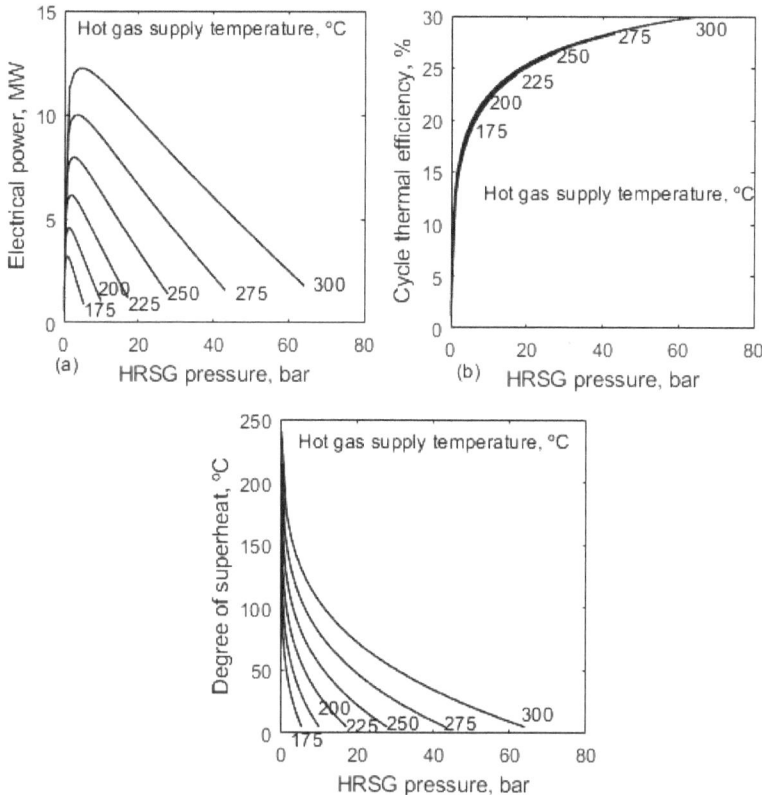

Fig. (4). Performance characteristics of SRC with of source temperature and HRSG pressure on **(a)** net power generation, **(b)** cycle energy conversion efficiency and **(c)** degree of superheat at the temperature of source below the critical temperature of working fluid.

The heat recovery decreases with increase in HRSG pressure. The pressure rise suppresses the steam capacity in the HRSG. Therefore, the power output decreases with further rise in pressure. But the rise in pressure increases the steam expansion in the turbine and power generation. These two effects of decreased fluid rate and increased power generates an optimum boiler pressure (Fig. **4a**). The decrease in heat supply with rise in pressure, favors the thermal efficiency. Therefore, the efficiency increases with increase in pressure in a diminishing rate. The degree of superheat changes inversely with pressure and directly with the temperature.

A non-linier equation is framed to find the optimum pressure at the gas supply temperature.

The optimum pressure,

$$P_{HRSG\ opt} = a_1 + a_2T + a_3T^2 \tag{29}$$

where T is the hot gas (source) temperature, °C.

The resulted coefficients for the above non-linear equation are:

$a_1 = 6.1234$, $a_2 = -0.0664$ and $a_3 = 0.0002$.

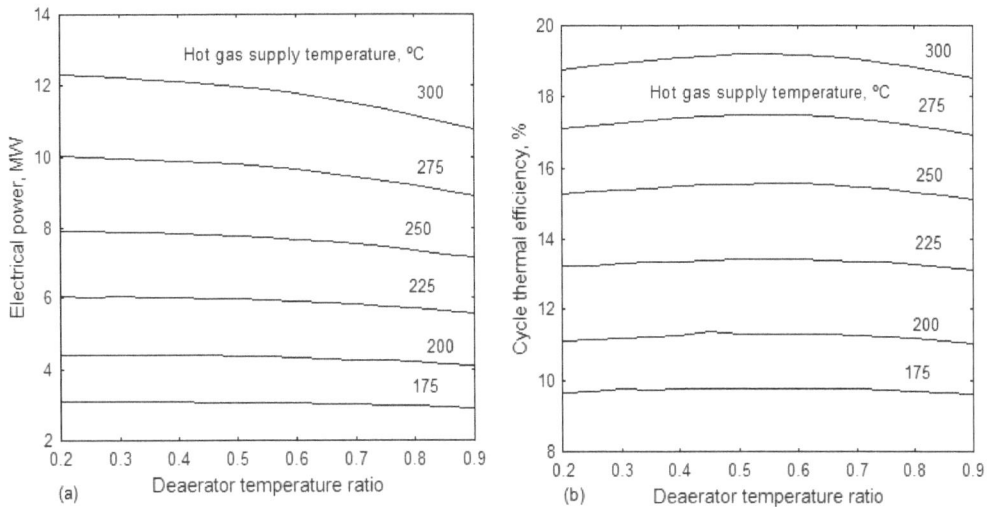

Fig. (5). Deaerator temperature ratio on **(a)** power generation and **(b)** thermal efficiency.

Deaerator has an optimum location for the maximum power and thermal efficiency. To find the optimum deaerator pressure, deaerator temperature ratio (Eq 3) has been defined and analyzed to study the maximum output and thermal efficiency (Fig. 5). If the deaerator temperature is equal to zero, the deaerator pressure is equal to condenser pressure. Else if the deaerator temperature is one, the deaerator temperature is equal to the HRVG saturation temperature. The deaerator temperature ratio is changed from 0.2 (above condenser pressure) to 0.9 (below HRSG pressure). A lower deaerator pressure favors the power as the turbine loss at the low pressure is not significant. But this state does not improve the feed water temperature hence less improvement in thermal efficiency. If the deaerator pressure is high and close to the HRSG pressure, more power loss from the turbine but higher feedwater temperature improves the efficiency. On overall view, a lower deaerator pressure favors the power and high deaerator pressure supports the efficiency. The analytical results show that the optimum deaerator temperature ratio for the maximum thermal efficiency is around 0.5. For maximum power conditions, the deaerator temperature ratio should be less than 0.5. In the further analysis, the deaerator temperature is considered as 0.3 due to opted maximum power condition.

Fig. (**6**) also shows the performance characters of SRC with supply temperature and HRSG with the source temperature above the critical temperature of water. The optimum HRSG pressure is increasing with increase in source temperature. The power is augmented from the source temperature below the critical temperature to the source temperature above the critical temperature of the working fluid. The effect of temperature on power is more than the effect of pressure. The efficiency variations are similar to the earlier analysis. The DSH with this temperature is higher than the source temperature below the critical temperature.

The resulted coefficients to find the optimum HRSG pressure at the source temperature above the critical temperature (Eq. 29),

$a_1 = 94.9900$, $a_2 = -0.4947$ and $a_3 = 0.0007$.

(Fig. **7**) analyzes the deaerator temperature ratio with source temperature which is above the critical temperature of water. The lower optimum deaerator temperature ratio is resulted for maximum pressure and 0.5 deaerator temperature ratio is resulted for the maximum thermal efficiency. The same optimum deaerator temperature ratio is continued at the source temperature above the critical temperature in further analysis.

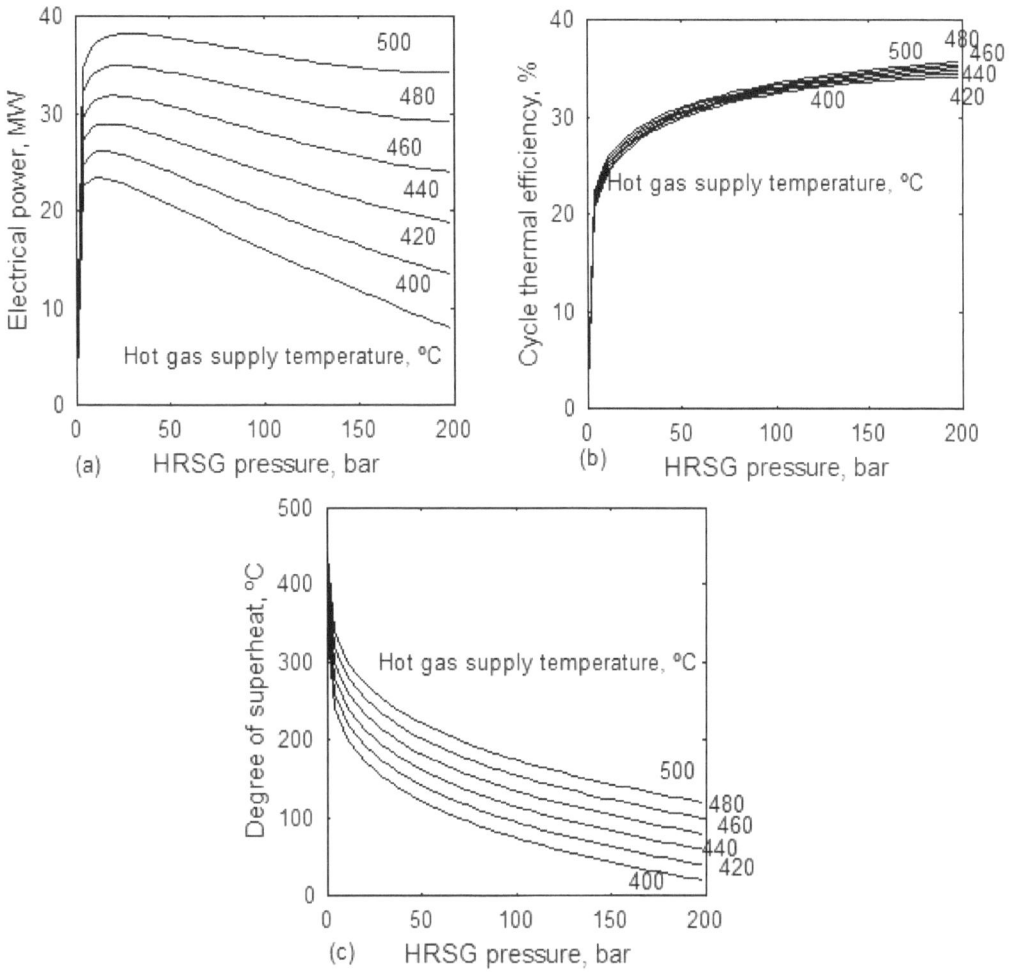

Fig. (6). Performance characteristics of SRC with of source temperature and HRSG pressure on **(a)** net power generation, **(b)** cycle energy conversion efficiency and **(c)** degree of superheat at the temperature of source above the critical temperature of fluid.

Fig. (**8**) shows the role of sink temperature with HRSG pressure on power and thermal efficiency. The HRSG pressure is varied between condenser pressure and HRSG pressure. Similar to the source temperature analysis with HRSG pressure, the sink temperature is also resulted low optimum pressure for maximum power and maximum thermal efficiency. The sink temperature is changed from 20 to 35 °C. There is no much variation in optimum pressure with the sink temperature. Therefore, HRSG optimum pressure is determined from the source temperature.

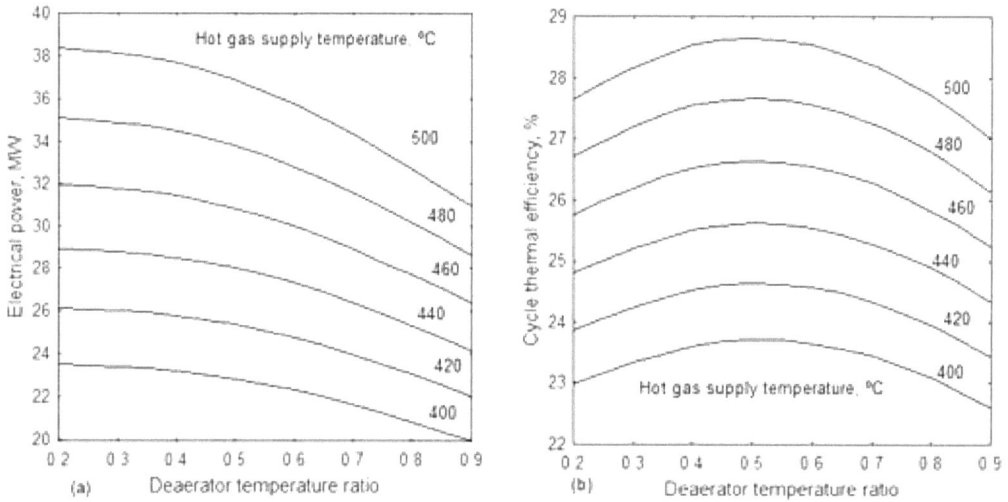

Fig. (7). Deaerator temperature ratio with turbine inlet temperature above the critical temperature.

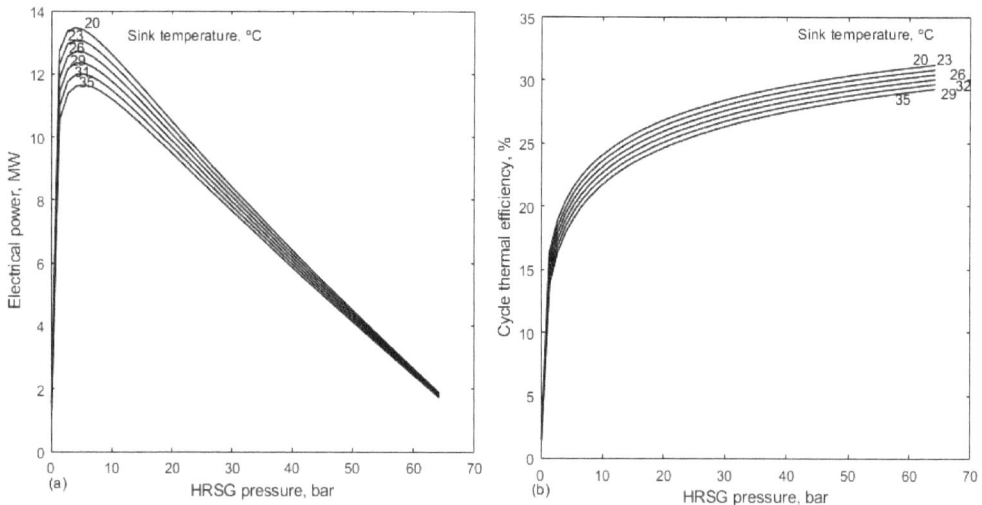

Fig. (8). Influence of sink temperature with HRSG pressure on **(a)** net power generation and **(b)** cycle energy conversion efficiency at the hot gas supply temperature of 300 °C.

Based on the above thermal analysis, at the optimized operational conditions, the results of thermodynamic evaluation have been tabulated in Table (**1**). The results are shown at the hot gas supply temperature of 300 °C. Three pressures in the plant are 4.20 bar, 0.33 bar and 0.07 bar respectively at HRSG, deaerator and condenser. The turbine inlet temperature at the hot gas supply temperature of 300 °C is 285 °C with the TTD of 15 K. The steam generated in HRSG is 22.82 kg/s at the hot gas flow of 368.9 kg/s. In water cooled condenser, the circulating water flow is 1487.76 kg/s.

Table 1. SRC material balance results at hot gas supply of 300 °C.

State	P, bar	t, °C	m, kg/s	h, kJ/kg	s, kJ/kg K
1.	4.20	285.00	22.82	3035.89	7.49
2.	0.33	78.97	1.18	2643.92	7.77
3.	0.07	40.00	21.64	2466.78	7.91
4.	0.07	40.00	21.64	167.45	0.57
5.	0.33	41.00	21.64	171.64	0.59
6.	0.33	71.63	22.82	299.78	0.97
7.	4.20	72.73	22.82	304.39	0.99
8.	4.20	140.42	22.82	590.88	1.74
9.	4.20	145.42	22.82	2739.84	6.88
10.	1.01	25.00	29.72	0.00	0.00
11.	1.01	35.00	339.21	4.09	0.01
12.	1.01	900.00	368.90	1006.40	1.54
13.	1.01	300.00	368.90	292.91	0.69
14.	1.01	283.38	368.90	274.60	0.66
15.	1.01	160.43	368.90	141.67	0.39
16.	1.01	143.77	368.90	123.95	0.35
17.	1.01	25.00	1487.76	0.00	0.00
18.	1.01	33.00	1487.76	33.44	0.11

The energy balance and exergy balance results are shown in Sankey diagram (Fig. **9a**) and Grassmann diagram (Fig. **9b**) respectively. The energy losses involved in the power plant are heat loses through exhaust gas and hot water from the condenser. The rest of the energy is converted into power (12.1%). Because of neglecting heat loss from HRSG, the energy loss in HRSG has not been shown. But in exergy analysis, the irreversibility is not zero in the heat exchanger as the heat transfers from high temperature to low temperature. Fig. (**9b**) shows the exergy losses in HRSG, turbine, deaerator, condenser, pump, exhaust and hot water. The majority of exergy losses observed from exhaust and HRSG followed by turbine. Deaerator, condenser, pump and hot water are resulted a minor exergy loss. After deducting these losses, the net available exergy is 38.5%.

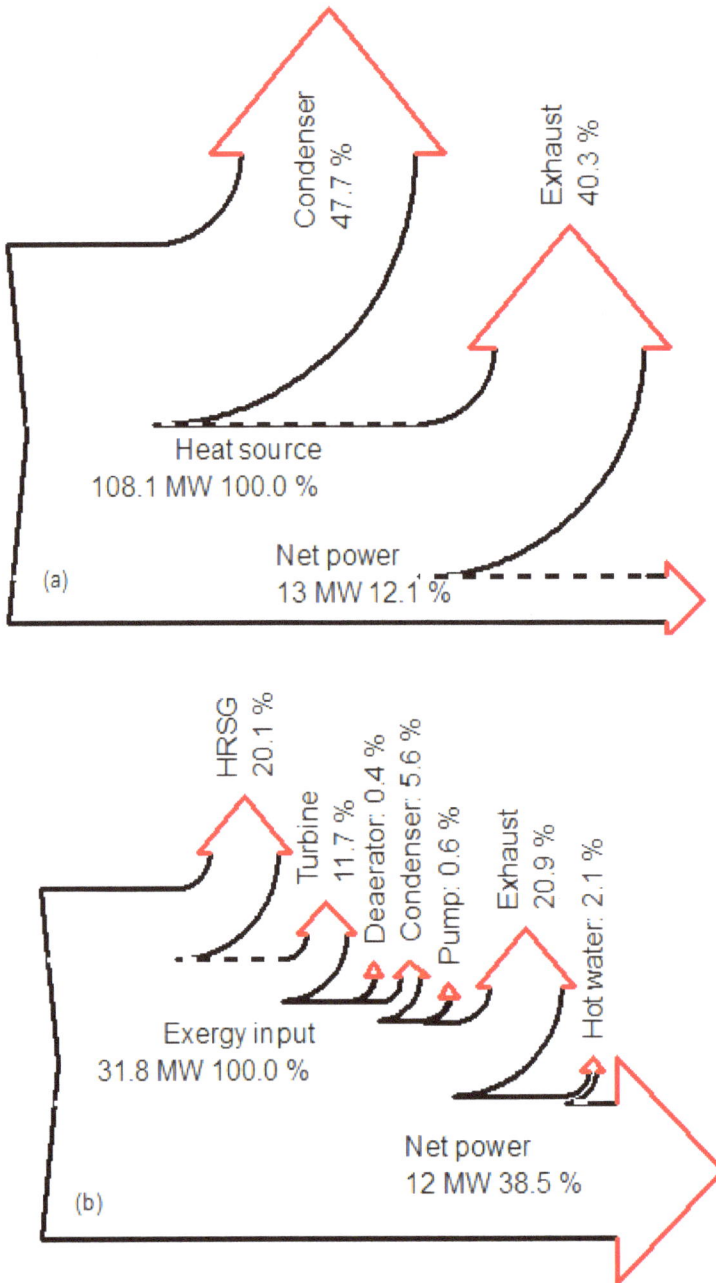

Fig. (9). (a) Sankey diagram and **(b)** Grassmann diagram of SRC.

Table **2** summarizes the specifications of SRC at the optimized conditions resulted from the analysis. The specific power output is 0.53 MW/kg of steam. The energy efficiency of cycle is determined from the heat supply to HRSG and energy efficiency of the plant is determined from heat of hot gas supplied. The exergy efficiency of the plant is determined from the exergy supply of hot gas at the inlet of HRSG. The total power output from the plant is 13.00 MW. The cycle and plant energy efficiencies are 18.95% and 12.10% respectively. The plant's exergy efficiency is 38.48%.

Table 2. The power plant specifications.

Working Fluid	Result
Hot fluid inlet temperature, °C	300.00
Turbine inlet temperature, °C	285.00
Optimized HRVG pressure, bar	4.20
Condenser pressure, bar	0.07
Specific power, MW/kg	0.53
Plant power, MW	13.00
Cycle thermal efficiency, %	18.95
Plant thermal efficiency, %	12.10
Plant exergy efficiency, %	38.48

CASE STUDY ON STEAM RANKINE CYCLE AT A CEMENT FACTORY

A case study has been conducted on a cement factory to identify the waste heat sources. The factory has two units. Fig. (**10**) shows the identified heat sources from the factory. Each unit has three hot gas sources which are not taped for the power generation through its recovery. Each unit is operated by the hot gas supplied from the coal fired furnace. The three gas streams are at the temperatures of 176 °C, 330 °C and 420 °C. This section is focused on identification of the potential of available heat and   conversion into to   useful   electricity generation.

A suitable power plant has developed to suit the waste heat of the cement factory. A single power plant is designed to the two units of factory get the benefit of high capacity steam turbine. The efficiency of the turbine increases with its size.

Fig. (**11**) shows the designed power plant layout suitable to the two units of a cement factory. For each unit two HRSG are configured *i.e.* total of four HRSGs. The four steam lines from two units are mixed and supplied to turbine for power generation. The selected power plant cycle to suit the heat is a basic Rankine cycle with a deaerator. The exhaust gas from the power plant is used in preheating of air and condensate preheating. The capacities of two plants are 1600 TPD (tons per day) and 5500 TPD with the total capacity of 7100 TPD. Coal is used in furnace to generate the process heat to cement manufacturing. The balanced heat is planed to the power plant. From the power plant still heat will reject through exhaust and condenser. The processes particulars are not furnished in the section as the current focus is on power generation. Out of the three heat sources identified in the cement factory, one source is at 176 °C which is not suitable to steam power cycle to generate the steam at the required pressure. To address this problem supplementary firing (SF) has been added to rise the temperature of gas from 176 °C to 420 °C by combustion and mixing of two gas streams. The SF has been iterated to result the required temperature. Now from each unit of cement factory, two hot gas streams are resulted at the temperature of 420 °C and 330 °C. Two HRSGs are configured to recover the heat from these two streams. Similarly, two superheated steams are generated from the second unit. The temperature of steam generated from 330 °C source is less than the temperature of steam from 420 °C source. These four steams from four HRSGs are mixed to form a single stream. The total steam is connected to a steam turbine where the power cycle repeats. The thermal power plant is a simple Rankine cycle with a deaerator. The condensed steam at saturated water state is pumped from the condenser pressure to deaerator pressure *via* condensate preheater. Again the water from the deaerator is pumped from the deaerator pressure to HRSG pressure and supplied to the four lines. The final exhaust gas is used in air preheater and condensate preheater. At the end, the exhaust gas is cleaned at electro static precipitator (ESP) before sending to atmosphere through chimney. The determined power from the factory's waste heat and SF is 15.41 MW against the factory demand of 15 MW. Therefore, sufficient power can be generated from the waste heat recovery as per the studied conditions.

The assumptions used in the solution of this case study are as follows. Atmospheric state is 1.01325 bar and 25 °C. The combustion temperature in main furnace and SF are 1000 °C and 900 °C respectively. These temperatures are below the adiabatic flame temperature of the solid fuel (coal). TTD between hot gas and superheater is 15 °C. PP in HRSG is 20 °C. Optimum HRSG pressure is used at the hot gas supply temperature. Deaerator temperature ratio is 0.3.

The isentropic efficiency of steam turbine is 80%. The isentropic efficiency for the pump is 75%. The mechanical efficiency for pump and turbine is considered as 96%. Electrical generator efficiency (η_{gen}) is taken as 98%. The pipe pressure and energy losses are neglected. The cement factory is results three hot streams at the temperatures of 420 °C, 330 °C and 176 °C. In this analysis, the hot has supply temperature, *i.e.* 420 °C is changed while the 330 °C is varied as 90% of hot gas supply temperature. The gas supply to SF, 176 °C is not changed in the analysis. The hot gas generation from the cement factory depends on its capacity. The specific mass of hot gas production to 330 °C supply is 1.6 Nm³/h per kg cement. From the capacity of cement factory unit, the total gas can be determined. The specific mass of hot gas supplied after mixing of main stream with SF stream is 1.25 Nm³/h per kg cement. After mixing the hot gas from SF with 176 °C stream, the resultant temperature is 420 °C. Therefore, the mixing of SF mixed gas and main stream happens at constant temperature of 420 °C. At this temperature, the mass ratio of SF mixed stream and main stream at 420 °C is 0.44:0.56.

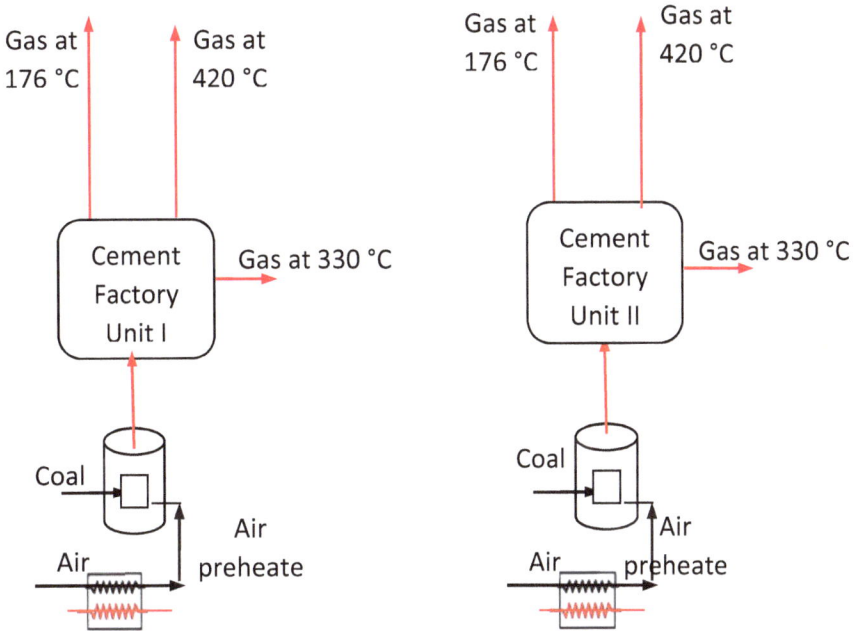

Fig. (10). Identified heat sources from a cement factory with two units.

The coal composition from the proximate analysis is:

C = 47, H = 3.17, O = 8.7, N = 1.5, S = 0.91 and ash = 33.

The moisture content is 11.7%. The determined higher heating value (HHV) of coal is 14871 kJ/kg coal. After solving the adiabatic combustion with stoichiometric air, the resulted air fuel ratio is 9.17. At this air fuel ratio, adiabatic flame temperature of coal is 1100 °C. Since the assumed combustion temperature is below the adiabatic flame temperature, the actual air fuel ratio is greater than the stoichiometric air fuel ratio.

In SF, coal is burned with air. The reaction in SF is,

$$\left(C_{a_1}H_{a_2}O_{a_3}N_{a_4}\right)_{coal} + \left(a_{5\ SF}(O_2 + 3.76N_2)\right)_{air} + \left(a_6 H_2O(l)\right)_{moisture\ content}$$
$$\Rightarrow \left(c_1 CO_2 + c_2 H_2O + c_3 O_2 + c_4 N_2\right)_{gas} \tag{30}$$

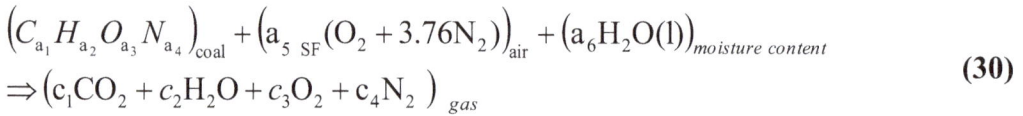

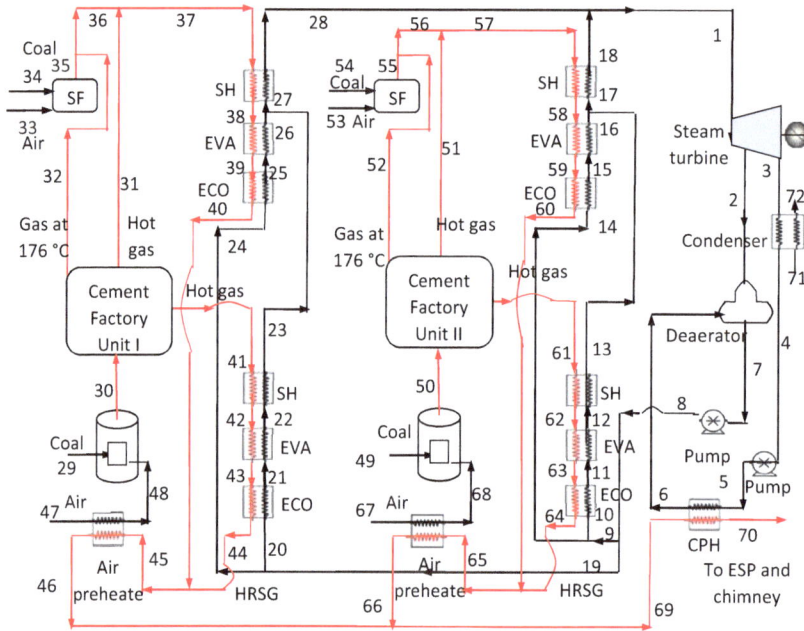

Fig. (11). Use of steam Rankine cycle for power generation using the waste heat from a cement factory.

The coefficients of reaction are determined from the C, H, O, N balance and energy balance equation. The reaction in SF with solved coefficients is,

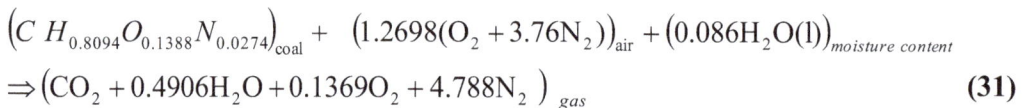

$$\left(C H_{0.8094}O_{0.1388}N_{0.0274}\right)_{coal} + \left(1.2698(O_2 + 3.76N_2)\right)_{air} + \left(0.086H_2O(l)\right)_{moisture\ content}$$
$$\Rightarrow \left(CO_2 + 0.4906H_2O + 0.1369O_2 + 4.788N_2\right)_{gas} \tag{31}$$

The actual air fuel ratio in SF is 10.28 which is more than the stoichiometric air fuel ratio.

At the outlet of SF, two hot gas streams are mixed to result the gas at 420 °C.

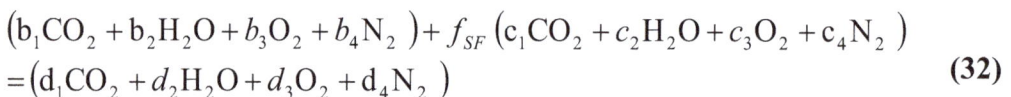

$$\left(b_1 CO_2 + b_2 H_2O + b_3 O_2 + b_4 N_2\right) + f_{SF}\left(c_1 CO_2 + c_2 H_2O + c_3 O_2 + c_4 N_2\right)$$
$$= \left(d_1 CO_2 + d_2 H_2O + d_3 O_2 + d_4 N_2\right) \tag{32}$$

The fraction, f_{SF} is SF gas ratio from the total mixed gas to result the required temperature.

The Eq. 32 has been solved with mass and energy balance formulations and the resultant reaction with the obtained coefficients is:

$$(CO_2 + 0.4906H_2O + 0.0.0796O_2 + 4.57N_2) + 0.4459 \begin{pmatrix} CO_2 + 0.4906H_2O \\ +0.1369O_2 + 4.788N_2 \end{pmatrix} \quad (33)$$
$$= (1.449CO_2 + 0.7094H_2O + 0.1406O_2 + 6.7076N_2)$$

The hot gas from the SF and the main gas will mix together to supply the thermal energy to the HRSG and then to power plant.

$$n_{31}(b_1CO_2 + b_2H_2O + b_3O_2 + bc_4N_2) + n_{36}(d_1CO_2 + d_2H_2O + d_3O_2 + d_4N_2) \quad (34)$$
$$= (e_{11}CO_2 + e_{21}H_2O + e_{31}O_2 + e_{41}N_2)$$

Eq 34 has been solved to result the coefficients of mixed gas.

$$0.0741(CO_2 + 0.4906H_2O + 0.0.0796O_2 + 4.57N_2) + 0.0397 \begin{pmatrix} 1.449CO_2 + 0.7094H_2O \\ +0.1406O_2 + 6.7076N_2 \end{pmatrix}$$
$$= (0.1316CO_2 + 0.0646H_2O + 0.0115O_2 + 0.6057N_2) \quad (35)$$

Two exhaust gases are mixed together after the HRSGs. The mixed gas is supplied to the APH and CPH.

$$(e_{11}CO_2 + e_{21}H_2O + e_{31}O_2 + e_{41}N_2) + n_{41}(b_1CO_2 + b_2H_2O + b_3O_2 + bc_4N_2)$$
$$= (f_{11}CO_2 + f_{21}H_2O + f_{31}O_2 + f_4N_2) \quad (36)$$

Eq 7.35 is solved for the mixing of two streams the the resulted coefficients are presented as follows.

$$(0.1316CO_2 + 0.0646H_2O + 0.0115O_2 + 0.6057N_2) + 0.2153 \begin{pmatrix} 1.449CO_2 + 0.7094H_2O \\ +0.1406O_2 + 6.7076N_2 \end{pmatrix}$$
$$= (0.347CO_2 + 0.1702H_2O + 0.0286O_2 + 1.59047N_2) \quad (37)$$

At the inlet of APH, also two hot exhaust gases are mixed together.

$$\left(f_{11}CO_2 + f_{21}H_2O + f_{31}O_2 + f_{41}N_2\right) + \left(f_{12}CO_2 + f_{22}H_2O + f_{32}O_2 + f_{42}N_2\right)$$
$$= \left(g_1CO_2 + g_2H_2O + g_3O_2 + g_4N_2\right) \tag{38}$$

The resulted coefficients in Eq. 38 are used in the following equation.

$$\left(0.347CO_2 + 0.1702\,H_2O + 0.0286\,O_2 + 1.5904\,7N_2\right) + \begin{pmatrix} 1.1927CO_2 + 0.5852\,H_2O \\ + 0.0984\,O_2 + 5.4671N_2 \end{pmatrix}$$
$$= \left(1.5397CO_2 + 0.7554\,H_2O + 0.127\,O_2 + 7.0575N_2\right) \tag{39}$$

The unit 2 of the cement factory has been solved with the similar approach applied to Unit 3.

The energy supplied from the hot gas to generate the steam are as follows.

$$Q_{process} = m_{30}h_{30} - m_{31}h_{31} - m_{32}h_{32} - m_{41}h_{41} + m_{50}h_{50} - m_{51}h_{51} - m_{52}h_{52} - m_{61}h_{61} \tag{40}$$

After cement process, the balance heat to power generation is.

$$Q_{supply\ cycle} = m_{37}(h_{37} - h_{40}) + m_{41}(h_{41}-h_{44}) + m_{57}(h_{57} - h_{60}) + m_{61}(h_{61}-h_{64})$$
$$+ m_{69}(h_{69}-h_{70}) \tag{41}$$

It can also be determined from power plant configuration as follows.

$$Q_{supply\ cycle} = m_1\,(h_1 - h_8) + (m_1 - m_2)\,(h_6 - h_5) \tag{42}$$

The net power output,

$$W_{net} = W_t - W_p =$$
$$W_{net} = W_t - W_p = \left(m_1(h_1 - h_2) + (m_1 - m_2)(h_2 - h_3)\right)\eta_t\eta_{eg} - \frac{m_5(h_5 - h_4)}{\eta_p} - \frac{m_8(h_8 - h_7)}{\eta_p} \tag{43}$$

The cycle thermal efficiency,

$$\eta_{cycle} = \left(\frac{W_{net}}{Q_{\text{sup}ply\, cycle}} \right) \times 100 \qquad \textbf{(44)}$$

The thermodynamic properties of power plant of this case study are developed with assumptions, mass balance and energy balance equations and tabulated in Table **3**.

The gas temperature is considered at 420 °C. The turbine inlet temperature with this hot gas is 405 °C. At this supply temperature the optimum HRSG pressure has been determined from the correlation equation (Eq. 28) with the coefficients developed above the critical temperature. The resulted optimized HRSG pressure at the source temperature is 4.13 bar. From this HRSG pressure, the selected deaerator pressure for maximum power is 0.33 bar. The gas supply from two sources of unit I are 24.27 kg/s and 39.47 kg/s. Similarly, unit 2 is giving 83.42 kg/s and 135.67 kg/s of hot gas respectively from 420 °C and 336 °C. The steam generating capacity in HRSG with the gas supply at four lines is 24 kg/s. The steam used for feedwater heater is 0.61 kg/s.

Table 3. The properties of proposed power plant (Fig. 11) after identifying the heat sources from a cement factory with two units.

State	P, bar	t, °C	m, kg/s	h, kJ/kg	State	P, bar	t, °C	m, kg/s	h, kJ/kg
1	4.13	405.00	24.00	3283.88	37	1.15	420.00	24.27	428.57
2	0.33	163.55	0.61	2807.64	38	1.14	353.60	24.27	353.18
3	0.07	56.20	23.39	2604.92	39	1.13	159.74	24.27	141.19
4	0.07	40.00	23.39	167.45	40	1.12	133.42	24.27	113.00
5	0.33	40.37	23.39	169.01	41	1.19	336.00	39.47	333.95
6	0.33	55.90	23.39	233.92	42	1.18	310.59	39.47	305.60
7	0.33	71.42	24.00	298.92	43	1.18	159.74	39.47	141.29
8	4.13	72.16	24.00	302.03	44	1.17	139.26	39.47	119.45
9	4.13	72.16	18.59	302.03	45	1.17	137.01	63.74	116.99
10	4.13	72.16	10.36	302.03	46	1.16	95.91	63.74	73.61
11	4.13	139.74	10.36	587.95	47	1.15	28.00	58.10	3.03
12	4.13	144.74	10.36	2739.00	48	1.15	75.00	58.10	50.62
13	4.13	321.00	10.36	3110.05	49	1.01	28.00	19.35	0.00
14	4.13	72.16	8.22	302.03	50	1.14	1000.00	219.04	1139.22
15	4.13	139.74	8.22	587.95	51	1.13	420.00	46.72	428.95

(Table 3) cont.....

State	P, bar	t, °C	m, kg/s	h, kJ/kg	State	P, bar	t, °C	m, kg/s	h, kJ/kg
16	4.13	144.74	8.22	2739.00	52	1.13	176.00	36.65	158.70
17	4.13	238.64	18.59	2945.51	53	1.12	30.00	0.05	3.62
18	4.13	405.00	18.59	3283.88	54	1.01	28.00	0.00	0.00
19	4.13	72.16	5.41	302.03	55	1.11	900.00	0.06	1006.66
20	4.13	72.16	3.01	302.03	56	1.10	420.00	36.71	428.08
21	4.13	139.74	3.01	587.95	57	1.10	420.00	83.42	428.57
22	4.13	144.74	3.01	2739.00	58	1.09	353.48	83.42	353.18
23	4.13	321.00	3.01	3110.05	59	1.09	159.74	83.42	141.19
24	4.13	72.16	2.39	302.03	60	1.08	133.29	83.42	113.00
25	4.13	139.74	2.39	587.95	61	1.08	336.00	135.67	333.95
26	4.13	144.74	2.39	2739.00	62	1.07	310.59	135.67	305.60
27	4.13	238.64	5.41	2945.51	63	1.07	159.74	135.67	141.29
28	4.13	405.00	5.41	3283.88	64	1.06	139.26	135.67	119.45
29	1.01	28.00	5.63	0.00	65	1.06	136.96	219.09	116.99
30	1.24	1000.00	63.72	1139.22	66	1.05	95.86	219.09	73.61
31	1.23	420.00	13.59	428.95	67	1.04	28.00	199.71	3.03
32	1.21	176.00	10.66	158.70	68	1.04	75.00	199.71	50.62
33	1.20	30.00	0.02	3.62	69	1.03	95.86	282.83	73.61
34	1.01	28.00	0.00	0.00	70	1.03	90.74	282.83	68.24
35	1.18	900.00	0.02	1006.66	71	1.15	28.00	1704.88	12.54
36	1.17	420.00	10.68	428.08	72	1.10	36.00	1704.88	45.98

Fig. (**12**) shows the variations in (a) power generating capacity, power cycle's thermal efficiency and exhaust gas temperature with change in hot gas supply temperature and HRSG pressure. The hot gas supply temperature is changed above the critical temperature of water. The HRSG pressure is varied between condenser pressure and HRSG pressure. The optimum HRSG pressure is increasing with increase in source temperature. The optimum HRSG pressure is low for maximum power condition and high for maximum thermal efficiency condition. A low pressure in HRSG increases the heat recovery and steam generating capacity. It results more power from the plant. At high pressure, the heat recovery decreases which favours the thermal efficiency with a loss in power. After the optimum steam pressure, the power is dropping with further increase in pressure. Therefore, the supercritical steam pressure is not suitable for the heat recovery operated power plants as it is not favouring the power. But the supercritical power plant is suitable for non-heat recovery power plants as they support the thermal efficiency.

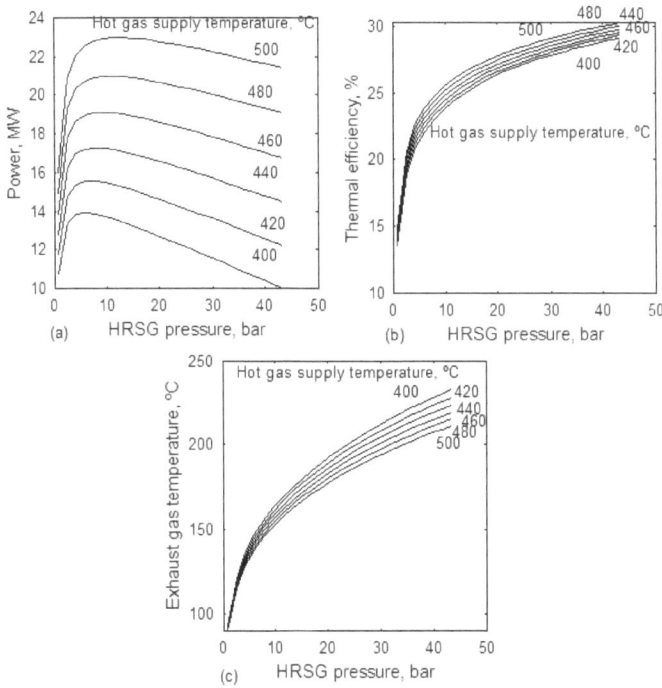

Fig. (12). **(a)** The power generation and **(b)** power cycle thermal efficiency with source temperature and steam pressure.

From the optimization of the power, the equation has been developed to find the HRSG pressure from the hot gas temperature.

The resulted coefficients for the above non-linear equation (Eq. 29) are:

$a_1 = 42.3032$, $a_2 = -0.2169$ and $a_3 = 0.0003$.

Fig. (**13**) analyzes the power plant capacity with cement factory capacity and the HRSG pressure. The size of the cement factory is expressed by the production of cement in tons per day (TPD). In this case study, the factory has two units producing 7100 TPD of total capacity. To analyze the relation between cement capacity and power plant capacity with HRSG pressure, the factory capacity has been changed from 5000 TPD to 10000 TPD. The steam pressure is changed between condenser pressure and critical pressure of water as subcritical power plant is suitable for heat recovery. Similar to the earlier analysis, optimum HRSG pressure has been resulted with the factory capacity. Since the hot gas supply temperature is kept at 420 °C, the optimum pressure is constant and independent of factory capacity. The cycle efficiency is mostly depending on thermal conditions of the power plant and nearly independent of factory capacity. Therefore, thermal efficiency has not been analyzed with the factory capacity.

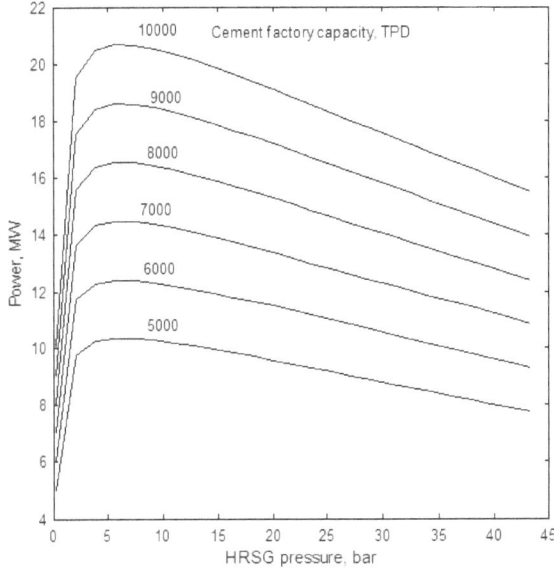

Fig. (13). Power plant capacity with cement factory size and steam pressure.

Table **4** consolidates the results of the case study conducted at the cement factory. The mass balance, energy balance and exergy balance formulation and its evaluation resulted the plant properties and performance of cycle and plant. The energy supply and outputs are presented at the heat sources of the cement factory. The air fuel ratio in main furnace and SF are 10.32 and 11.35 respectively. The coal consumption for process heat and power generation is nearly 90 t/h. The steam generating capacity is 19.47 t/h and 66.92 t/h respectively with 1600 TPD and 5500 TPD capacity units. The electricity generation from Khattak *et al.* [56] study is 29 kWh/t. The current cement factory's power plant generates 42 kWh/t. This value is higher than the reported results because of power plant layout with minimum heat losses. The gross energy consumption of Madloola *et al.* [57] report is 3.64 GJ/t and the current work results 4.74 GJ/t for process heat and power generation. The same report showed 25% cycle thermal efficiency. Since the current work is focused on power production than thermal efficiency, the resulted cycle thermal efficiency is 21.09%. The case study results are satisfactorily agreed with the literature reported results.

Table 4. Specifications of power plant operating from the waste heat of cement factory obtained from the case study.

S. No.	Description	Result
1.	Coal consumption, t/h	89.92
2.	Air used for combustion, Nm³/h	720630.00
3.	Exhaust gas, Nm³/h	764390.00
4.	Air fuel ratio at combustion, kg/kg coal	10.32
5.	Air fuel ratio at SF, kg/kg coal	11.35
6.	Exhaust gas temperature, °C	133.00
7.	Steam capacity at Unit I, t/h	19.47
8.	Steam capacity at Unit II, t/h	66.92
9.	Total steam capacity, kg/h	86.39
10.	Circulating water at condenser, t/h	6137.60
11.	Fuel energy, MW	396.10
12.	Process heat, MW	230.26
13.	Heat supply to power plant, MW	73.07
14.	APH, MW	12.27
15.	CPH, MW	1.51
16.	Condenser, MW	57.01
17.	Exhaust gas, MW	19.30
18.	Total heat losses, MW	150.22
19.	Power, MW	15.41
20.	Specific heat consumption, GJ/t	4.74
21.	Cycle thermal efficiency, %	21.09
22.	Cycle exergy efficiency, %	45.62

SUMMARY

The study recommends a low pressure for maximum power with a heat recovery. The SRC has a great potential with the ideal properties of the water compared to the other power plant cycles. Study has been conducted with the source temperature below the critical temperature and above the critical temperature of the water. Curve fitting equations are developed to find the optimum HRSG pressure from the source temperature for general power plant layout and a specially designed layout to suit the cement factory. The heat recovery requires low pressure for maximum power production but high pressure for high thermal efficiency. From the case study, it can be understood that the power plant configuration needs to modify with supplementary firing and branches of heat recoveries to suit the source temperature and flow. The heat recovery based power plant demands only subcritical operation. The designed power plant configuration and its evaluation to a case study results 15 MW of electrical output from the waste heat recovery.

CHAPTER 8

Steam Flash Cycle

Abstract: Similar to the organic flash cycle (OFC), steam flash cycle (SFC) consists of flasher to expand the liquid into liquid-vapour mixture from high pressure to low pressure. It has a significant role in power augmentation through heat recovery. This chapter reports the formulation, analysis, and performance characteristics of SFC. The performance changes of OFC with single flash and double flash are studied and compared. The source temperature is changed below and above the critical temperature of the water. Correlations are developed to find the optimum heat recovery steam generator's (HRSG's) pressure as a function of source temperature. The location of a single flash and double flash units are optimized at maximum power condition. The work is extended to generalize the SFC formulation to solve with 'n' number of flashers to simplify the complex formulae. This generalization can be used to optimize the number of flashers. A case study on SFC has been presented to understand the power generation processes with heat recovery.

Keywords: Optimum boiler pressure, Power augmentation, Steam flash, Thermal efficiency, Thermal power plant.

INTRODUCTION

Use of heat recovery in place of fuel firing in a thermal power plant has benefits *viz.* fuel-free, economic, no emissions, and no ash and dust. The industry wasting thermal energy can use it and generate its own electricity demand without depending on the grid. These power plants are also suitable for decentralized power generation to meet the load of small colonies, villages, *etc.* This chapter is focused on the steam flash cycle (SFC) for augmented power compared to the conventional steam Rankine cycle (SRC). The waste heat from a typical cement factory is considered as a heat source to the SFC. The study of the cement factory is not the scope of the current chapter. This chapter is focused on the power generation unit to maximize heat recovery and power. In the combustion chamber, fuel (coal) and air are supplied to generate heat used for cement production, and the rest of the waste heat is supplied to the steam-generating unit. Since steam is generated in the boiler from heat recovery, the boiler is known as a heat recovery steam generator (HRSG). In a regular steam power plant with a deaerator, steam has been generated at single pressure in three units *i.e.* economizer (ECO), evaporator (EVA) and superheater (SH). The local hot gas temperatures are determined using pinch point (PP) and terminal temperature difference (TTD). The condensed steam is pumped to deaerator and boiler using two pumps. In SRC, steam is consumed for feedwater heater and it drops the power production.

Flashing of pressurized hot water differ from geothermal power plant's flasher to OFC. In geothermal plant, total fluid undergoes flashing process. But in SFC, a small amount of pressurized water from the economizer of heat recovery steam generator (HRSG) is used in a flasher. The balanced fluid undergoes the regular SRC processes. SFC combines the features of SRC and flash cycle. The heat load in economizer increases compared to SRC due to the handling of excess fluid. Therefore, thermal efficiency increases with an increase in heat supply. Since the main objective of the heat recovery is to maximize power generation, the drop in efficiency is not the issue in the flash cycle. After flashing, the separated liquid goes to the boiler and the vapour expands in the turbine. The added vapour in the turbine increases the power with a penalty in efficiency. The process in a regenerative steam Rankine cycle is reversed compared to a flash cycle. In a feedwater heating system, steam from the turbine is used to preheat the water to save the heat supply to the plant. So, feedwater heating system supports the thermal efficiency but cuts the power from the turbine. In SFC, flasher supports the power but not the efficiency. In the multi flashing plant, the balanced liquid from the first flasher is supplied to the subsequent flasher. The increased economizer load drops the exhaust gas temperature. The work is focused on thermodynamic development, optimization of boiler pressure with source temperature, performance characteristics, generalization of SFC, and case study on SFC.

STEAM FLASH CYCLE

Fig. (**1**) shows the plant layout of SFC with a single flasher. The concerned thermodynamic cycle has been depicted in Fig. (**2**) . In addition to the components used in a basic Rankine cycle, flasher with separator is used as an extra component. The flash cycle works on three pressures *viz.*, HRSG pressure, flash pressure, and condenser pressure with the flasher. The superheated steam expands from the high pressure to flash pressure. At the exit of first stage expansion, the vapour is mixed with the separated vapour from the flasher unit. The mixed vapour is then expanded from the flash pressure to condenser pressure. The expanded steam as usual condensed in the condenser as shown in water cooled condenser. The condensate is pumped from condenser (low) pressure to flash pressure. At this intermediate pressure, the pumped fluid and liquid part from the flashing unit are mixed and followed by pumping from flash pressure to HRSG (high) pressure. The pressurized water is supplied to the economizer section of HRSG. In the HRSG the temperature of water is reached closed to saturation temperature corresponding to HRSG pressure. This is below the saturation point by the approach point (AP). The heat recovery consists of hot gas from the industry, such as a typical cement factory. The hot gas temperature is decreased in superheater, evaporator and economizer with a constraint of pinch point (PP).

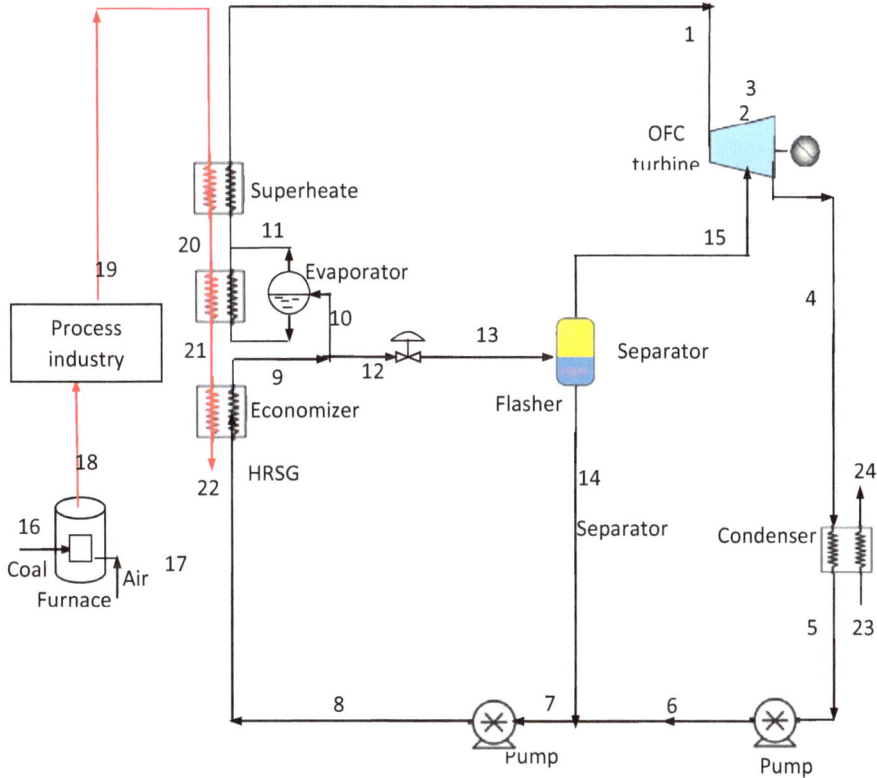

Fig. (1). SFC with single flasher.

The hot gas solution and its formulation are furnished in the earlier chapters. In the single flash SFC, the performance has been evaluated from the mass and energy formulation. For this assumptions made are as follows:

The hot gas flow rate is 1000000 m³/h. The combustion temperature is 900 °C. The temperature of circulating water in the condenser is 30 °C. The mechanical efficiency of pump and turbine is 98%. Similarly, the electrical efficiency of generator is 98%. The isentropic efficiency of pump and turbine are 75% and 80% respectively. Approach point is 5 °C. Terminal temperature difference (TTD) in condenser is 10 °C. Initially, the power cycle has been solved with unit mass flow at the turbine inlet. Later, steam generation is evaluated from the energy balance of evaporator and economizer sections.

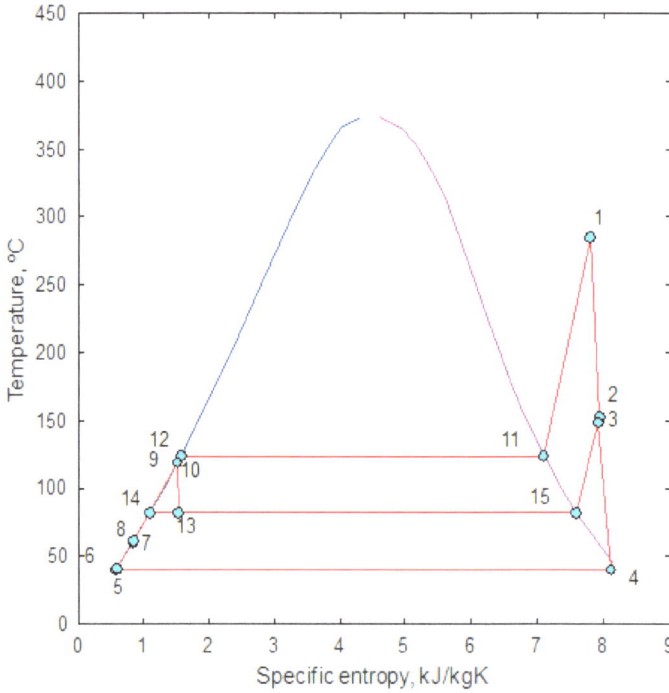

Fig. (2). Temperature-specific entropy diagram of a thermal power cycle with single flasher.

The flash pressure is maintained between condenser pressure and HRSG pressure. The location of flasher plays a key role in performance variations. To find the optimum flash pressure, let θ_F is the temperature difference ratio in flasher as follows:

$$\theta_{flash} = \frac{T_{flash} - T_{HRSG}}{T_{HRSG} - T_{cond}} \tag{1}$$

In Eq. 8.1, T_{HRSG} and T_{cond} are respectively boiler and condenser saturation temperatures. With the initial assumption of θ_{flash}, the saturation temperature of flasher is determined with the simplification of the above equation.

$$T_{flash} = T_{cond} + \theta_{flash} (T_{HRSG} - T_{cond}) \tag{2}$$

Eq. 8.2 shows that the flash temperature is the function of boiler temperature, condenser temperature, and flasher temperature difference ratio. From the determined flash temperature, flash pressure is determined.

The HRSG's steam capacity can be determined from the energy balance of the evaporator and superheater.

$$m_{10} = \frac{m_{19}(h_{19} - h_{21})}{h_1 - h_{10}} \tag{3}$$

In the above equation, h_{21} is determined from the constraint of PP. Flashing is a process of expansion of pressurized saturated hot liquid into liquid-vapor mixture. The enthalpy of saturated liquid (12) before flashing is equal to the total enthalpy of liquid-vapor mixture after the flashing (13). The flashed mixture (13) is separated into liquid (14) and vapour (15).

$$h_{13} = (1 - d_{13})\, h_{14} + d_{13} h_{15} \tag{4}$$

In Eq. 8.4, 'd' is the dryness fraction of steam.

The simplification of Eq. 8.4,

$$h_{13} = h_{14} - d_{13}\, h_{14} + d_{13} h_{15}$$

$$h_{13} = h_{14} + d_{13}\,(h_{15} - h_{14}) \tag{5}$$

The dryness fraction of steam, d after the expansion in the flasher at state 13 can be determined from Eq. 8.5.

Let FMR is the flash mass ratio defined as the mass ratio of liquid used for flash to total liquid mass in economizer.

$$FMR = \frac{m_{12}}{m_9} = \frac{m_{12}}{m_{10} + m_{12}} \tag{6}$$

The simplification of Eq. 8.6 to determine the vapour from FMR and,

$$m_{10}\, FMR + m_{12}\, FMR = m_{12}$$

$$m_{10}\, FMR = m_{12}\,(1 - FMR)$$

Since, the fluid used in evaporator and superheater, m_{10} is equal to the flow at the steam turbine inlet, m_1,

$$m_{12} = m_1 \frac{FMR}{(1 - FMR)} \tag{7}$$

The steam generating capacity in evaporator and supherheater sections,

$$m_1 = \frac{m_{19}(h_{19} - h_{21})}{(h_1 - h_{10})}$$ (8)

From the energy balance equation in HRSG's economizer,

$$h_{20} = h_{19} - \frac{m_1(h_1 - h_{11})}{m_{19}}$$ (9)

Eq. 8.9 is iterated to result T_{20} from h_{20}.

The net power from the plant with single flasher,

$$W_{net} = W_t - W_p = m_1(h_1 - h_2) + m_3(h_3 - h_4) - m_5(h_6 - h_5) - m_7(h_8 - h_7)$$ (10)

The thermal efficiency of cycle,

$$\eta_1 = \frac{W_{net}}{m_8(h_9 - h_8) + m_1(h_1 - h_{10})} \times 100$$ (11)

The mass balance and energy balance of these equations are solved and the resulted thermodynamic properties are furnished in Table 1. In the plant, the three pressures are 5.77 bar, 0.97 bar and 0.07 bar at the hot gas supply temperature of 300 °C. In the subsequent sections, the optimum boiler pressure at a source temperature has been analyzed. At the hot gas supply temperature of 300 °C, the turbine inlet temperature is 285 °C with the TTD. The total fluid heated at the economizer is 29.62 kg/s. Out of this fluid, 25% has been used for flashing, *i.e.* 7.4 kg/s. The balance fluid, 22.21 kg/s is superheated for the turbine. The flash mass of 7.4 kg/s is sum of 6.65 kg/s of liquid and 0.75 kg/s of vapour. Therefore, with this single flash unit, 0.75 kg/s of fluid is available for the power.

Table 1. Thermodynamic properties of SFC with single flash and hot gas supply temperature at 300 °C.

State	P, bar	t, °C	m, Kg/s	h, KJ/kg	s, KJ/kg K
1	5.77	285.00	22.21	3031.64	7.34
2	0.97	129.26	22.21	2735.33	7.53
3	0.97	128.25	22.96	2733.32	7.53

(Table 1) cont.....

State	P, bar	t, °C	m, Kg/s	h, KJ/kg	s, KJ/kg K
4	0.07	40.00	22.96	2422.50	7.77
5	0.07	40.00	22.96	167.45	0.57
6	0.97	41.00	22.96	171.64	0.59
7	0.97	54.06	29.62	226.31	0.76
8	5.77	55.11	29.62	230.61	0.77
9	5.77	152.34	29.62	642.23	1.87
10	5.77	152.34	22.21	642.23	1.87
11	5.77	157.34	22.21	2753.80	6.77
12	5.77	152.34	7.40	642.23	1.87
13	0.97	98.67	7.40	642.23	1.91
14	0.97	98.67	6.65	413.45	1.29
15	0.97	98.67	0.75	2673.93	7.37
16	1.01	25.00	29.72	0.00	0.00
17	1.01	25.00	339.21	4.09	0.01
18	1.01	900.00	368.90	1006.40	1.54
19	1.01	300.00	368.90	292.91	0.69
20	1.01	284.82	368.90	276.18	0.66
21	1.01	167.35	368.90	149.04	0.41
22	1.01	136.27	368.90	116.00	0.33
23	1.01	25.00	1548.42	0.00	0.00
24	1.01	33.00	1548.42	33.44	0.11

Performance characteristics are developed with the source temperature below the critical temperature and above the critical temperature of the water. Fig. (3) is used to formulate the correlation for optimum HRSG or boiler pressure at the source temperature below the critical temperature. Since the source is a heat recovery with the constraints of AP, PP and TTD, the optimum HRSG pressure is different compared to the fuel fired power plants. The hot gas temperature is varied from 175 °C to 300 °C. The HRSG pressure is changed from the condenser pressure to the saturation pressure of source temperature. The increase in HRSG pressure permits more expansion and generated more power. But it also limits the steam generating capacity due to restriction with pinch point. Therefore, a rise and fall of power can be observed with increase in pressure. The optimum HRSG pressure is increasing with increase in source temperature. The pressure rise decreases the heat supply to the plant. Therefore, initially the thermal efficiency is increasing with increase in pressure. Efficiency is increasing with diminishing rate with the limit in steam capacity. The degree of superheat is analyzed with pressure and temperature. It increases with increase in temperature but decreases with increase in pressure. At the hot gas supply temperature of 300 °C, the resulted maximum power is approximately 13 MW at 5.8 bar.

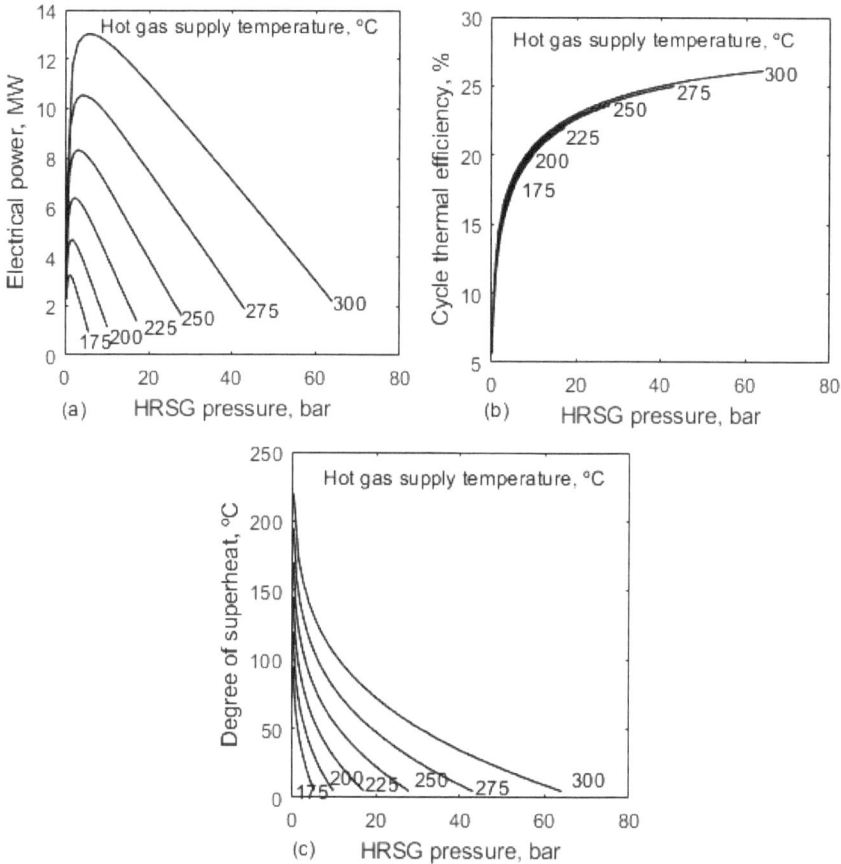

Fig. (3). Performance characteristics of SFC with single flasher with the source temperature below the critical temperature.

Following equation is developed for SFC with single flasher to evaluate the optimum HRSG's pressure from source temperature at maximum power.

$$P_{HRSG\ opt} = a_1 + a_2T + a_3T^2 \qquad (12)$$

where T is the hot gas (source) temperature in °C.

The resulted coefficients for the above non-linear equation at the source temperature below the critical temperature are.

$a_1 = 5.2036$, $a_2 = -0.0581$ and $a_3 = 0.0002$.

After solving the optimum HRSG pressure, next question is where to locate the flasher. To answer this question, Fig. (**4**) analyzed the flash temperature ratio with supply temperature. At each temperature, optimum HRSG pressure has been selected in this analysis. Eq. 8.1 defines the flash temperature ratio varied between condenser and HRSG. Similar to the HRSG pressure, the flash condition also different for maximum power and maximum thermal efficiency. The optimum flash temperature ratio is 0.5 for maximum power generation. The results are showing that optimum flash temperature ratio is mostly independent of the operational conditions. Therefore, there is no need further iteration of optimum HRSG pressure with flash temperature ratio. The optimum flash temperature ratio of 0.5 indicates that the flash temperature is the mid temperature of condenser temperature and HRSG's evaporator temperature. The optimum flash temperature ratio for maximum thermal efficiency is 0.8. It indicates high pressure flasher is required for maximum efficiency compared to the maximum power.

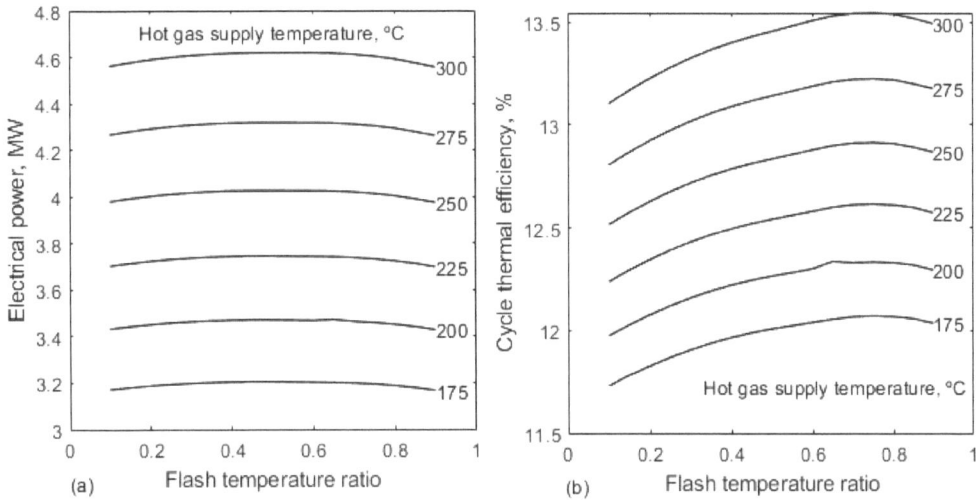

Fig. (4). Performance variations with flasher temperature ratio and source temperature below the critical temperature.

The optimum HRSG pressure with the source temperature above the critical temperature has been furnished in Fig. (**5**). The optimum pressure is increasing with increase in source temperature. The optimum pressure with source temperature above the critical temperature is more than the optimum pressure with source temperature below the critical temperature. At 400 °C of source temperature, the power production is 24.2 MW at the optimum pressure of 8.7 bar. Compared to the source temperature at 300 °C, this power is high. Therefore, SFC is highly recommended, if the source temperature is above the critical temperature of the water. The thermal efficiency also improves with the source temperature above the critical temperature.

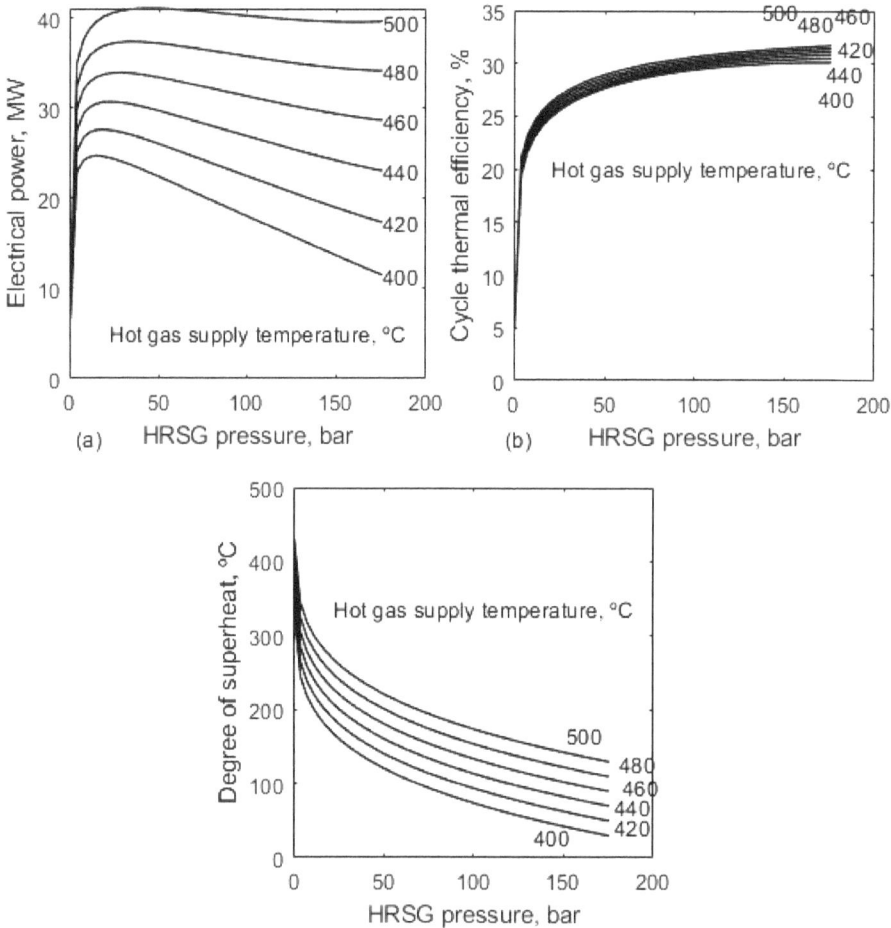

Fig. (5). Performance characteristics of SFC with single flasher with the source temperature above the critical temperature.

The correlated coefficients to find the optimum pressure for maximum power are as follows at above the critical temperature of source temperature with single flasher,

$a_1 = 185.8540$, $a_2 = -1.0054$ and $a_3 = 0.0014$.

Fig. (**6**) shows the flow diagram of SFC with two flashers. The thermodynamic cycle of SFC with two flashers has been depicted in temperature-entropy diagram Fig. (**7**). The hot gas from the process industry is supplied to HRSG where superheated steam is generated for power plant. In fuel fired thermal power plant, the source temperature is high and above 1000 °C. Therefore, steam pressure is high, either closed to critical pressure (221 bar) or above it. From the earlier section, it can be noticed that a low pressure is sufficient with the heat recovery power plant as the source temperature is very low compared to the firing option. The double flash in SFC results four pressures in the plant. In earlier case of single flash, we noticed three pressures. The four pressures with two flashers allows three stage expansion in the turbine. First the steam expands from high pressure (HRSG) to HP flash pressure (first flasher). In second stage, steam expands from HP flash pressure to LP flash pressure (second flasher). Finally, the steam expands from the LP flash pressure to condenser pressure. At the end of the turbine, steam condensed to the saturated liquid in a water cooled condenser. Two pumps are used to supply the water to HRSG. In the first pump, the pressure of condensate is raised from condenser pressure to LP flash pressure. At the exit of this pump, the liquid from LP flasher and pumped fluid are mixed. The mixed fluid is again pumped from the LP flash pressure to HRSG pressure. The HRSG has three sections *viz.* economizer, evaporator and superheater. The hot gas flow in counter flow direction from superheater to economiser. In this analysis, 25% of the pumped fluid has been used for flashing. This fluid is supplied to the HP flasher. The balance fluid is supplied to evaporator and superheater for power production. The fluid is flashed in the HP flasher from HRSG pressure to HP flash pressure. After flashing, the liquid and vapour are separated. The vapour is rooted to turbine and the liquid is connected to the LP flasher. In the LP flasher, the liquid is flashed from HP flash pressure to LP flash pressure. After this LP flasher, again the liquid and vapour are separated. The vapour goes to turbine and the liquid is mixed with the pumped fluid. So this cycle repeats for continuous power production. The distinctiveness of SFC is addition of extra steam to turbine. It is opposite phenomena of removal of steam from the turbine in a regular thermal power plant with a feedwater heater. These two processes are opposite in nature. Therefore, SFC supports the power and regular thermal power plant supports the efficiency. For heat recovery power plant, SFC is more suitable than the conventional power plant.

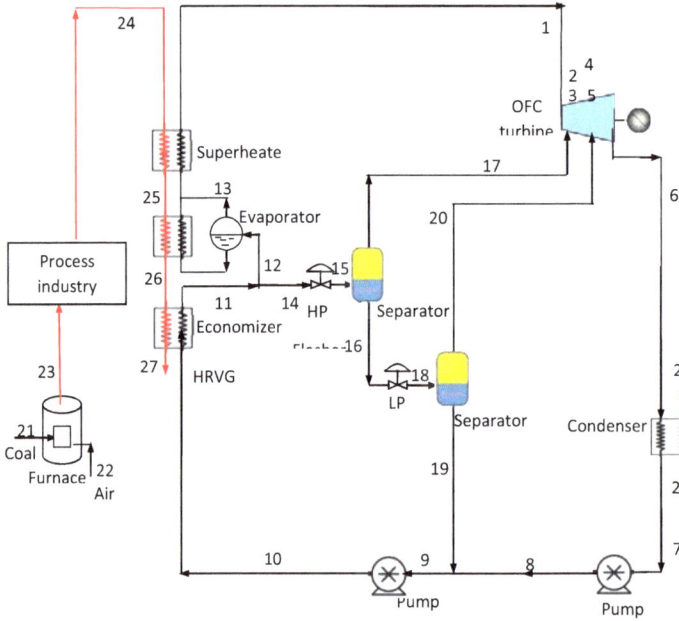

Fig. (6). SFC with two flashers.

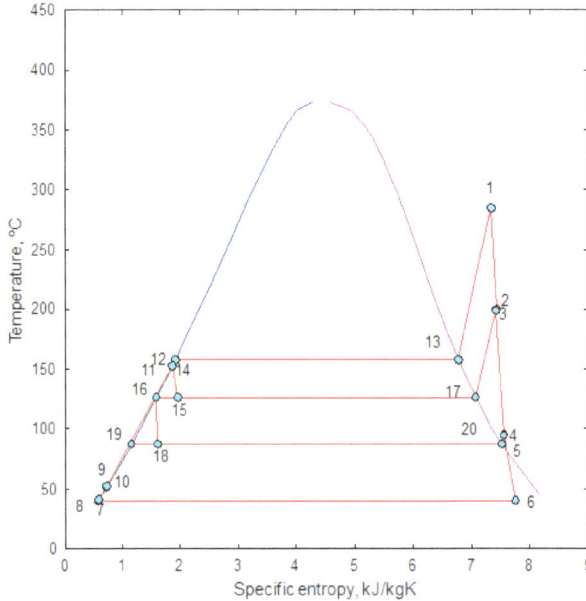

Fig. (7). Temperature-specific entropy diagram for SFC with two flashers.

In Fig. (**7**), the flash process in HP flasher and LP flasher has been shown as a vertical line, which expander from one pressure to another. The vertical line is bending right as entropy generated during the throttling (flashing) which is an irreversible process. After the flashing, the separation has been shown as a horizontal line, the left state shows the separated portion of liquid and the right side indicates the vapour state. As per the lever rule, the length toward the saturated liquid state is proportional to the vapour mass. This small length reveals that the vapour separated after the flash is a small amount. Even though it is little, a considerable improvement in the power can be observed in the subsequent performance characteristics.

The thermodynamic work for two flash SFC is similar to the single flasher.

The steam capacity in the boiler is determined from the energy balance applied to evaporator and superheater of HRSG.

$$m_1 = m_{12} = \frac{m_{24}(h_{24} - h_{27})}{h_1 - h_{12}} \tag{13}$$

Flashing is an irreversible expansion process from high pressure to low pressure into liquid-vapour mixture. It is an isenthalpic process (throttling). Therefore, the enthalpy before and after the flashing is same. The entropy increases due to irreversibility.

The enthalpy balance in a HP flasher,

$$h_{14} = h_{15} = h_{16} + d_{15}(h_{17} - h_{16}) \tag{14}$$

Similarly, the enthalpy balance in a LP flasher,

$$h_{16} = h_{18} = h_{19} + d_{18}(h_{20} - h_{19}) \tag{15}$$

The dryness fraction of steam, 'd' after flash is determined from Eq. 8.14 and Eq. 8.15 respectively for HP flasher and LP flasher.

For double flash SFC, let FMR is the flash mass ratio as per the following,

$$FMR = \frac{m_{14}}{m_{11}} = \frac{m_{14}}{m_{12} + m_{14}} \tag{16}$$

The simplification of Eq. 8.16 as follows to find the flash mass, m_{14},

$$m_{14} = FMR\, m_{12} + m_{14}\, FMR$$

$$m_{14}(1 - FMR) = FMR \, m_1$$

$$m_{14} = m_1 \frac{FMR}{(1 - FMR)} \tag{17}$$

The energy balance in the economizer results the exhaust gas temperature as follows:

$$h_{27} = h_{26} - \frac{m_{11}(h_{11} - h_{10})}{m_{26}} \tag{18}$$

For optimum location of HP flasher and LP flasher, the flasher position is defined as a temperature ratio. Let the temperature ratio in HP flasher and LP flasher are θ_{HPF} and θ_{LPF} respectively.

$$\theta_{HPF} = \frac{T_{HPF} - T_{LPF}}{T_{HRSG} - T_{LPF}} \tag{19}$$

$$\theta_{LPF} = \frac{T_{LPF} - T_{cond}}{T_{HPF} - T_{cond}} \tag{20}$$

The temperatures, T_{HRSG} and T_{cond} are HRSG evaporator temperature and condenser temperature respectively.

From the θ_{HPF} and θ_{LPF} the HP flash temperature and LP flash temperature are determined. These two temperatures are the saturation temperatures are used to find the HP flash pressure and LP flash pressure from the simplification of Eq. 8.19 and Eq. 8.20.

In Eq. 8.19, T_{LPF} is eliminated as follows:

$$\theta_{HPF} = \frac{T_{HPF} - \left(T_{cond} + \theta_{LPF}(T_{HPF} - T_{cond})\right)}{T_{HRSG} - \left(T_{cond} + \theta_{LPF}(T_{HPF} - T_{cond})\right)}$$

$$= \frac{T_{HPF} - T_{cond} - \theta_{LPF}T_{HPF} + \theta_{LPF}T_{cond}}{T_{HRSG} - T_{cond} - \theta_{LPF}T_{HPF} + \theta_{LPF}T_{cond}}$$

$$(1 - \theta_{LPF} + \theta_{HPF}\theta_{LPF})T_{HPF} = \theta_{HPF}T_{HRSG} + (-\theta_{HPF} + \theta_{HPF}\theta_{LPF} - \theta_{LPF} + 1)T_{cond}$$

$$T_{HPF} = \frac{T_{HRSG}\theta_{HPF} + T_{cond}(1 - \theta_{HPF} - \theta_{LPF} + \theta_{HPF}\theta_{LPF})}{(1 - \theta_{LPF} + \theta_{HPF}\theta_{LPF})} \qquad (21)$$

In Eq. 8.20, T_{HPF} is eliminated as follows:

$$\theta_{LPF} = \frac{T_{LPF} - T_{cond}}{T_{LPF} + \theta_{HPF}(T_{HRSG} - T_{LPF}) - T_{cond}}$$

$$= \frac{T_{LPF} - T_{cond}}{T_{LPF} + \theta_{HPF}T_{HRSG} - \theta_{HPF}T_{LPF} - T_{cond}}$$

$$\theta_{HPF}\theta_{LPF}T_{HRSG} + (1 - \theta_{LPF})T_{cond} = (1 - \theta_{LPF} + \theta_{HPF}\theta_{LPF})T_{LPF}$$

$$T_{LPF} = \frac{T_{HRSG}\theta_{HPF}\theta_{LPF} + T_{cond}(1 - \theta_{LPF})}{(1 - \theta_{LPF} + \theta_{HPF}\theta_{LPF})} \qquad (22)$$

From Eq.8.21 and Eq.8.22, the HP flash temperature and LP flash temperature are the function of HRSG's evaporator temperature, condenser temperature, temperature different ratio in HP flasher and temperature difference in LP flasher.

The net power from SFC with double flasher,

$$W_{net} = W_t - W_p$$

$$= m_1(h_1 - h_2) + m_3(h_3 - h_4) + m_5(h_5 - h_6) - m_7(h_8 - h_7) - m_9(h_{10} - h_9) \qquad (23)$$

The exergy balance at each component of SFC with two flashers gives the irreversibility of the component/process.

$$\varepsilon_{in} = \varepsilon_{out} + I_{component} \qquad (24)$$

The total irreversibility of the SFC,

$$\sum I_{components} = I_{HRSG} + I_t + I_{HP\,Flasher} + I_{LP\,Flasher} + I_{cond} + I_p + I_{ex} + I_{hw} \qquad (25)$$

Thermal efficiency is the ratio of power to heat supply to the cycle. The heat supply is the energy supply for superheat steam generation from feed water.

Thermal efficiency of SFC cycle,

$$\eta_1 = \frac{W_{net}}{m_{10}(h_{11} - h_{10}) + m_1(h_1 - h_{12})} \times 100 \tag{26}$$

The exergy efficiency of SFC cycle,

$$\eta_2 = \frac{\varepsilon_{hot\,gas} - \sum I_{components}}{\varepsilon_{hot\,gas}} \times 100 \tag{27}$$

Table **2** presents the results of the above formulation applied to the SFC with two flashers at the source temperature of 300 °C. The HRSG pressure, HP flasher pressure, LP flasher pressure and condenser pressure are respectively 4.29 bar, 1.47 bar, 0.39 bar and 0.07 bar. The exhaust gas temperature is 125 °C. By increasing the flash mass, the exhaust gas temperature can be decreased. The liquid in economizer is 31.38 kg/s. Out of this, 8.85 kg/s of liquid is used for the flashing. The balanced fluid, 23.54 is supplied to evaporator and superheater. From the 7.85 kg/s of flashed mass, the fluid available for LP flasher is 7.39 kg/s. The vapour supplied to steam turbine from HP flasher and LP flasher is 0.46 kg/s and 0.47 kg/s respectively.

Fig. (**8**) analyzes the SFC performance with hot gas supply temperature and HRSG pressure. The hot gas supply is below the critical temperature of water. The optimum HRSG pressure is increasing with increase in source temperature. At the source temperature of 300°C, the maximum power is 13 MW. The comparison of performance from single flasher to double flasher with source temperature below the critical temperature, a minor improvement only observed. The performance comparison with source temperature above the critical temperature is different and discussed in the next sections.

The correlated coefficients to find the optimum HRSG pressure of SFC with double flasher and temperature below the critical temperature are:

$a_1 = 6.1499$, $a_2 = -0.0662$ and $a_3 = 0.0002$.

Fig. (**9**) studies the location of HP flasher and LP flasher in SFC plant at the source temperature of 300 °C. The optimum HRSG pressure at this source temperature has been selected to develop the optimum condition for HP flasher and LP flasher. The location of HP flasher and LP flasher is expressed as a temperature ratio, defined in Eq. 8.19 and Eq. 8.20 respectively. The LP flash temperature ratio is changed from 0.1 to 0.9 and the HP flash temperature ratio is varied from 0.2 to 0.8. The HP flash temperature is changed between HP flash temperature and HRSG saturation temperature. Similarly, the LP flash temperature is changed between LP flash temperature and HP flash temperature. For maximum power, the optimum LP flash temperature ratio is decreasing with increase in HP flash temperature ratio. Therefore, the combination of low LP flash temperature with high HP flash temperature or high LP flash temperature with low HP flash temperature result the maximum power. A mid temperature of HP flasher and LP flasher results a great power. The maximum efficiency condition demands the higher side of the HP flash temperature and LP flash temperature. The exhaust gas temperature is increasing with increase in LP flash temperature and HP flash temperature.

Table 2 . Thermodynamic properties of SFC with two flashers and hot gas supply temperature at 300 °C.

State	P, bar	t, °C	m, kg/s	h, kJ/kg	s, kJ/kg K
1	4.29	285.00	23.54	3035.66	7.48
2	1.47	184.78	23.54	2842.67	7.59
3	1.47	183.37	23.99	2839.81	7.58
4	0.39	82.54	23.99	2649.97	7.72
5	0.39	82.40	24.47	2649.70	7.72
6	0.07	40.00	24.47	2453.78	7.87
7	0.07	40.00	24.47	167.45	0.57
8	0.39	41.00	24.47	171.64	0.59
9	0.39	48.65	31.38	203.62	0.69
10	4.29	49.68	31.38	207.90	0.70
11	4.29	141.16	31.38	594.07	1.75
12	4.29	141.16	23.54	594.07	1.75
13	4.29	146.16	23.54	2740.75	6.87
14	4.29	141.16	7.85	594.07	1.75
15	1.47	110.77	7.85	594.07	1.85
16	1.47	110.77	7.39	464.58	1.43
17	1.47	110.77	0.46	2692.46	7.23
18	0.39	75.39	7.39	464.58	1.45
19	0.39	75.39	6.91	315.55	1.02
20	0.39	75.39	0.47	2636.04	7.68
21	1.01	25.00	29.72	0.00	0.00
22	1.01	308.15	339.21	4.09	0.01

(Table 2) cont.....

State	P, bar	t, °C	m, kg/s	h, kJ/kg	s, kJ/kg K
23	1.01	900.00	368.90	1006.40	1.54
24	1.01	300.00	368.90	292.91	0.69
25	1.01	282.92	368.90	274.10	0.66
26	1.01	156.17	368.90	137.14	0.38
27	1.01	125.19	368.90	104.29	0.30
28	1.01	25.00	1672.82	0.00	0.00
29	1.01	33.00	1672.82	33.44	0.11

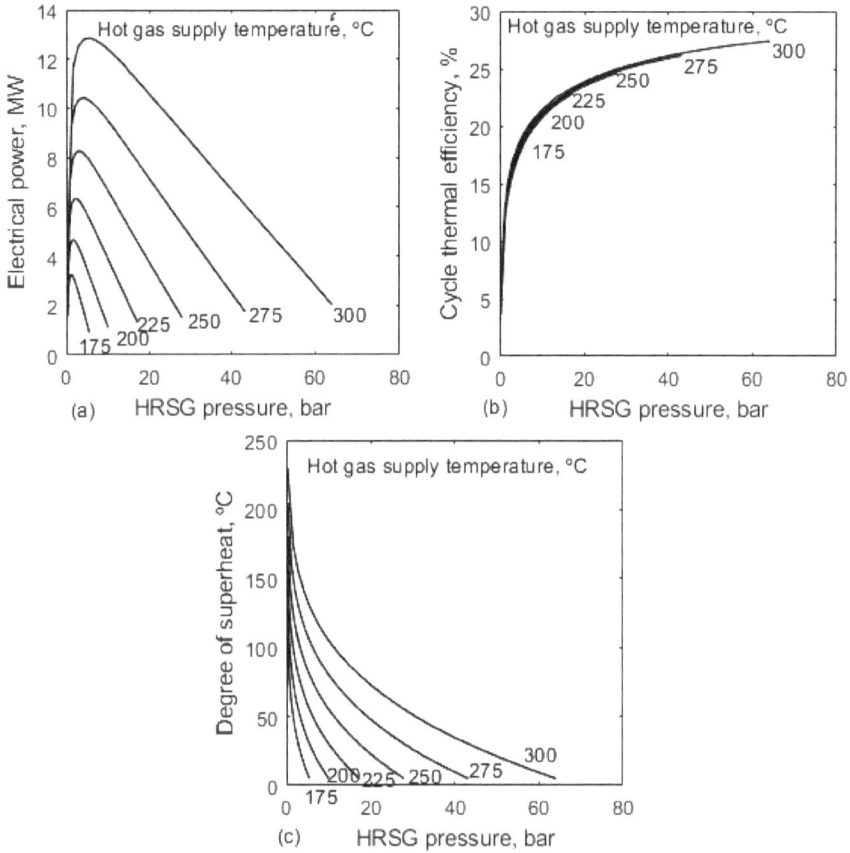

Fig. (8). Performance characteristics of SFC with two flashers with turbine inlet temperature below the critical temperature.

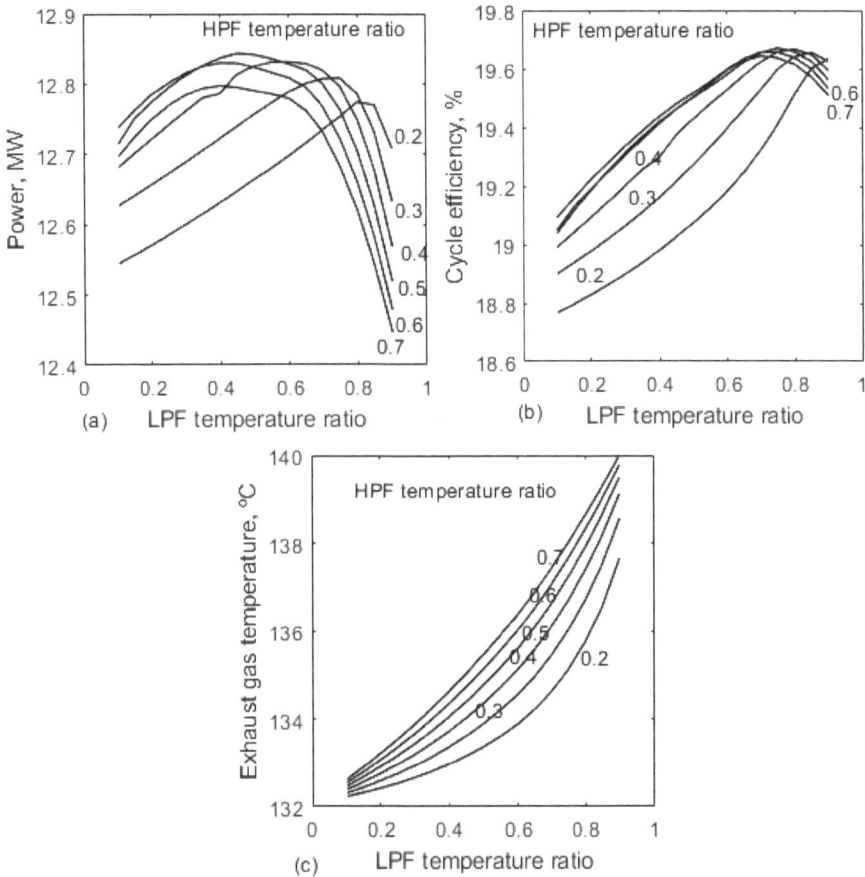

Fig. (9). Optimum location for low pressure flasher and high pressure flasher in a double flash SFC plant with source temperature below the critical temperature (300 °C).

Fig. (**10**) depicts the role of hot gas supply temperature above the critical temperature. Similar to single flash results, the temperature above the critical state boosting the performance. The optimum HRSG pressure is increasing with increase in HRSG pressure. The power augmentation from single flasher to double flasher is also at the satisfactory level with source temperature above the critical temperature.

To find the optimum HRSG pressure with the source temperature above the critical temperature in a SFC with two flashers, the following coefficients are developed.

$a_1 = 400.7293$, $a_2 = -1.9710$ and $a_3 = 0.0025$.

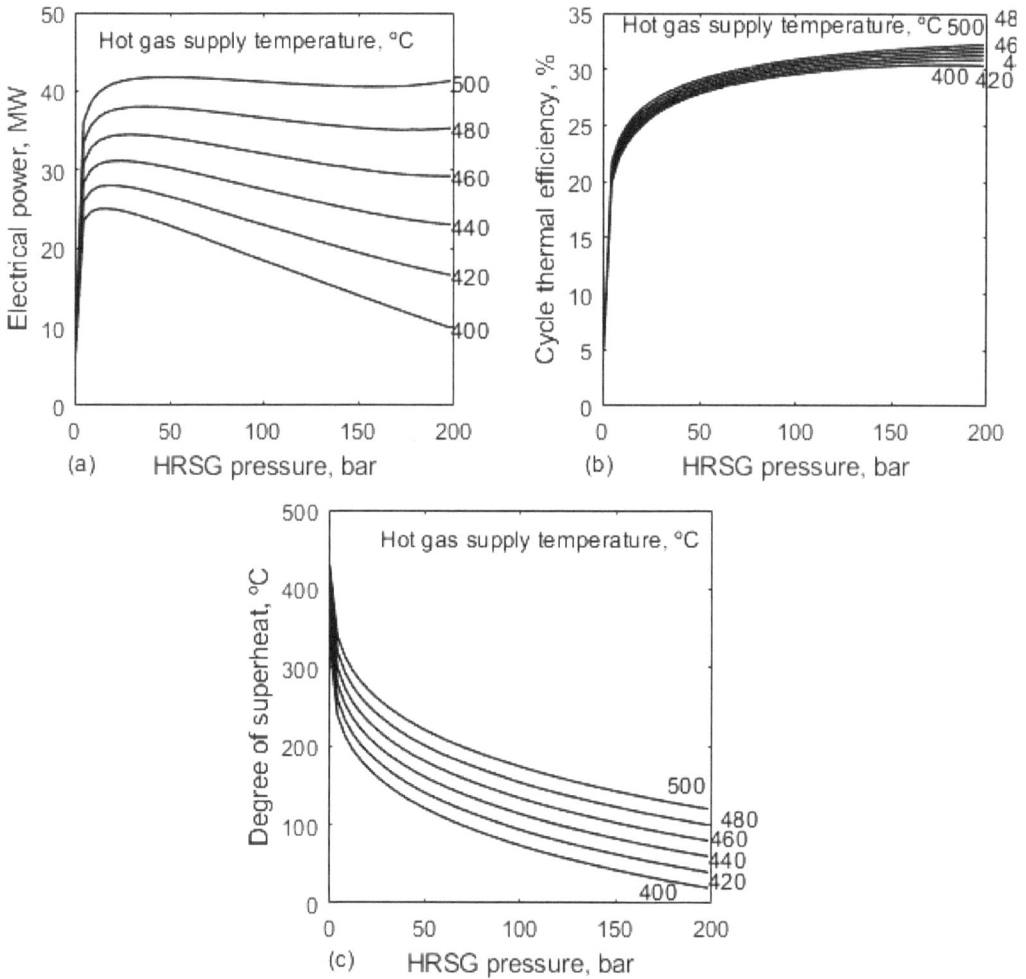

Fig. (10). Performance characteristics of SFC with two flashers with turbine inlet temperature above the critical temperature.

Fig. (**11**) analyzes the place of HP flasher and LP flasher at the source temperature of 400 °C.

Fig. (**12**) shows the role of flash mass ratio with source temperature on the SFC performance with two flashers. These results are developed with the source temperature below the critical temperature. The studied results are (a) SFC power with two flashers (b) thermal efficiency of SFC cycle and (c) exhaust gas temperature. Flash mass ratio is expressed as the mass ratio of liquid used for flashing to the total mass in economizer. Optimum HRSG pressure is used at the source temperature. The power is increased with increase in source temperature and flash mass ratio. It is in increasing order with increase in mass ratio. Since, the flasher will not favor the thermal efficiency, the efficiency is decreasing with increase in mass ratio. The exhaust gas temperature also decreasing with increase in mass ratio. Even though, a considerable power boost with increase in flash mass, it should be in the limits, keeping in mind on exhaust gas temperature.

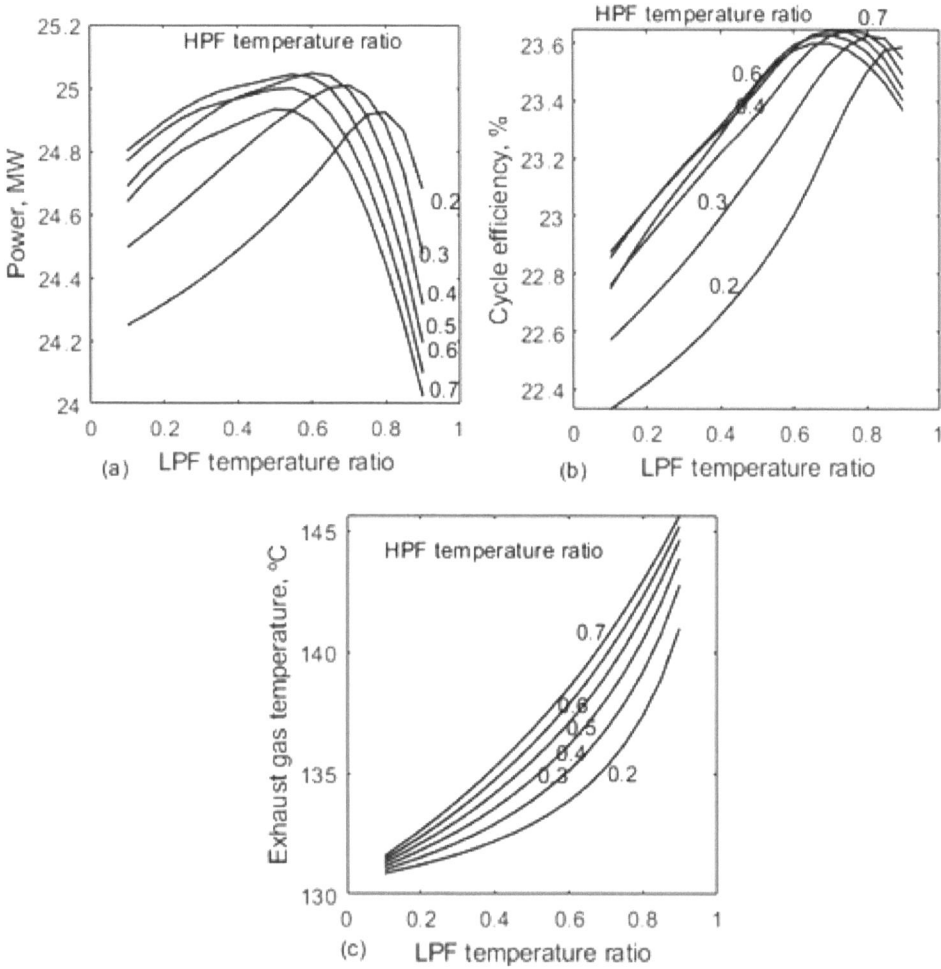

Fig. (11). Optimum location for low pressure flasher and high pressure flasher in double flash SFC plant with source temperature above the critical temperature (400 °C).

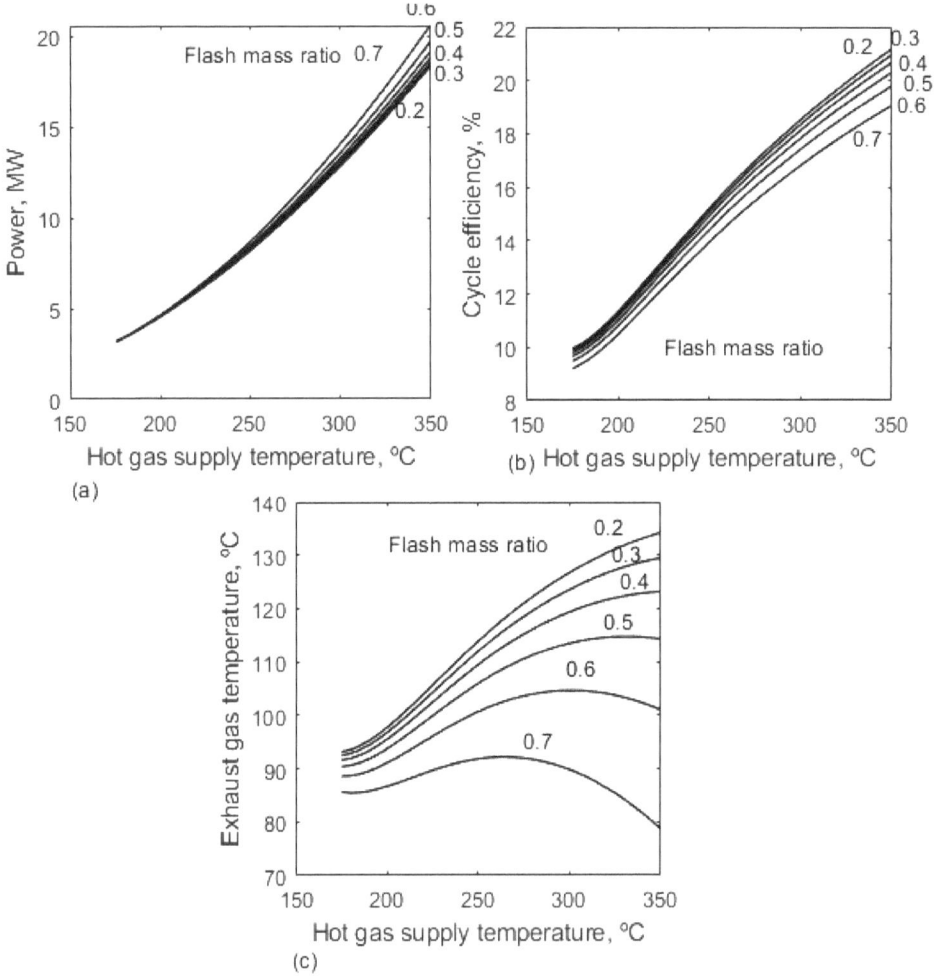

Fig. (12). Influence of quantity of flashed fluid in a double flash SFC plant with a change in source temperature below the critical temperature.

Fig. **(13)** compares the exergy losses (decay) in a regular thermal power plant (SRC) and SFC with two flashers at the source temperature of 300 °C. The exergy losses are expressed in percent of exergy of hot gas. The exergy losses in HRSG, turbine, condenser, pump, exhaust, hot water from condenser and deaerator (SRC)/flasher (SFC) are presented. The major portion of exergy loss is associated with exhaust in a SRC plant. The heat recovery from hot gas is increased with the use of flasher. Therefore, the exergy loss in exhaust is decreased with SFC. By properly choosing the flash fluid from the economizer, the exhaust gas temperature can be controlled at minimum level. So the exergy loss through exhaust also can be minimized with SFC. Since the working fluid is a single pure substance, the exergy losses in HRSG has significant influence. Compared to single fluid system, dual fluid cycle, such as Kalina power plant results lower exergy loss in heat recovery. The exergy losses in pump, hot water, flasher and deaerator are minor compared to the other components. The deaerator exergy loss is too small compared to the exergy with flashing. After the exhaust and HRSG, a considerable exergy loss has been observed in the turbine. The power is expressed as a fractional output from the exergy of hot gas. Therefore, it is the exergy efficiency of the cycle. The power has been enhanced with SFC over SRC. Nearly 2% improvement in exergy efficiency with SFC has been observed. The improvement is with the drop in exhaust losses by 4%.

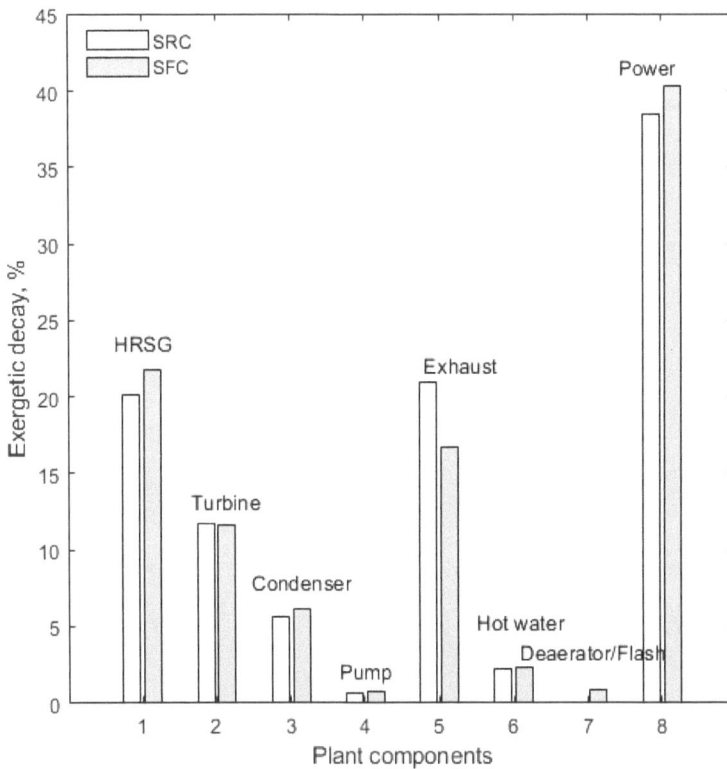

Fig. (13). Exergy analysis of steam Rankine cycle and double flashed SFC.

Fig. **(14)** consolidates the energy balance results in (a) Sankey diagram and exergy balance results in (b) Grassmann diagram for a SFC with two flashers. The energy supply and exergy supply to the SFC are determined from the flow, gas composition and temperature of hot gas. The energy supply to the SFC is 108.1 MW. From this supply of energy, the major energy loss is 55.93 MW (51.8%) at the condenser and the balanced reject is 38.47 MW (35.6%) with the exhaust gas.

The power production is 14 MW which is 12.6% of energy supply (thermal efficiency). The exergy losses in all the components are shown in the Grossman diagram (Fig. **14b**). The exergy supply with hot gas is 31.8 MW. The major exergy losses are at HRSG (6.92 MW, 21.8%), exhaust (5.32 MW, 16.7%), and turbine (3.69 MW, 11.6%). In contract to the energy results, the exergy loss in condenser is 1.94 MW (6.1%). It is due to the closeness of heat rejection to the sink. The other heat losses in pump, flasher and hot water from the condenser are minor.

Fig. (14). (a) Sankey diagram and **(b)** Grassmann diagram of SFC with two flashers.

GENERALIZED SOLUTION FOR MULTI FLASHERS

This section is focused on solution of SFC with multiple flashers. The evaluation of SFC with more number of flashers is a tedious process and consumes more time. To simplify this complex problem, generalized solution has been developed for multi flashed SFC plant. This work is useful for optimization of number of flashing units in SFC.

Fig. (**15**) shows the schematic layout of SFC with four flashers. The hot gas is connected to HRSG where superheated steam is generated in HRSG. Turbine receives the superheated steam from the HRSG and saturated steam from the flashers. The expanded steam is completely condensed into saturated liquid and followed by pumping to HRSG. The designer may easily substitute the number of flashers to generate the SFC results from the developed common formulae by substituting the total number of flashers. If 'n' is the number of flashers, n + 1 pressure levels exist in the cycle. It includes the HRSG pressure and condenser pressure. The partial amount of pressurized hot water from the economizer section of HRSG is used for flashing. The flashed fluid or wet steam consists of water and saturated vapor. The saturated vapor is supplied to the turbine from subsequent stages of flashers to the turbine. The liquid part of wet steam is supplied to the next flasher to continue the flashing to the next pressure.

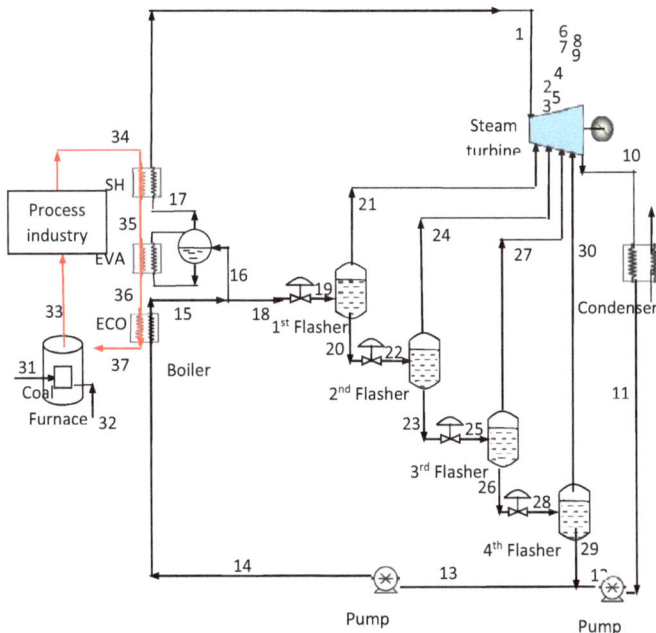

Fig. (15). Arrangement of four flashers in SFC.

Fig. (**16**) shows the temperature-entropy diagrams for (a) regular steam power plant and (b) SFC with multi flashers (four in this plot). The arrow lines indicate the direction of steam flow into the turbine and/or from the turbine. In a deaerator configuration, one steam line is connected to the open feed water heater. In Fig. (**16b**), four steam lines are connected to the turbine from four flashers in addition to HRSG steam. The heat supply to evaporator and superheater is same for the two plants under the comparison. Therefore, the steam supply from HRSG is same. The flashers add the extra steam to turbine. In SFC, the economizer load increases over the regular plant.

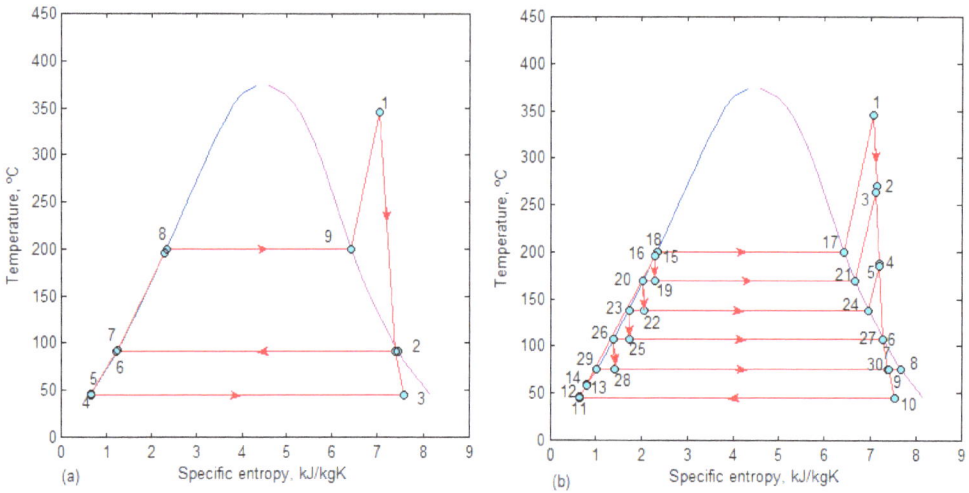

Fig. (16). Temperature-entropy diagram of (**a**) regular steam power plant with a deaerator (SRC) and (**b**) SFC with four flashers.

Fig. (**17**) shows the concept of generalization of 'n' flashers in SFC. It is used to develop the mathematical formulate with n flashers and i[th] (instantaneous) flasher. After studying the regular changes in the state points, all the state properties are formulated as a function of 'n' and 'i'. It will help the designer to find the performance of SFC with 'n' flashers without repeating the mass and heat balance equations. It also saves the computational time and simplifies the procedure.

The energy balance in HRSG with evaporator and superheater results the steam capacity.

$$m_{2(n+4)} = \frac{m_{14+5n}(h_{14+5n} - h_{16+5n})}{h_1 - h_{2(n+4)}} \qquad (28)$$

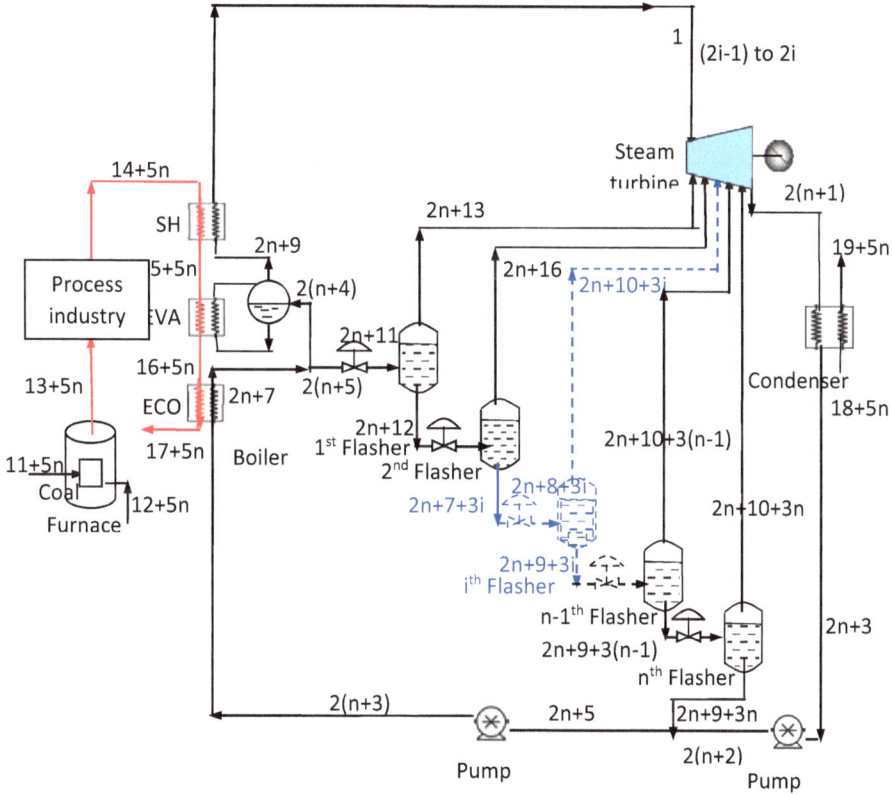

Fig. (17). Generalization of 'n' number of flashers in SFC plant, n: total number of flashers, i: i[th] flasher.

Flashing is a process of expansion of pressurized liquid into liquid and vapor mixture. The enthalpy of saturated liquid before flashing is equal to the total enthalpy of mixture after flashing.

$$h_{2n+7+3i} = h_{2n+8+3i} = h_{2n+9+3i} + d_{2n+8+3i}(h_{2n+10+3i} - h_{2n+9+3i}) \qquad (29)$$

The dryness fraction of steam at state, $2n + 8 + 3i$ can be determined from Eq. 8.29.

$$FMR = \frac{m_{2(n+5)}}{m_{2n+5}} = \frac{m_{2(n+5)}}{m_1 + m_{2(n+5)}} \qquad (30)$$

after the simplification of Eq. 8.30,

$$m_{2(n+5)} = m_1 \frac{FMR}{1-FMR} \tag{31}$$

From the energy balance equation in HRSG' economizer,

$$T_{14+5n} = T_{13+5n} - \frac{m_{2(n+3)}(h_{2n+7} - h_{2(n+3)})}{m_{13+5n}(h_{13+5n} - h_{14+5n})} \tag{32}$$

The saturation temperature of wet steam at the exit of flasher,

$$T_{2n+8+3i} = \frac{n+1-i}{n+1} \times \frac{T_{2n+8+3i} - T_{2n+3}}{T_{2n+9} - T_{2n+3}} \tag{33}$$

$T_{2n+8+3i}$ is the flasher temperature, T_{2n+9} is the boiler saturation temperature and T_{2n+3} is the condenser temperature.

The net power from the plant with HP flasher and LP flasher is.

$$W_{turbine} = \sum_{i=1}^{n+1} m_{2i-1}(h_{2i-1} - h_{2i}) \tag{34}$$

$$W_{pump} = m_{2n+3}(h_{2(n+2)} - h_{2n+3}) + m_{2n+5}(h_{2(n+3)} - h_{2n+5}) \tag{35}$$

$$W_{net} = W_{st} - W_p \tag{36}$$

The energy (thermal) efficiency of cycle,

$$\eta_1 = \frac{W_{net}}{m_{2(n+3)}(h_{2n+7} - h_{2(n+3)}) + m_{2(n+4)}(h_1 - h_{2(n+4)})} \times 100 \tag{37}$$

The exergy (second law) efficiency of cycle,

$$\eta_2 = \frac{E_{hot\,gas} - \sum I_{components}}{E_{hot\,gas}} \times 100 \tag{38}$$

In this section, the influence of FMR, hot gas temperature and HRSG pressure are studied with number of flashers. The exergy decays (irreversibilities) in the plant's components are compared to find the strong and weak areas. The focused results are power, energy efficiency, and dryness fraction of steam and exhaust gas temperature.

Fig. (**18**) studies the effect of FMR with number of flashers on (a) power, (b) energy efficiency, (c) dryness fraction of steam at turbine exit and (d) exhaust gas temperature. The results show that there is a considerable change in operational and performance results up to two flashers. The results also compare with and without flashing in a steam power plant. There is no much significant improvement with more than two flashers. The augmentation in power is increasing with increase in FMR but the exhaust gas temperature is decreasing. So the FMR is restricted according to the required exhaust gas temperature. Since the current focus is on maximum heat recovery, the results are focused on maximum power. The lowest exhaust gas temperature is 100 °C at 0.6 FMR. A further increase in mass ratio drops the gas temperature below the 100 °C and not is recommendable due to thermal head needed to expel the gas through the chimney or stack.

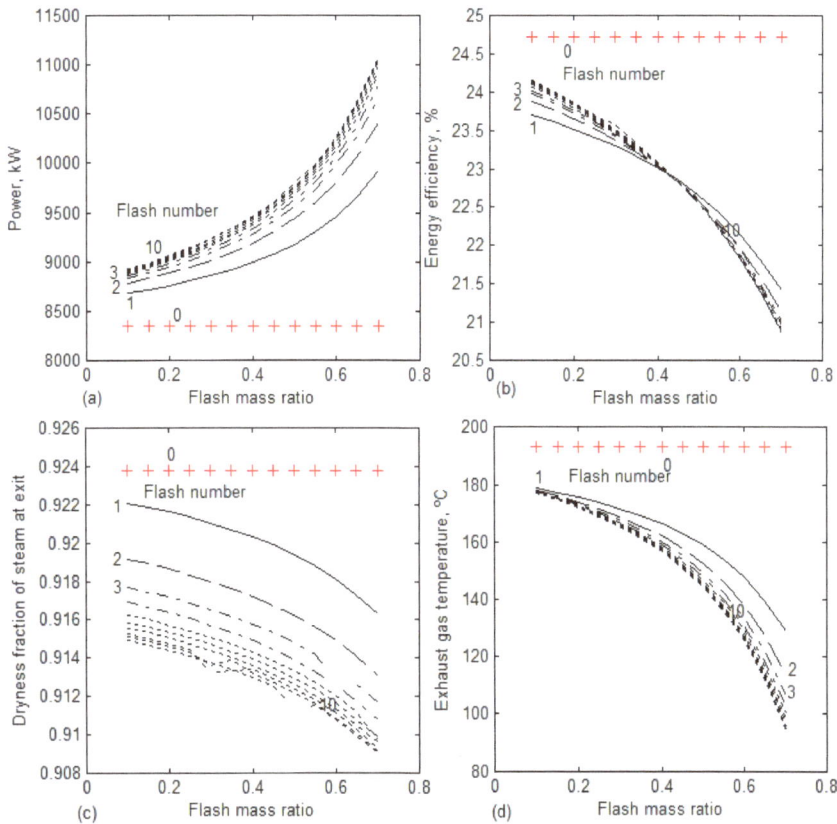

Fig. (18). Influence of number of flashers and FMR on **(a)** power generation, **(b)** energy efficiency, **(c)** dryness fraction of steam at turbine exit and **(d)** exhaust gas temperature.

Fig. (**18b**) shows loss in energy efficiency with flashers and also compares with the efficiency of regular steam power plant. The SFC results are compared with the SRC plant. The improvement in power and drop in efficiency of the plant are shown in Fig. (**18a**) and Fig. (**18b**) respectively. The adding flasher in SFC enhances the efficiency up to with 0.4 FMR. Beyond this, the efficiency is decreasing with increase in number. For the current system, exhaust gas temperature limits the flash fluid. Dryness fraction of steam at exit of the turbine should not fall below 0.85 to save the life of the turbine blades. The results show that the dryness fraction decreases with increase in flashers and FMR. But there is no much drop in dryness fraction of steam with the above changes. The dryness fraction of steam is limited to 0.91 which is safe in condition in the specified range. So the dryness fraction is not the big issue with use of flashers.

Fig. (**19**) shows the influence of heat source temperature to power system with a change in flashers. It also compares the details of plant operational conditions with and without flasher. The augmented power has been founded with the flashers compared to the regular plant (zero flash) in Fig. (**19a**). The efficiency is increasing with an increase in source temperature. A significant increment has been found with the two flashers and the increasing rate is diminishing with the increase in number. Since the flashing system demands more heat recovery, the energy efficiency will drops drastically as shown in Fig. (**19b**). The dryness fraction of steam at exit of turbine decreases with increase in flashers. The drop in dryness fraction increases with increase in flashers and hot gas inlet temperature. The minimum dryness fraction within the range of operation conditions is approximately 0.89. It is in safe limit. Since the flashing increases heat recovery, the exhaust gas temperature is decreasing with increase in flash points and hot gas inlet temperature.

Fig. (**20**) shows the influence of HRSG pressure with flasher's number on power, energy efficiency, steam's dryness fraction and exhaust gas temperature. There is a considerable improvement in power has been seen up to the double flasher. The augmentation of power is in diminishing rate with increase in the number of flashers. The power is dropping with increase in HRSG pressure as the pinch point restricts the heat recovery. There is a considerable scope of heat recovery for more power at the low pressure. Since the selected FMR (0.5) is nearer to the critical flash mass ratio (0.4), there is no much changes in the energy efficiency with flashers. The dryness fraction is decreasing and exhausts gas temperature is increasing with increase in HRSG pressure. Therefore, a low pressure in HRSG has been recommended to result more power and dry steam at the exit of turbine.

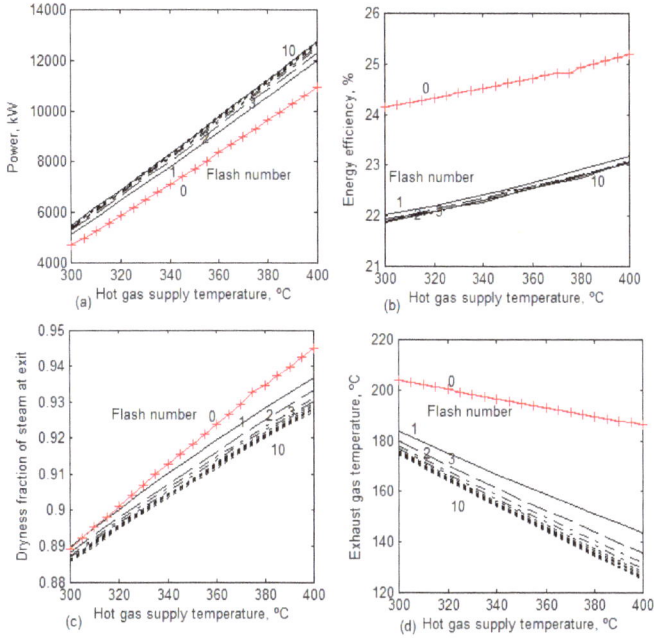

Fig. (19). Influence of hot gas supply temperature with number of flashers on **(a)** power generation, **(b)** energy efficiency, **(c)** dryness fraction of steam at turbine exit and **(d)** exhaust gas temperature.

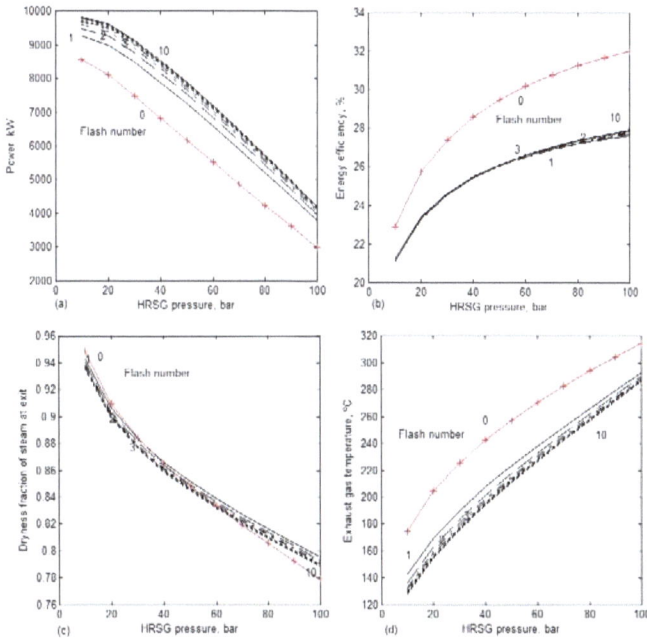

Fig. (20). Influence of flashers and HRSG pressure on **(a)** power generation, **(b)** energy efficiency, **(c)** dryness fraction of steam at turbine exit and **(d)** exhaust gas temperature.

Fig. (**21**) shows the influence of number of flashers on exergy decay of plant components. It has been noticed that the major exergetic loss is with the loss of exhaust gas and the minimum is with the flashers and pump. It also shows that the exergy decay with exhaust can be controlled by adopting the flashing. The HRSG loss is increasing with increase in flashes as it increases the waste heat recovery.

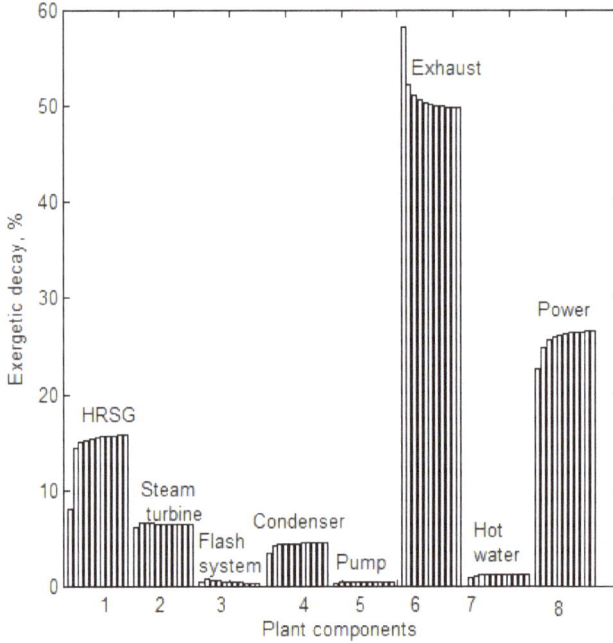

Fig. (21). Effect of increase in flashers (0-10) on component's exergetic decays.

The condenser and steam turbine losses in SFC are significant. The increment in power is shown as exergy efficiency with the flashing.

Fig. (**22**) compares the key parameters and performance details of SFC from zero flash (non-SFC) to 10 flashers. These results are generated at a fixed hot gas flow of 500000 Nm³/h and 360 °C. A FMR, 0.5 indicates that half of the preheated water from the economizer is supplied to flashers for the power augmentation. The maximum dryness fraction in a regular plant is 0.924 and minimum is 0.911 with 10 flashers. The exhaust gas temperature is 193 °C in a regular plant due to improper heat recovery. It is 144 °C with 10 flashers. The maximum drop in exhaust gas temperature is observed with two flashers *i.e.* from 193 °C to 152 °C. Due to addition of steam from the flashers to turbine, the condenser load increases with increase in flash number. The steam flow in the condenser is HRSG steam minus the deaerator steam in a regular power plant. The condensate is increasing from 11.33 kg/s without flashing to 15.21 kg/s with 10 flashers. Similar to exhaust gas temperature, the heat recovery changed from 33.76 MW to 43.4 MW from zero to 10 flashers. The power change is from 8.3 MW to 9.8 MW with flash number (0-10). Nearly 10% power augmentation has been obtained with the first flasher and an additional 3% with second flasher. But a maximum of 18% increase in power has been found with 10 flashers. Similarly, the improvement in exergy efficiency is shown with power boost at fixed exergy supply. Not much considerable change in thermal efficiency (energy efficiency) is found with increase in flashers. The adoption of flashers increases the size of heat recovery, condenser and pumps per unit output. The additional equipment cost returns after the payback period due to generation of excess power from waste heat instead of depending on fuel firing.

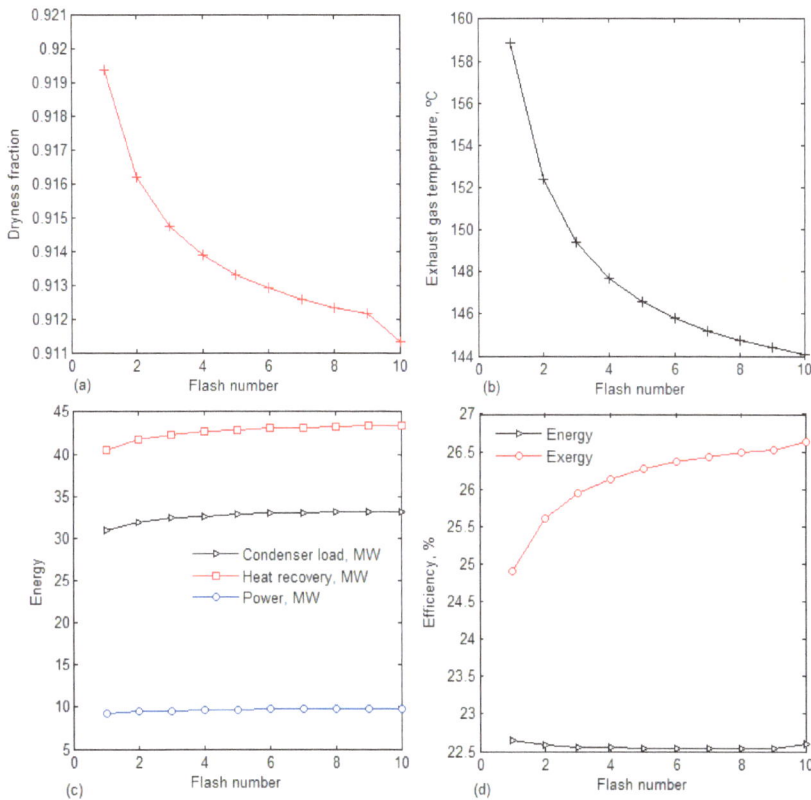

Fig. (22). Influence of adding flashers on operating conditions (turbine dryness fraction, exhaust gas temperature, condenser load and HRSG load) and performance (power, energy efficiency and exergy efficiency) conditions.

CASE STUDY ON SFC

A case study on SFC has been conducted to understand the actual power plant process and compare with the existing power generation system. A cement factory located at Telangana, India is generating its own electricity by recovering the heat from the furnace which is used for cement production. The power plant is working on SFC mode with two flashers. The plant operational data has been used to validate the current thermodynamic work. The plant results are also compared with the existing conventional thermal power plant.

Fig. (**23**) is the regular thermal power plant working on Rankine cycle. The heat source has been arranged as per the power plant layout adopted from case study. This layout is prepared to compare the performance of case study's plant with the existing plant. The power cycle in Fig. (**23**) consists of two HRSGs, steam turbine, water cooled condenser, two pumps and deaerator.

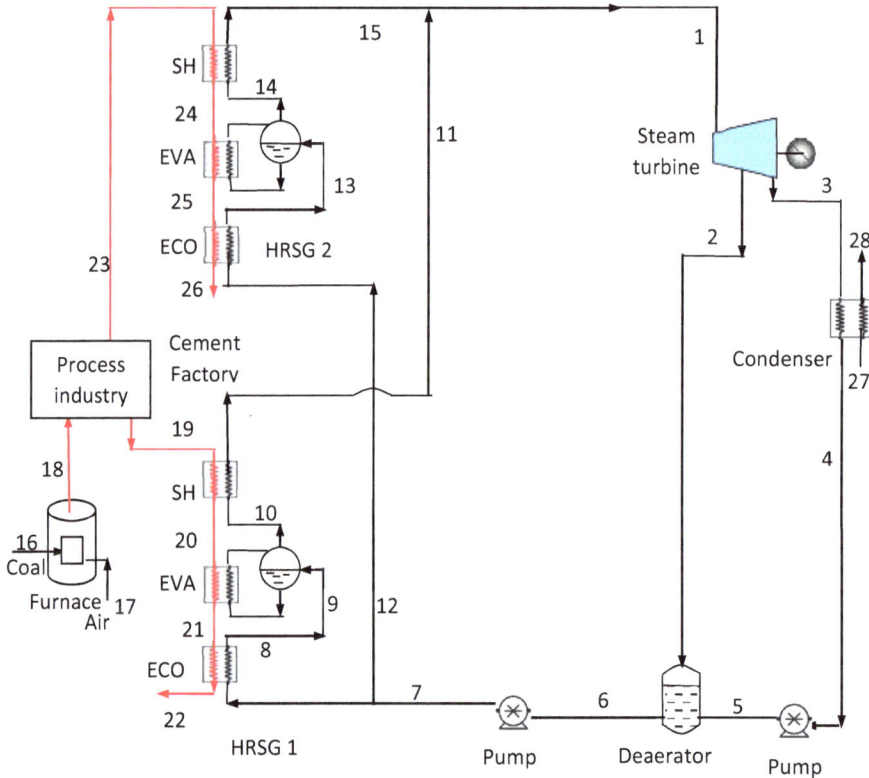

Fig. (23). Power plant layout for SRC suitable to a cement factory prepared for comparative purpose.

Fig. (**24**) is the power plant layout considered from the case study conducted at a cement factory. The cement factory is generating electricity by generating steam from the factory's thermal energy. The temperature-entropy diagram for both regular SRC and SFC from the case study has been depicted at Fig. (**25**).

Fig. (24). Power plant layout of SFC with two HRSGs and two flashers in a cement factory developed from the case study.

Fig. (25). Temperature-entropy diagram of (**a**) SRC and (**b**) SFC with two flashers.

Fig. (**24**) is plant layout of SFC which consists of two HRSGs, steam turbine, water cooled condenser, two pumps, HP flasher, LP flasher and two separators. There are two gas streams resulted from the process industry at different temperatures. Therefore, two HRSGs are connected to these two gas streams without adopting any supplementary firing. The same operational data has been used to solve the plant conditions and performance. The hot gas flow in HRSG 1 and HRSG 2 are 191600 Nm³/h and 360000 Nm³/h respectively. The temperatures of these two gases are 340 °C and 360 °C respectively in HRSG 1 and HRSG 2. Superheated steams are generated in these two HRSGs. The two streams of generated superheated steams are mixed before supplying to a steam turbine. Instead of connecting two steam turbines, the use of one steam turbine offers the high steam turbine efficiency as the steam turbine efficiency increases with increase in its size. Single pressure HRSGs are considered in this power plant due to small capacity of cement factory. For example, in a typical combined cycle power plant, the bottoming steam power plant consists of multi pressure HRSG due to the high capacity plant. Each HRSG consists of three sections for steam generation *viz.*, economizer (ECO), evaporator (EVA) and superheater (SH). A steam separating drum has been used at each HRSG. Pinch point, approach point and terminal temperature difference are used to find the local temperature of hot gas in HRSG. The energy balance in respective HRSG results the steam generating capacity. The generated steam from two HRSGs are mixed, expanded in a turbine, condensed and pumped to HRSG. At the exit of two economizers, predetermined amount of fluid has been taken for the flasher. These two liquids are mixed before supplying to the HP flasher. The economizer load in a steam power plant with flasher is more than the load in a regular plant. But the dryness fraction of steam at the exit of turbine decreases due to mixing of saturated steam in turbine. The drop in dryness fraction is small and it is above the safe limit of 85%. Two hot gas streams are supplied to HRSG 1 and HRSG 2. In a thermal power plant operated by fuel, the steam pressure is high due to high temperature of the source. Therefore, the plant layout consists of reheater, multi feedwater heaters *etc.* But the heat recovery power plant conditions are different compared to the conventional grid connected power plants. Since the source temperature is low, low pressure is sufficient for the steam generation. A simple power plant layout without too many sub systems is suitable to the low temperature source. Therefore, single Rankine cycle is sufficient for heat recovery operated power plant. During the expansion, the turbine receives extra vapour from the two flashers. The steam expands in three stages. It expands from high pressure to HP flasher pressure in first stage. At the end of the first stage, steam receives the saturated vapour from the HP flasher. After mixing of these two streams, the steam expands from the HP flasher to LP flasher. Again at the end of the LP flasher, steam receives the vapour from the LP flasher. The mixed steam further expands from the LP flash pressure to condenser pressure. Therefore, during the journey of the expansion, the steam receives flash vapour two times. In HRSG, the supply hot gas temperature decreases from supehreater to economizer *via* evaporator. The working fluid temperature increases from economizer to superheater *via* evaporator. In HP flasher, the preheated water from the economizer is flashed into wet steam. Later, the wet steam is separated into saturated water and saturated vapour. The saturated vapour is used in steam turbine and the saturated water is forwarded for further flashing in LP flasher. In LP flasher, the process is similar to HP flasher and repeats.

Fig. (**26**) shows the temperature-heat transferred diagram of a (a) SRC and (b) SFC with two flashers. The heat recovery consists of two parts *viz*. HRSG 1 and HRSG 2. The adoption of flasher in a steam power plant increases the economizer load and generates more steam. The load in first part of heat recovery (HRSG 1) is increased from 13 MW to 15 MW with the adoption of two flashers in place of a regular steam power plant. Similarly, the heat load in second part of heat recovery (HRSG 2) is increased from 21 MW to 26 MW from SRC to SFC. The same steam pressure in HRSG 1 and HRSG 2 results approximately same exhaust gas temperature.

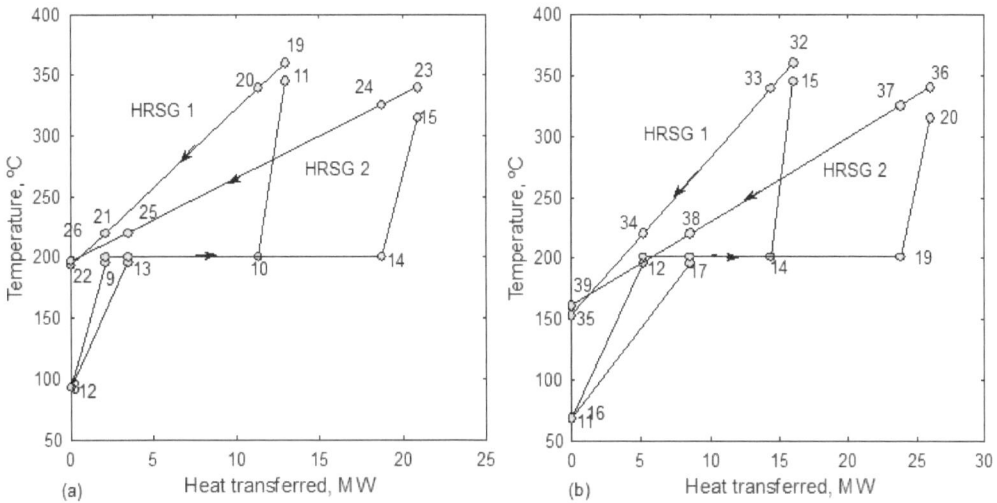

Fig. (26). Temperature-heat transferred diagram for (**a**) SRC and (**b**) SFC with double flashers.

For validation of the theoretical work, a case study has been conducted at a cement factory generating its own electricity from its heat at Telangana, India. Table **3** validates the theoretical results with the observation of the case study. The power plant has two flashers and two HRSGs. HRSG 1 has three sections of heat exchangers *viz*., economizer, evaporator and superheater. HRSG 2 has only evaporator and superheater. A minor change in HRSG 2 is made from the plant information over the layout shown in Fig. (24) . The hot gas flow in HRSG 1 and HRSG 2 respectively are 191600 Nm³/h and 360000 Nm³/h. The hot gas inlet and exit temperatures at HRSG 1 are 340 °C and 90 °C respectively. They are 360 °C and 230 °C respectively at HRSG 2. The comparison is carried out under the similar operational conditions and supply. The exhaust gas temperature from HRSG 1 is differed from the original by 5 °C due to fixed pinch point in the heat recovery. The steam capacity is deviated from 3 to 8%. The error in the estimation of power is 2.86%. The assumption in PP results deviation in HRSG 1's exhaust gas temperature by 5 K. The steam generation in boilers is deviated from 3 to 8%.

Table 3. Validation of plant simulation with the results obtained from the SFC case study.

S. No.	Description	Plant Simulation	Existing Plant Readings	Deviation, %
1.	HRSG 1, exhaust gas temperature, °C	95.00	90.00	-5.56
2.	HRSG 2, exhaust gas temperature, °C	220.00	230.00	4.35
3.	Fluid to HP flasher, kg/s	3.71	3.88	4.38
4.	Fluid to LP flasher, kg/s	3.42	3.26	-4.91
5.	HP flash pressure, bar	5.96	5.60	-6.43
6.	LP flash pressure, bar	1.32	1.30	-1.54
7.	HRSG 1 steam capacity, kg/s	4.19	4.06	-3.20
8.	HRSG 2 steam capacity, kg/s	7.20	7.82	-7.93
9.	Net power generation, MW	7.92	7.70	-2.86

Fig. (27) is developed to find the optimum HRSG pressure with a change in source temperature below the source temperature adopted to the case study power plant configuration. The comparison of SRC and SFC characteristics shows that the optimum HRSG pressure of SFC is greater than the pressure in SRC.

The resulted curve fitting coefficients to find the optimum HRSG pressure as a function of source temperature are:

$a_1 = 8.0641$, $a_2 = -0.0739$ and $a_3 = 0.0002$.

Thermodynamic properties of SFC, determined from the mass balance and energy balance are furnished in Table 4. The optimum HRSG pressure is determined at the source temperature. The optimum HRSG pressure resulted at the source temperature of 360 °C is 7.38 bar. The HP flash pressure and LP flash pressure are 2.3 bar and 0.52 bar respectively. The condenser pressure is 0.07 at the circulating water supply temperature of 30 °C. After mixing of two streams, the turbine inlet temperature is 319.42 °C. The generated total steam in two HRSG sis 14.04 kg/s. The vapour collected from LP flasher and HP flasher is 0.34 kg/s from each flasher. The water used to flashers out of the total liquid from the condenser is 25%. The condensed fluid is 18.72 kg/s and out of this 4.68 kg/s of liquid is used in LP flasher and HP flasher.

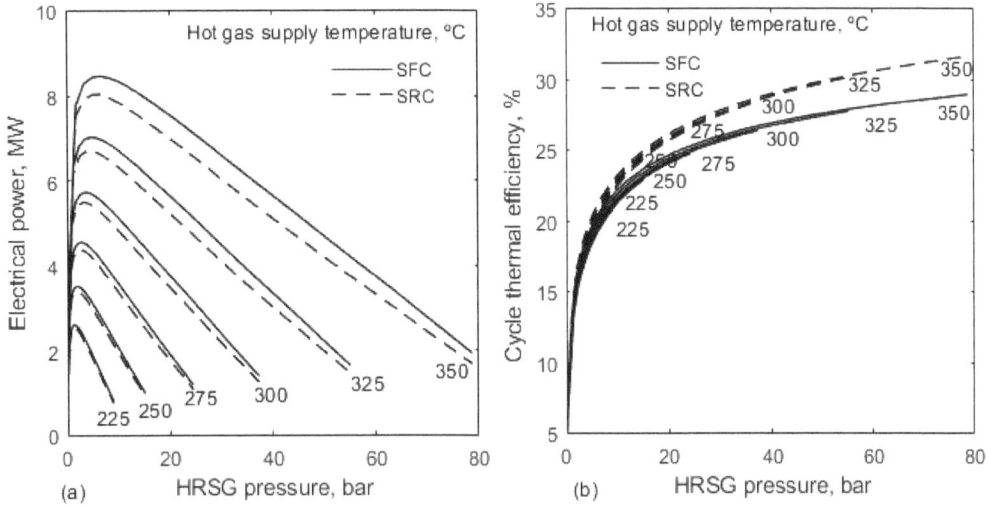

Fig. (27). Optimum HRSG pressure for **(a)** maximum power and **(b)** maximum thermal efficiency with a change in source temperature below the critical temperature applied to the case study with the comparison of SRC and SFC.

Table 4. Material flow properties of a flashed steam power plant shown in Fig. (24).

State	P, bar	t,°C	m, kg/s	h, kJ/kg	s, kJ/kg K	State	P, bar	t,°C	m, kg/s	h, kJ/kg	s, kJ/kg K
1	8.38	319.42	14.04	3099.57	7.34	21	7.38	162.10	1.81	684.57	1.96
2	2.30	205.24	14.04	2879.65	7.46	22	7.38	162.10	2.87	684.57	1.96
3	2.30	203.28	14.38	2875.66	7.45	23	2.30	124.73	4.68	684.57	1.98
4	0.52	87.19	14.38	2657.17	7.61	24	2.30	124.73	4.34	523.83	1.58
5	0.52	87.08	14.72	2656.95	7.61	25	2.30	124.73	0.34	2712.65	7.08
6	0.07	40.00	14.72	2427.66	7.79	26	0.52	82.37	4.34	523.83	1.61
7	0.07	40.00	14.72	167.45	0.57	27	0.52	82.37	4.00	344.85	1.10
8	0.52	41.00	14.72	171.64	0.59	28	0.52	82.37	0.34	2647.67	7.58
9	0.52	49.91	18.72	208.93	0.70	29	1.01	25.00	16.42	0.00	0.00
10	7.38	50.95	18.72	213.22	0.72	30	1.01	308.15	187.36	7.20	0.02
11	7.38	50.95	7.23	213.22	0.72	31	1.01	900.00	203.76	1006.40	1.54

(Table 4) cont.....

State	P, bar	t,°C	m, kg/s	h, kJ/kg	s, kJ/kg K	State	P, bar	t,°C	m, kg/s	h, kJ/kg	s, kJ/kg K
12	7.38	162.10	7.23	684.57	1.96	32	1.01	360.00	70.68	359.60	0.80
13	7.38	162.10	5.42	684.57	1.96	33	1.01	333.28	70.68	329.75	0.76
14	7.38	167.10	5.42	2764.17	6.69	34	1.01	187.10	70.68	170.16	0.46
15	7.38	345.00	5.42	3153.12	7.43	35	1.01	141.87	70.68	121.93	0.34
16	7.38	50.95	11.48	213.22	0.72	36	1.01	328.34	133.08	324.28	0.75
17	7.38	162.10	11.48	684.57	1.96	37	1.01	310.75	133.08	304.76	0.71
18	7.38	162.10	8.61	684.57	1.96	38	1.01	187.10	133.08	170.16	0.46
19	7.38	167.10	8.61	2764.17	6.69	39	1.01	148.99	133.08	129.48	0.36
20	7.38	303.34	8.61	3065.85	7.29	40	1.01	25.00	994.81	0.00	0.00
--	--	--	--	--	--	41	1.01	33.00	994.81	33.44	0.11

Table **5** compares the SRC and SFC on the bases of operational conditions and performance. The optimum HRSG pressure is determined from the source temperature. The steam capacity is determined from the evaporator and superheater section with PP constraint. The resulted optimum HRSG pressure for SRC is less than the optimum pressure of SFC. The different steam pressure results different steam capacity in HRSG of SRC and SFC. The energy supply and source state is same to both the power plants under this comparison. The hot gas supply is 70.68 kg/s and 133.08 kg/s respectively to HRSG 1 and HRSG 2. The air fuel ratio in fuel combustion is 11.41. The amount of working fluid in the economizer section is different due to flashers in SFC. This increases the condenser load in the SFC compared to SRC. With SFC, the power is augmented from 8.39 MW to 9.08 MW (8.22% rise). The same SFC influenced in decrease in thermal efficiency from 23.9% to 21.25% (11,08% drop). As power is augmented, the exergy efficiency is increased from 38.15% to 41.28% (8.2% rise). Therefore, the SFC increases heat recovery, power, and exergy efficiency with a loss in thermal efficiency.

Table 5. Comparison of SRC and SFC with two flashers at the respective optimized HRSG pressures.

S. No.	Description	SRC	SFC
1.	Fuel (coal) supply, kg/s	16.41	16.41
2.	Air supply to furnace, kg/s	187.36	187.36
3.	Hot gas to HRSG 1, kg/s	70.68	70.68
4.	Hot gas to HRSG 2, kg/s	133.08	133.08
5.	Total hot gas supply to power plant, kg/s	203.76	203.76
6.	Air fuel ratio, kg/kg coal	11.41	11.41

(Table 5) cont.....

S. No.	Description	SRC	SFC
7.	HRSG 1, exhaust gas temperature, °C	178.00	141.87
8.	HRSG 2, exhaust gas temperature, °C	183.00	149.00
9.	HRSG 1, steam capacity, kg/s	5.03	5.42
10.	HRSG 2, steam capacity, kg/s	7.80	8.61
11.	Circulating water to condenser, kg/s	788.00	994.90
12.	Power, MW	8.39	9.08
13.	Cycle thermal efficiency, %	23.90	21.25
14.	Cycle exergy efficiency, %	38.15	41.28

SUMMARY

In SRC, saturated steams from the flashers are mixed with the turbine flowing steam. It drops the dryness fraction of steam at the exit of the turbine compared to the SRC but it is within the safe limits. The power augmentation with two flashers compared to the single flasher is more. The subsequent addition of flashers increases the power in a diminishing rate. The optimum HRSG pressure determined at the source temperature is high in SFC compared to the SRC. The optimum pressure resulted in SFC is greater than the optimum pressure of SRC. The major exergetic loss occurs in exhaust gas can be controlled by the flashers in the plant. The influence of flashers is more at the higher temperature than low supply temperature. Maximum power condition demands lower HRSG pressure and maximum energy efficiency condition needs higher HRSG pressure. A double flash SFC with low HRSG pressure results 90% of dry steam at the exit of turbine. Adoption of flashing in steam power plant increases power from 8.34 MW to 9.43 MW. The efficiency of the SFC is increasing with number of flashers up to 0.4 flash mass ratio (FMR) and above this, efficiency is inversely relating to FMR.

CHAPTER 9

Comparison of Thermal Power Cycles

Abstract: This chapter summarizes the book by the comparison of all the thermal power cycles under a common platform. Some outcomes of the book have been drawn through this comparative analysis. All the thermal cycles of heat recovery power plants are consolidated in this comparative analysis. The organic Rankine cycle (ORC), organic flash cycle (OFC), steam Rankine cycle (SRC) and steam flash cycle (SFC) are compared and analyzed below and above the critical temperature of working fluid. The six working fluids used in ORC and OFC are R123, R124, R134a, R245fa, R717 and R407C. The performance of Kalina cycle (KC) is also added to differentiate with the ORC and OFC characteristics. R123 in OFC has been recommended to result in higher power compared to the other fluids. After the low temperature cycles (ORC, OFC and KC), the next level of source temperature has been analyzed with steam power cycles (SRC and SFC). The SRC, SFC with one flasher (SFC 1), and SFC with two flashers (SFC 2) are compared for the power and efficiency variations with a change in temperature and pressure. From the steam cycles, SFC 2 has been recommended to waste heat transfer for power augmentation. The analysis shows that the R134a and R245fa are not able to augment the power with OFC plant. OFC- R407C is only escalating the thermal efficiency in addition to the power augmentation due to the use of zeotropic working fluid. R717 exhibits maximum specific power with ORC and OFC with higher optimized heat recovery pressure.

Keywords: Comparative analysis, Energy efficiency, Power augmentation, Thermal power plants, Waste heat recovery.

INTRODUCTION

In the earlier chapters, the individual power plant cycles are detailed *viz.* organic Rankine cycle (ORC), organic flash cycle (OFC), Kalina cycle (KC), steam Rankine cycle (SRC) and steam flash cycle (SFC). The studied working fluids in ORC and OFC are R123, R124, R134a, R245fa, R717 and R407C. Out of these working fluids, R407C is the zeotropic mixture. KC also works with zeotropic mixture but the cycle undergoes binary fluid processes such as absorption and separation. Zeotropic mixture has an advantage of glide in boiling and condensation. This glide helps to match the temperature profile of heat source. Two types of steam cycles are selected in the thermal power cycle *viz.*, SRC and SFC. The main focus of this chapter is the comparative analysis of all these thermal power cycles to make some useful recommendations.

COMPARISON OF ORC, OFC, AND KC

Each chapter in this book is focused on a thermal cycle of heat recovery power plant. Performance of each thermal cycle has been studied with the selected working fluids. The recommendations in the earlier chapters are made from the comparison of same category of layout with a change in working fluid. This chapter compares all the categories under common ground. A wider outcome can be expected from this overall comparison. Figs. (**1 - 6**) compares the ORC and OFC with the six working fluids for power and efficiency changes.

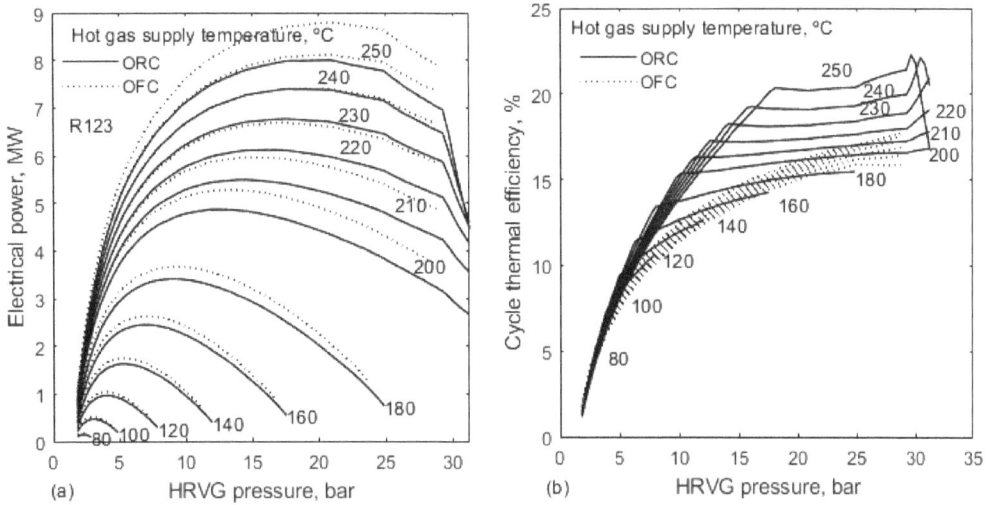

Fig. (1). Comparison of **(a)** power and **(b)** thermal efficiency of ORC and OFC with source temperature below and above the critical temperature of R123.

Figs. (**1a** and **1b**) shows the power and thermal efficiency characteristics developed for ORC and OFC with R123. The augmentation in power with OFC increases with increase in source temperature. The performance characteristics below and above the critical temperature are plotted. A greater improvement in power has been observed from the source temperature, below the critical temperature to above the critical temperature of the fluid. This Figure also shows that the optimum pressure increases with an increase in the source temperature. Thermal efficiency increases with an increase in source temperature at a diminishing rate. The increase in source temperature increases the optimum HRVG pressure. It limits the vapor generation with a limit in the pinch point. Therefore, the efficiency increases at a diminishing rate with a rise in the temperature.

Fig. (**2**) outlines the performance characteristics of R124. A favorable power augmentation is observed with OFC. In the majority of cases, the optimum OFC pressure is more than the optimum ORC pressure. R717 shows a lower OFC pressure compared to ORC pressure. It is due to the higher pump supply with high HRVG pressure. The power and thermal efficiency decrease with OFC-R134a. The deviation between ORC and OFC characteristics is more with R134a. A similar deviation in power and efficiency has been observed with R245fa (Fig. **4**).

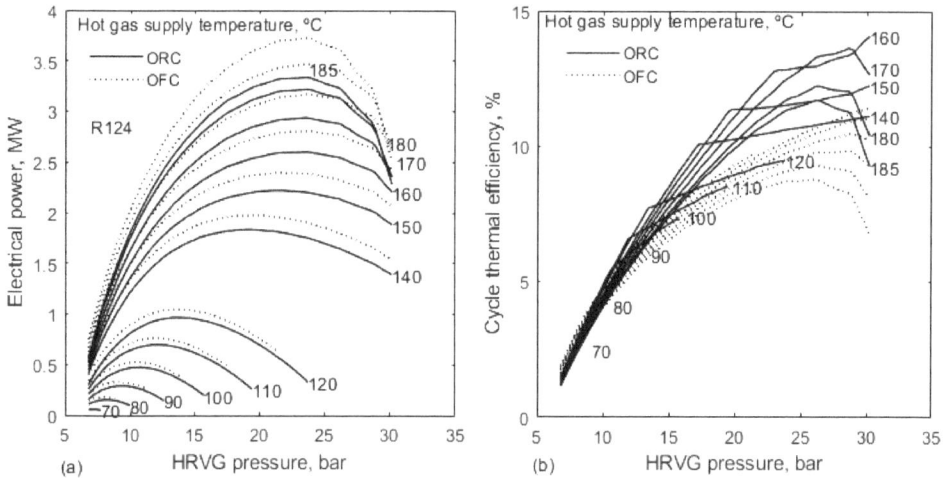

Fig. (2). Comparison of (**a**) power and (**b**) thermal efficiency of ORC and OFC with source temperature below and above the critical temperature of R124.

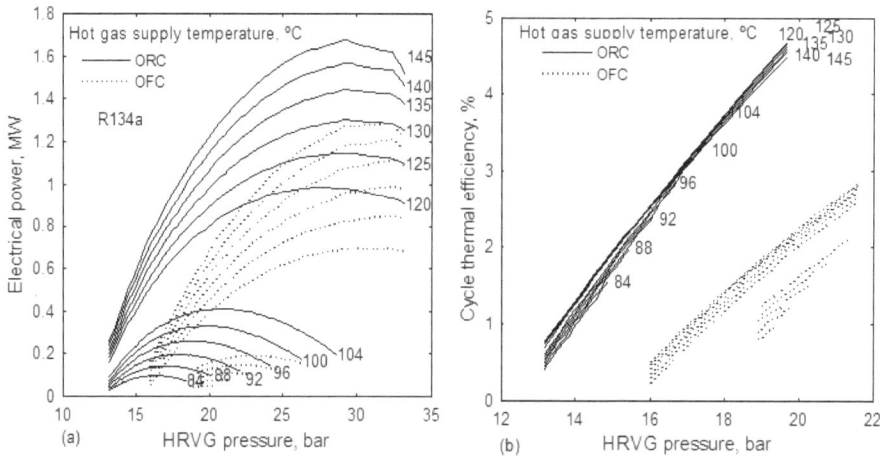

Fig. (3). Comparison of (**a**) power and (**b**) thermal efficiency of ORC and OFC with source temperature below and above the critical temperature of R134a.

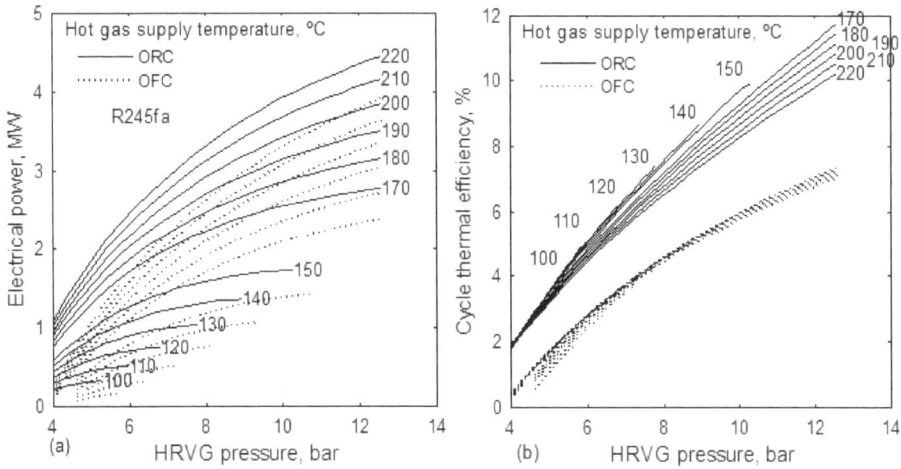

Fig. (4). Comparison of **(a)** power and **(b)** thermal efficiency of ORC and OFC with source temperature below and above the critical temperature of R245fa.

Fig. (**5**) shows the power and thermal efficiency characteristics of R717 for ORC and OFC. A satisfactory power boost has been found with OFC over the ORC with this fluid. Thermal efficiency decay has been observed with OFC and the efficiency is increased with increase in source temperature. R407C is a zeotropic fluid, hence its performance characteristics differs compared to others. Both the power and efficiency are improved with R407C in OFC with the comparison of ORC.

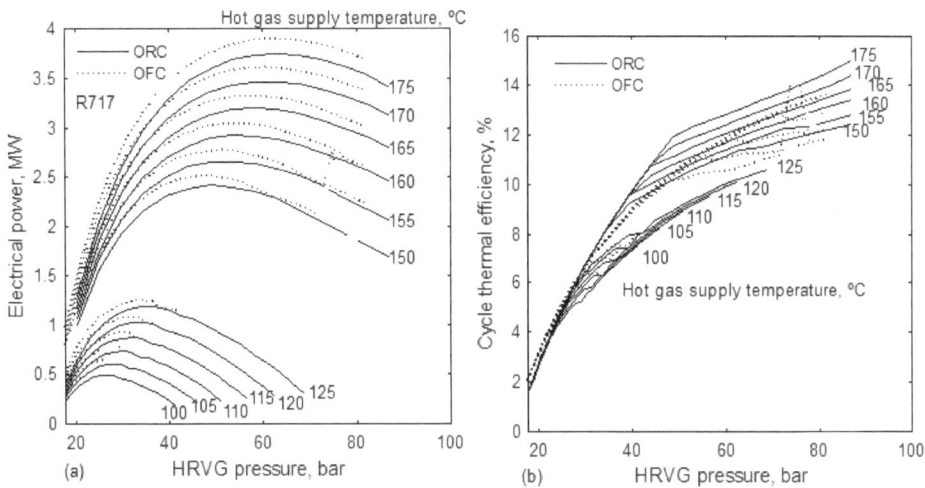

Fig. (5). Comparison of **(a)** power and **(b)** thermal efficiency of ORC and OFC with source temperature below and above the critical temperature of R717.

Fig. (**7**) compares the power generation from ORC and OFC with six fluids. The first five working fluids are pure substances, and the last fluid is zeotropic fluid. The source temperature is increased from below the critical temperature to above the critical temperature of the fluid. In the earlier chapters, separate curve fitting equations are developed for the optimum HRVG pressure below and above the critical temperature. At each temperature, an optimum heat recovery pressure has been determined to evaluate the power and efficiency. The comparative study shows that the highest power is found with the R123 working fluid and the lowest with R407C. The results show that the OFC is boosting the power over the ORC. The augmentation of power with OFC increases with an increase in temperature. The high source temperature can be observed with R123, R124, and R245fa as per the critical temperature of working fluid. R407C results in more power augmentation with OFC over the others. Similarly, the power augmentation at high temperature with R134a fluid is high.

Similar to power, the thermal efficiency of the power plant with the six working fluids has been analyzed in Fig. (**8**). The ORC and OFC plants are compared with a change in source temperature below and above the critical temperature. Since the excess fluid increases the heat supply, the cycle's thermal efficiency decreases from ORC to OFC. The power increment with OFC is not justified to thermal efficiency. The efficiency loss is more at high temperatures. Below the critical temperature, OFC-R407 augments the thermal efficiency compared to ORC-R407C.

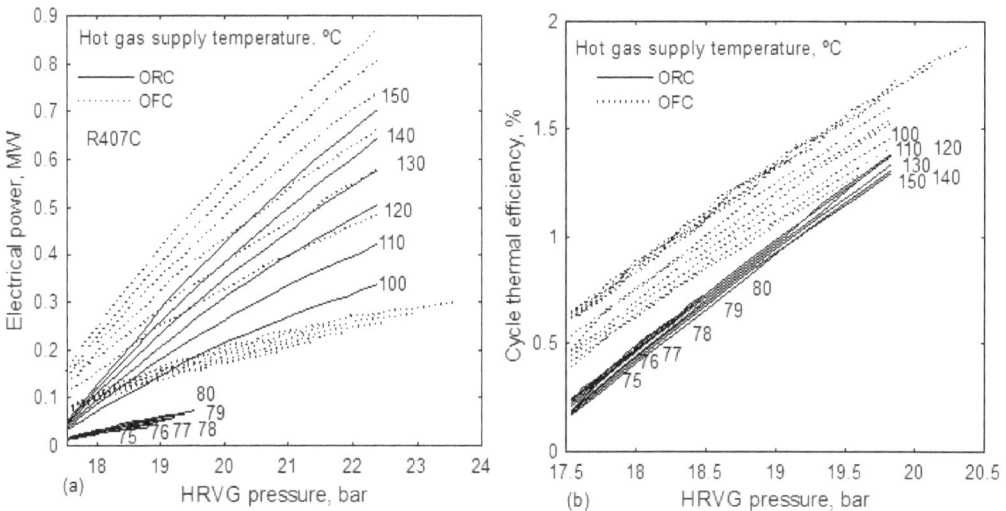

Fig. (6). Comparison of (**a**) power and (**b**) thermal efficiency of ORC and OFC with source temperature below and above the critical temperature of R407C.

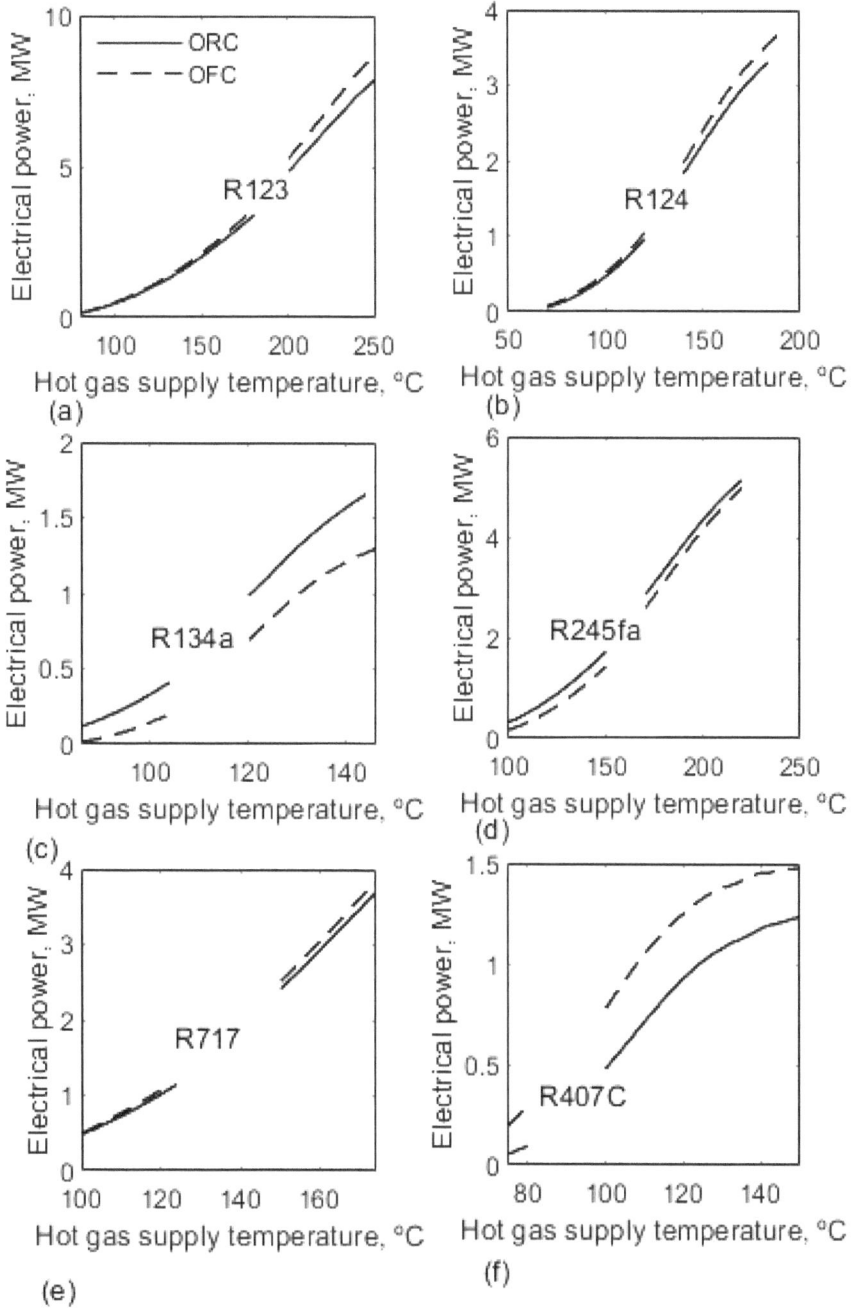

Fig. (7). Comparison of power from ORC and OFC with the source temperature below and above the critical temperature of working fluid and optimized HRVG pressure.

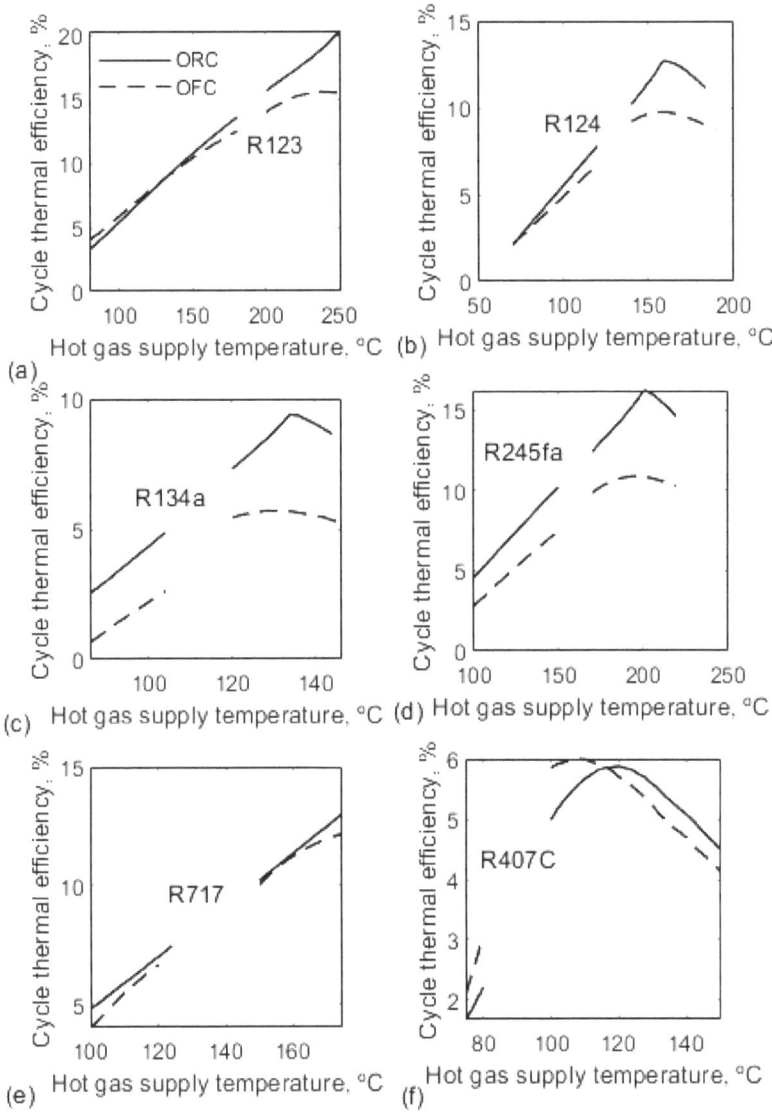

Fig. (8). Comparison of thermal efficiency of ORC and OFC with the source temperature below and above the critical temperature of working fluid and optimized HRVG pressure.

(Figs. **9a** and **b**) are focused on KC power and KC thermal efficiency, respectively. These results can be compared with the earlier results of ORC and OFC characteristics. The power and thermal efficiency increase with an increase in source temperature. These results are generated at the optimum process conditions discussed in the earlier chapter.

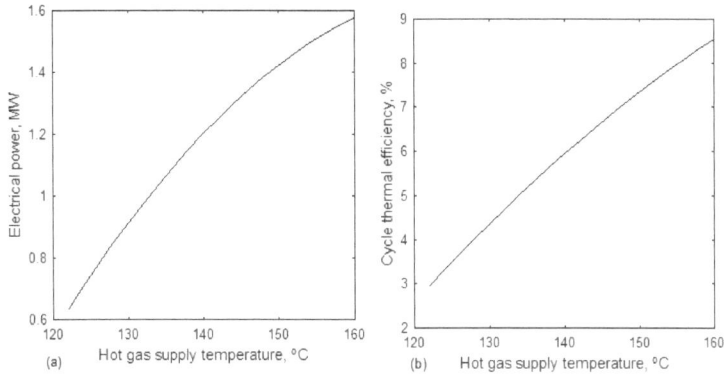

Fig. (9). Power and thermal efficiency of Kalina cycle.

COMPARISON OF SRC AND SFC

Fig. (**10**) summarizes the SRC, SFC with single flasher (SFC 1), and SFC with two flashers (SFC 2) on power and thermal efficiency. A large power augmentation has been observed from characteristics below the critical temperature to above the critical temperature. The power augmentation also increases with an increase in source temperature. The HRSG pressure is changed between condenser pressure and critical pressure. The optimum pressure increases with an increase in source temperature. Since the flash cycle is not favoring thermal efficiency, we may have understood that SFC 1 efficiency is greater than the SFC 2 efficiency. The results show that the thermal efficiency of SFC 2 is greater than the SFC 1 with the influence of power augmentation. However, the efficiency of SFC 1 and SFC 2 is always below the efficiency of SRC. Thermal efficiency increases with an increase in the source temperature.

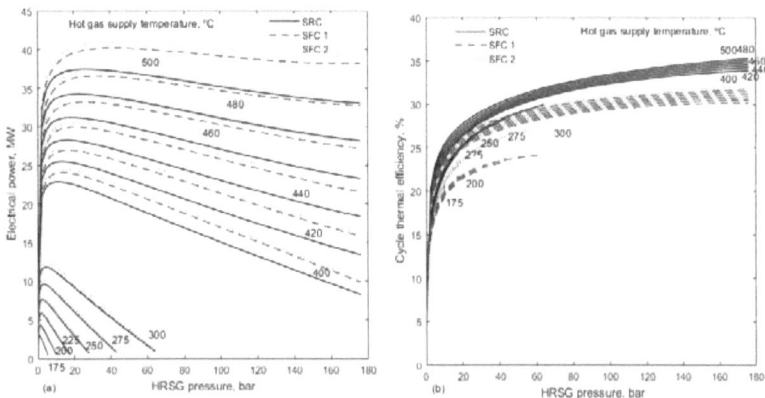

Fig. (10). Comparison of SRC, SFC 1 and SFC 2 at below and above the critical temperature of steam.

Fig. (**11**) compares the three thermal power cycles *viz.*, SRC, SFC 1, and SFC 2 with a change in source temperature. The power has been studied below the critical temperature of water (Fig. **11a**) and above the critical temperature of the water (Fig. **11b**). The power curves are close at low source temperature below the critical temperature. These curves are wider at the temperature above the critical state. The power generation in SFC 2 is greater than the SFC1 and SRC.

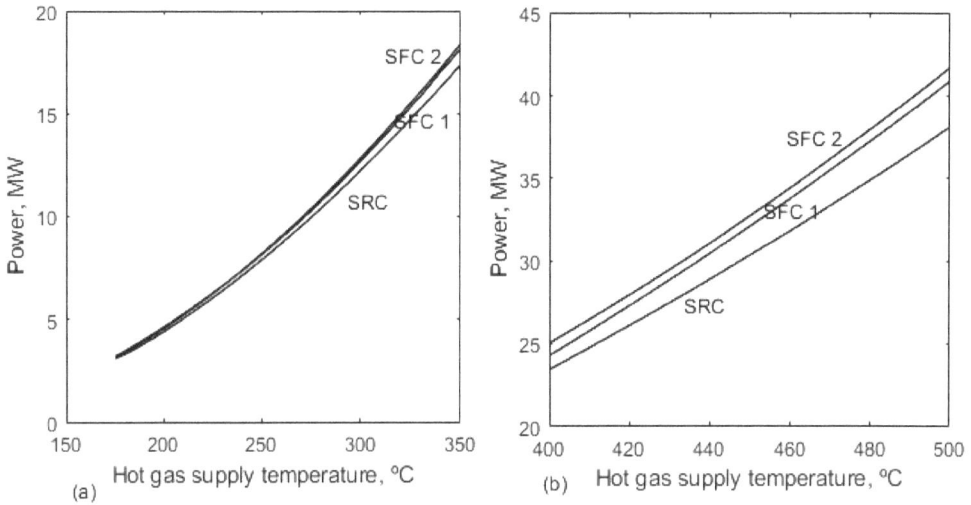

Fig. (11). Comparison of power from SRC, SFC 1 and SFC 2 plants with source temperature (**a**) below the critical temperature and (**b**) above the critical temperature.

Fig. (**12**) compares the thermal efficiency of power cycles with the source temperature below and above the critical state of water. SFC 1 results in high efficiency compared to SRC and SFC 2 at the source temperature below the critical temperature. At this low temperature, efficiency loss is not predominant compare to the power augmentation with SFC 1. The efficiency of two flash cycles (SFC 1 and SFC 2) is lower than the SRC at the temperature above the critical temperature. Finally, from 400 °C onwards, SFC 2 efficiency dominates the other two.

Table **1** summarizes the performance of ORC, OFC, KC, SRC and SFC with the source temperature and/or turbine inlet temperature below the critical temperature of the working fluid. The power plants operating with industrial waste heat have been compared with the difference working fluids stated in the earlier sections. To refer the turbine supply temperature and pressure with the critical temperature and critical pressure data is furnished. In ORC and OFC power plants, R123 has high critical temperature among the considered fluids. Similarly, R717 has highest critical pressure among the fluids. Therefore, the turbine inlet temperature with R123 is 165 °C after the TTD between the gas and working fluid. Since the critical temperature and the turbine inlet temperature with R123 is high, the resulted cycle thermal efficiency is 13.59% which is the highest among ORCs. The critical temperature of R407C is low, therefore, a minimum thermal efficiency of 2.00% results at 63 °C. After R407C, the second low temperature plant is with R134a fluid. The optimum HRSG pressures at the source temperatures are presented. The HRVG pressure is increasing from ORC to OFC. But with R717, the optimum pressure is decreasing from ORC to OFC. The power generation is high with R717 with high critical pressure. The power augmentation has been observed from ORC to OFC except with the fluids R134a and R245fa. HRVG pressure is high and its declines the steam generating capacity. Therefore, these two fluids are not augmenting the power. The other fluids not influencing the power much as the pressure is nearly same.

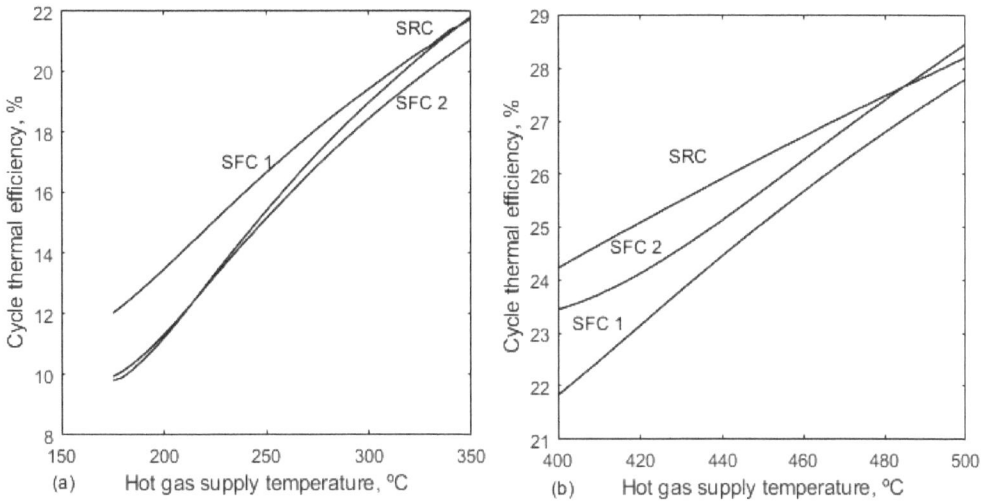

Fig. (12). Comparison of cycle thermal efficiency of SRC, SFC 1 and SFC 2 with source temperature **(a)** below the critical temperature and **(b)** above the critical temperature.

Table 1. Summary of ORC, OFC, KC, SRC and SFC power plants with the working fluids at the source temperature below and the hot gas flow rate of 100000 Nm³/h, CT: critical temperature; CP: critical pressure; TIT: turbine inlet temperature; 1: energy; 2: exergy.

Thermal Cycle	CT, °C	CP, bar	TIT, °C	P_b, bar	T_{ex}, °C	W_{net}, MW	$\eta_{cycle,\,1}$, %	$\eta_{plant,\,2}$, %
ORC-R123	183.68	36.68	165.00	9.23	118.72	3.41	13.59	29.02
ORC-R124	122.47	36.34	105.00	13.67	89.27	0.96	7.78	20.00
ORC-R134a	101.08	40.60	85.00	20.17	81.27	0.33	4.33	10.56

(Table 1) cont.....

Thermal Cycle	CT, °C	CP, bar	TIT, °C	P_b, bar	T_{ex}, °C	W_{net}, MW	$\eta_{cycle, 1}$, %	$\eta_{plant, 2}$, %
ORC-R245fa	154.10	36.40	135.00	11.55	106.96	1.72	10.21	21.65
ORC-R717	132.40	113.40	110.00	36.48	86.52	1.18	8.03	22.25
ORC-R407C	86.03	46.30	63.00	21.42	70.48	0.08	2.00	4.92
OFC-R123	183.68	36.68	165.00	10.58	108.80	3.63	12.50	30.96
OFC-R124	122.47	36.34	105.00	13.30	82.38	1.05	6.67	21.68
OFC-R134a	101.08	40.60	85.00	22.07	86.00	0.14	2.20	4.62
OFC-R245fa	154.10	36.40	135.00	12.79	107.20	1.43	7.44	18.01
OFC-R717	132.40	113.40	110.00	33.23	84.82	1.24	7.17	23.39
OFC-R407C	86.03	46.30	63.00	22.47	53.25	0.25	2.68	15.79
SRC	374.00	221.2	285.00	4.20	143.77	11.82	18.96	37.18
SFC	374.00	221.2	285.00	4.29	131.28	12.38	18.43	38.96
KC	Binary fluid		135.00	13.78	100.23	1.42	7.34	17.83

Table **2** quantifies the ORC, OFC, SRC and SFC with the source temperature and/or turbine inlet temperature above the critical temperature of the working fluid. From ORC to OFC, the optimum pressure is increased with R124, R134a, R245fa and R407C. The optimum pressure is decreasing with the fluid R123 and R717. The power augmentation has been observed from ORC to OFC except in R134a and R245fa. The thermal efficiency is decreasing from ORC to OFC. The exergy efficiency is improved from ORC to OFC with power augmentation.

Table 2. Summary of ORC, OFC, SRC and SFC power plants with the working fluids at the source temperature above the critical temperature and the hot gas flow rate of 100000 N^3/h, CT: critical temperature; CP: critical pressure; TIT: turbine inlet temperature; 1: energy; 2: exergy.

Thermal Cycle	CT, °C	CP, bar	TIT, °C	P_b, bar	T_{ex}, °C	W_{net}, MW	$\eta_{cycle, 1}$, %	$\eta_{plant, 2}$, %
ORC-R123	183.68	36.68	235	22.95	149.31	7.87	20.32	34.87
ORC-R124	122.47	36.34	165	23.30	106.20	3.21	11.55	27.37
ORC-R134a	101.08	40.60	130.00	28.97	91.25	1.67	8.52	22.60
ORC-R245fa	154.10	36.40	205.00	19.75	136.08	5.13	14.62	29.12
ORC-R717	132.40	113.40	160.00	64.04	101.86	3.74	13.12	33.77
ORC-R407C	86.03	46.30	105.00	35.59	78.99	0.93	5.89	19.27
OFC-R123	183.68	36.68	235.00	22.45	101.41	8.73	15.46	38.66
OFC-R124	122.47	36.34	165.00	24.64	81.65	3.45	9.17	29.34
OFC-R134a	101.08	40.60	115.00	32.15	84.55	0.99	5.68	16.99
OFC-R245fa	154.10	36.40	205.00	23.63	100.42	4.98	10.29	28.27
OFC-R717	132.40	113.40	160.00	62.92	93.99	3.89	12.20	35.15

(Table 2) cont.....

Thermal Cycle	CT, °C	CP, bar	TIT, °C	P_b, bar	T_{ex}, °C	W_{net}, MW	$\eta_{cycle, 1}$, %	$\eta_{plant, 2}$, %
OFC-R407C	86.03	46.30	105.00	36.36	63.06	1.25	5.70	25.83
SRC	374.00	221.2	385.00	11.56	167.00	22.87	24.22	42.95
SFC	374.00	221.2	385.00	12.33	142.13	23.44	24.42	45.87

SUMMARY

In this chapter, all the thermal power plants with heat recovery are compared to understand the relative performance changes. The performance of thermal power cycles has been compared with a change in temperature and pressure. The overall comparison gave more insight on the power plant categories over the individual comparison. The qualitative changes and the performance quantities of all the thermal power cycles have been critically examined. OFC and SFC are boosting the power compared of ORC and SRC respectively with a penalty in thermal efficiency. OFCs with R134a and R245fa are not capable of power augmentation over ORC. At low temperatures, R407C is augmenting the thermal efficiency with OFC. R717 has highest specific power compared to the other fluids. Recommended R123 for good performance as its critical temperature is high compared to the other fluids. Supercritical power plants (operating at super critical pressure) are not suitable for the waste heat recovery as the optimum pressure is low to maximize the heat recovery. Recommended to select the source temperature above the critical temperature as it performs well compare to the source temperature below the critical temperature. The case studies at thermal industries are resulting a satisfactory result for self-generation of power to meet its own load.

References

[1]   S. Quoilin, M. Van Den Broek, S. Declaye, P. Dewallef, and V. Lemort, "Techno-economic survey of Organic Rankine Cycle (ORC) systems", *Renew. Sustain. Energy Rev.,* vol. 22, pp. 168-186, 2013. [http://dx.doi.org/ 10.1016/j.rser.2013.01.028]

[2]   T. Engin, and V. Ari, "Energy auditing and recovery for dry type cement rotary kiln systems—A case study", *Energy Convers. Manage.,* vol. 46, no. 4, pp. 551-562, 2005. [http://dx.doi.org/ 10.1016/j.enconman.2004.04.007]

[3]   K. Chandra, and J.N. Palley, "Cement plant cogeneration opportunities", *IEEE Trans. Ind. Appl.,* no. 1, pp. 37-42, 1983. [http://dx.doi.org/ 10.1109/TIA.1983.4504153]

[4]   H. Chen, D.Y. Goswami, and E.K. Stefanakos, "A review of thermodynamic cycles and working fluids for the conversion of low-grade heat", *Renew. Sustain. Energy Rev.,* vol. 14, no. 9, pp. 3059-3067, 2010. [http://dx.doi.org/ 10.1016/j.rser.2010.07.006]

[5]   J. Bao, and L. Zhao, "A review of working fluid and expander selections for organic Rankine cycle", *Renew. Sustain. Energy Rev.,* vol. 24, pp. 325-342, 2013. [http://dx.doi.org/ 10.1016/j.rser.2013.03.040]

[6]   P.J. Mago, L.M. Chamra, K. Srinivasan, and C. Somayaji, "An examination of regenerative organic Rankine cycles using dry fluids", *Appl. Therm. Eng.,* vol. 28, no. 8-9, pp. 998-1007, 2008. [http://dx.doi.org/ 10.1016/j.applthermaleng.2007.06.025]

[7]   J. Sun, and W. Li, "Operation optimization of an organic Rankine cycle (ORC) heat recovery power plant", *Appl. Therm. Eng.,* vol. 31, no. 11-12, pp. 2032-2041, 2011. [http://dx.doi.org/ 10.1016/j.applthermaleng.2011.03.012]

[8]   J. Frutiger, J. Andreasen, W. Liu, H. Spliethoff, F. Haglind, J. Abildskov, and G. Sin, "Working fluid selection for organic Rankine cycles–Impact of uncertainty of fluid properties", *Energy,* vol. 109, pp. 987-997, 2016. [http://dx.doi.org/ 10.1016/j.energy.2016.05.010]

[9]   T. Yamamoto, T. Furuhata, N. Arai, and K. Mori, "Design and testing of the organic Rankine cycle", *Energy,* vol. 26, no. 3, pp. 239-251, 2001. [http://dx.doi.org/ 10.1016/S0360-5442(00)00063-3]

[10] B. Lei, W. Wang, Y.T. Wu, C.F. Ma, J.F. Wang, L. Zhang, C. Li, Y.K. Zhao, and R.P. Zhi, "Development and experimental study on a single screw expander integrated into an Organic Rankine Cycle", *Energy,* vol. 116, pp. 43-52, 2016. [http://dx.doi.org/ 10.1016/j.energy.2016.09.089]

[11] D.S. Lee, T.C. Hung, J.R. Lin, and J. Zhao, "Experimental investigations on solar chimney for optimal heat collection to be utilized in organic Rankine cycle", *Appl. Energy,* vol. 154, pp. 651-662, 2015. [http://dx.doi.org/ 10.1016/j.apenergy.2015.05.079]

[12] K. Hu, J. Zhu, W. Zhang, K. Liu, and X. Lu, "Effects of evaporator superheat on system operation stability of an organic Rankine cycle", *Appl. Therm. Eng.,* vol. 111, pp. 793-801, 2017. [http://dx.doi.org/ 10.1016/j.applthermaleng.2016.09.177]

[13] M. White, and A.I. Sayma, "Improving the economy-of-scale of small organic Rankine cycle systems through appropriate working fluid selection", *Appl. Energy,* vol. 183, pp. 1227-1239, 2016. [http://dx.doi.org/ 10.1016/j.apenergy.2016.09.055]

[14] D. Wei, X. Lu, Z. Lu, and J. Gu, "Performance analysis and optimization of organic Rankine cycle (ORC) for waste heat recovery", *Energy Convers. Manage.,* vol. 48, no. 4, pp. 1113-1119, 2007.

[http://dx.doi.org/ 10.1016/j.enconman.2006.10.020]

[15]   N. Yamada, Y. Tominaga, and T. Yoshida, "Demonstration of 10-Wp micro organic Rankine cycle generator for low-grade heat recovery", *Energy,* vol. 78, pp. 806-813, 2014. [http://dx.doi.org/ 10.1016/j.energy.2014.10.075]

[16]   H. Yamaguchi, X.R. Zhang, K. Fujima, M. Enomoto, and N. Sawada, "Solar energy powered Rankine cycle using supercritical CO2", *Appl. Therm. Eng.,* vol. 26, no. 17-18, pp. 2345-2354, 2006. [http://dx.doi.org/ 10.1016/j.applthermaleng.2006.02.029]

[17]   Y. Chen, P. Lundqvist, A. Johansson, and P. Platell, "A comparative study of the carbon dioxide transcritical power cycle compared with an organic Rankine cycle with R123 as working fluid in waste heat recovery", *Appl. Therm. Eng.,* vol. 26, no. 17-18, pp. 2142-2147, 2006. [http://dx.doi.org/ 10.1016/j.applthermaleng.2006.04.009]

[18]   J.C. Hsieh, B.R. Fu, T.W. Wang, Y. Cheng, Y.R. Lee, and J.C. Chang, "Design and preliminary results of a 20-kW transcritical organic Rankine cycle with a screw expander for low-grade waste heat recovery", *Appl. Therm. Eng.,* vol. 110, pp. 1120-1127, 2017. [http://dx.doi.org/ 10.1016/j.applthermaleng.2016.09.047]

[19]   J. Galindo, S. Ruiz, V. Dolz, and L. Royo-Pascual, "Advanced exergy analysis for a bottoming organic Rankine cycle coupled to an internal combustion engine", *Energy Convers. Manage.,* vol. 126, pp. 217-227, 2016. [http://dx.doi.org/ 10.1016/j.enconman.2016.07.080]

[20]   J. Lu, J. Zhang, S. Chen, and Y. Pu, "Analysis of organic Rankine cycles using zeotropic mixtures as working fluids under different restrictive conditions", *Energy Convers. Manage.,* vol. 126, pp. 704-716, 2016. [http://dx.doi.org/ 10.1016/j.enconman.2016.08.056]

[21]   A. Dagdas, "Performance analysis and optimization of double-flash geothermal power plants", *ASME J of Energy Resources Technology,* vol. 129, pp. 25-133, 2007. [http://dx.doi.org/ 10.1115/1.2719204]

[22]   N.A. Lai, and J. Fischer, "Efficiencies of power flash cycles", *Energy,* vol. 44, no. 1, pp. 1017-1027, 2012. [http://dx.doi.org/ 10.1016/j.energy.2012.04.046]

[23]   G.P. Varma, and T. Srinivas, "Power generation from low temperature heat recovery", *Renew. Sustain. Energy Rev.,* vol. 75, pp. 402-414, 2017. [http://dx.doi.org/ 10.1016/j.rser.2016.11.005]

[24]   U. Muhammad, M. Imran, D.H. Lee, and B.S. Park, "Design and experimental investigation of a 1 kW organic Rankine cycle system using R245fa as working fluid for low-grade waste heat recovery from steam", *Energy Convers. Manage.,* vol. 103, pp. 1089-1100, 2015. [http://dx.doi.org/ 10.1016/j.enconman.2015.07.045]

[25]   Ł. Witanowski, P. Klonowicz, P. Lampart, T. Suchocki, Ł. Jędrzejewski, D. Zaniewski, and P. Klimaszewski, "Optimization of an axial turbine for a small scale ORC waste heat recovery system", *Energy,* vol. 205, no. C, 2020. [http://dx.doi.org/ 10.1016/j.energy.2020.118059]

[26]   A.I. Kalina, "Combined cycle and waste heat recovery power systems based on a novel thermodynamic energy cycle utilizing low-temperature heat for power generation", *In Turbo Expo: Power for Land, Sea, and Air,* vol. 79368, , 1983p. V001T02A003

[27]   M. Mirolli, H. Hjartarson, H.A. Mlcak, and M. Ralph, "Testing and operating experience of the 2 MW Kalina cycle geothermal power plant in Húsavík", *In Proceedings,* Iceland, 2002.

[28]   H. Mlcak, M. Mirolli, H. Hjartason, O. Húsavíkur, and M. Ralph, "Notes from the North: a report on the debut year of the 2 MW Kalina Cycle Geothermal power plant in Húsavik", *Trans. Geotherm. Resour. Counc,* Iceland. Germany, pp. 715-718, 2002.

[29]   F. Marcuccilli, and S. Zouaghi, "Radial inflow turbines for Kalina and organic Rankine cycles system", *Proceedings European Geothermal Congress,* 2007pp. 1-7 Unterhaching, Germany

[30]   P.A. Lolos, and E.D. Rogdakis, "A Kalina power cycle driven by renewable energy sources", *Energy,* vol. 34, no. 4, pp. 457-464, 2009.
[http://dx.doi.org/ 10.1016/j.energy.2008.12.011]

[31]   C. Dejfors, E. Thorin, and G. Svedberg, "Ammonia–water power cycles for direct-fired cogeneration applications", *Energy Convers. Manage.,* vol. 39, no. 16-18, pp. 1675-1681, 1998.
[http://dx.doi.org/ 10.1016/S0196-8904(98)00087-9]

[32]   O. Bai, M. Nakamura, Y. Ikegami, and H. Uehara, "A simulation model for hot spring thermal energy conversion plant with working fluid of binary mixtures", *J. Eng. Gas Turbine. Power,* vol. 126, no. 3, pp. 445-454, 2004.
[http://dx.doi.org/ 10.1115/1.1760526]

[33]   V.A. Prisyazhniuk, "Alternative trends in development of thermal power plants", *Appl. Therm. Eng.,* vol. 28, no. 2-3, pp. 190-194, 2008.
[http://dx.doi.org/ 10.1016/j.applthermaleng.2007.03.025]

[34]   R.G. Bloomquist, "Integrating small power plants into agricultural projects", *Geothermics,* vol. 32, no. 4-6, pp. 475-485, 2003.
[http://dx.doi.org/ 10.1016/S0375-6505(03)00048-8]

[35]   E. Wang, and Z. Yu, "A numerical analysis of a composition-adjustable Kalina cycle power plant for power generation from low-temperature geothermal sources", *Appl. Energy,* vol. 180, pp. 834-848, 2016.
[http://dx.doi.org/ 10.1016/j.apenergy.2016.08.032]

[36]   X. Zhang, M. He, and Y. Zhang, "A review of research on the Kalina cycle", *Renew. Sustain. Energy Rev.,* vol. 16, no. 7, pp. 5309-5318, 2012.
[http://dx.doi.org/ 10.1016/j.rser.2012.05.040]

[37]   H. Esen, M. Inalli, M. Esen, and K. Pihtili, "Energy and exergy analysis of a ground-coupled heat pump system with two horizontal ground heat exchangers", *Build. Environ.,* vol. 42, no. 10, pp. 3606-3615, 2007.
[http://dx.doi.org/ 10.1016/j.buildenv.2006.10.014]

[38]   N.S. Ganesh, and T. Srinivas, "Design and modeling of low temperature solar thermal power station", *Appl. Energy,* vol. 91, no. 1, pp. 180-186, 2012.
[http://dx.doi.org/ 10.1016/j.apenergy.2011.09.021]

[39]   N. Shankar Ganesh, and T. Srinivas, "Power augmentation in a Kalina power station for medium temperature low grade heat", *J. Sol. Energy Eng.,* vol. 135, no. 3, 2013.
[http://dx.doi.org/ 10.1115/1.4023559]

[40]   T. Srinivas, N.S. Ganesh, and R. Shankar, *Flexible Kalina Cycle Systems.* CRC Press, 2019.
[http://dx.doi.org/ 10.1201/9780429487774]

[41]   R. Shankar, and T. Srinivas, "Performance investigation of Kalina cooling cogeneration cycles", *Int. J. Refrig.,* vol. 86, pp. 163-185, 2018.
[http://dx.doi.org/ 10.1016/j.ijrefrig.2017.11.019]

[42]   R. Sharkar, and T. Srinivas, "Novel cooling augmented cogeneration cycle", *Int. J. Refrig.,* vol. 91, pp. 146-157, 2018.
[http://dx.doi.org/ 10.1016/j.ijrefrig.2018.05.026]

[43]   S. Khurana, R. Banerjee, and U. Gaitonde, "Energy balance and cogeneration for a cement plant", *Appl. Therm. Eng.,* vol. 22, no. 5, pp. 485-494, 2002.
[http://dx.doi.org/ 10.1016/S1359-4311(01)00128-4]

[44]   G.K. Alexis, "Performance parameters for the design of a combined refrigeration and electrical power

cogeneration system", *Int. J. Refrig.,* vol. 30, no. 6, pp. 1097-1103, 2007.
[http://dx.doi.org/ 10.1016/j.ijrefrig.2006.12.013]

[45]   T. Srinivas, A.V.S.S.K.S. Gupta, and B.V. Reddy, *Thermodynamic modeling and optimization of multi-pressure heat recovery steam generator in combined power cycle.,* vol. 67, no. 10, pp. 827-834, 2008.

[46]   S.C. Kamate, and P.B. Gangavati, "Exergy analysis of cogeneration power plants in sugar industries", *Appl. Therm. Eng.,* vol. 29, no. 5-6, pp. 1187-1194, 2009.
[http://dx.doi.org/ 10.1016/j.applthermaleng.2008.06.016]

[47]   S. Karellas, A.D. Leontaritis, G. Panousis, E. Bellos, and E. Kakaras, "Energetic and exergetic analysis of waste heat recovery systems in the cement industry", *Energy,* vol. 58, pp. 147-156, 2013.
[http://dx.doi.org/ 10.1016/j.energy.2013.03.097]

[48]   D. Sen, R. Panua, P. Sen, and D. Das, "Thermodynamic analysis and cogeneration of a cement plant in India-A case study", *International Conference on Energy Efficient Technologies for Sustainability, ,* 2013pp. 641-646
[http://dx.doi.org/ 10.1109/ICEETS.2013.6533459]

[49]   M. Sharma, and O. Singh, "Exergy analysis of the dual pressure HRSG for varying physical parameters", *Appl. Therm. Eng.,* vol. 114, pp. 993-1001, 2017.
[http://dx.doi.org/ 10.1016/j.applthermaleng.2016.12.042]

[50]   G.P. Varma, and T. Srinivas, "Design and analysis of a cogeneration plant using heat recovery of a cement factory", *Case Studies in Thermal Engineering,* vol. 5, pp. 24-31, 2015.
[http://dx.doi.org/ 10.1016/j.csite.2014.12.002]

[51]   T. Srinivas, A.V.S.S.K.S. Gupta, and B.V. Reddy, "Generalized thermodynamic analysis of steam power cycles with 'n'number of feedwater heaters", *Int. J. Thermodyn.,* vol. 10, no. 4, pp. 177-185, 2007.

[52]   T. Srinivas, and A.V.S.S.K.S. Gupta, ""Thermodynamic analysis of Rankine cycle with generalisation of feed water heaters", J Institution of Engineers (India), Part MC", *Mechanical Engineering Division,* vol. 87, pp. 56-63, 2007.

[53]   J. Wang, Y. Dai, and L. Gao, "Exergy analyses and parametric optimizations for different cogeneration power plants in cement industry", *Appl. Energy,* vol. 86, no. 6, pp. 941-948, 2009.
[http://dx.doi.org/ 10.1016/j.apenergy.2008.09.001]

[54]   G.V. Pradeep Varma, and T. Srinivas, "Power-augmented steam power plant in a cogeneration cement factory", *J. Energy Eng.,* vol. 143, no. 1, p. 04016020, 2017.
[http://dx.doi.org/ 10.1061/(ASCE)EY.1943-7897.0000374]

[55]   T.J. Kotas, *The exergy method of thermal plant analysis.* Elsevier, 2013.

[56]   Z. Khattak, J. Ahmad Khan, A. Ahmad, S. Shah, and S. Masaud, "Co-generation of power through waste heat recovery–a cement plant case study", In: *International Conference on Future Electrical Power and Energy Systems (ICFEPES 2012).* Sanya: China, 2012, pp. 21-22.

[57]   N.A. Madlool, R. Saidur, M.S. Hossain, and N.A. Rahim, "A critical review on energy use and savings in the cement industries", *Renew. Sustain. Energy Rev.,* vol. 15, no. 4, pp. 2042-2060, 2011.
[http://dx.doi.org/ 10.1016/j.rser.2011.01.005]

SUBJECT INDEX

www.ingramcontent.com/pod-product-compliance
Lightning Source LLC
Chambersburg PA
CBHW050816220326
41598CB00006B/223